Magnesium-Based Nanocomposites

Advances and applications

Magnesium-Based Nanocomposites

Advances and applications

Sankaranarayanan Seetharaman
Department of Mechanical Engineering, National University of Singapore, Singapore

Sravya Tekumalla
School of Mechanical and Aerospace Engineering, Nanyang Technological University, Singapore

Manoj Gupta
Department of Mechanical Engineering, National University of Singapore, Singapore

IOP Publishing, Bristol, UK

ISBN 978-0-7503-3535-5 (ebook)
ISBN 978-0-7503-3533-1 (print)
ISBN 978-0-7503-3536-2 (myPrint)
ISBN 978-0-7503-3534-8 (mobi)

DOI 10.1088/978-0-7503-3535-5

Version: 20201101

IOP ebooks

British Library Cataloguing-in-Publication Data: A catalogue record for this book is available from the British Library.

Published by IOP Publishing, wholly owned by The Institute of Physics, London

IOP Publishing, Temple Circus, Temple Way, Bristol, BS1 6HG, UK

US Office: IOP Publishing, Inc., 190 North Independence Mall West, Suite 601, Philadelphia, PA 19106, USA

Cover image: From the top (centre) in a clockwise direction the symbols represent material characteristics such as mechanically durable, biocompatible, light-weight, flexible, electromagnetic shield, recyclable.

The supreme force that guides each of us & our families and friends…

Contents

Preface

Due to their light weight and biocompatibility, magnesium-based materials are potential candidates for use in a range of applications in the engineering and biomedical sectors. The driving aim is sustainability and keeping bodies of water, air, and land clean. As they are one of the most promising metallic materials of the future, this book looks into the recent advancements in the processing and properties of magnesium-based nanocomposites, a category of magnesium-based materials with an enhanced combination of properties. This book comprehensively summarizes the room temperature and high temperature tensile/compressive response, tribological characteristics, dynamic behaviour, fatigue and creep response, ignition, oxidation, corrosion, and biocompatibility of magnesium-based nanocomposites with particular emphasis on the effects of different types of nano-scale reinforcements. The contents of this book will enable researchers and engineers, irrespective of their field, to better exploit the benefits of magnesium nanocomposites in a wide array of applications.

Acknowledgements

We would like to take this opportunity to express our heartfelt thanks to all the people who have helped and contributed to the publication of this book; in particular to our families for their continual support and understanding, and to co-workers, friends and fellow students for their assistance and encouragement. The authors also acknowledge the research funding received from the Ministry of Education, Singapore (Grant number #R265000622112).

Author biography

Sankaranarayanan Seetharaman

Dr Sankaranarayanan Seetharaman is currently a Research Fellow in the Department of Mechanical Engineering at the National University of Singapore (NUS). His primary research interest is the exploration of process–structure–property relationships in metallic materials. He completed his BE in Mechanical Engineering at Anna University, India, in 2007 and his PhD at the NUS in 2014. During his PhD, he was also a visiting researcher at the Indian Institute of Science, Bangalore (2012–13), and Helmholz Zentrum Geesthacht, Germany (2014). He was also awarded the NUS President Graduate Fellowship in 2013 for his exceptional research accomplishments as a doctoral candidate. After his PhD, he worked as a Development Scientist at the Advanced Remanufacturing Technology Centre, Singapore (2015–18) and led multiple industry sponsored research projects on the metallurgical aspects of metal additive manufacturing. He also worked briefly as a Senior Application Engineer at ANSYS Inc., India (2018–20), where he provided technical support to additive manufacturing simulation activities. He is also a professional member of The Minerals, Metals and Materials Society and actively contributes to the Light Metals Division. He has published over 50 research articles in various peer-reviewed journals and conference proceedings. His current h-index and RG index are 19 and 28, respectively.

Dr Sravya Tekumalla

Dr Sravya Tekumalla is a materials scientist, who is passionate about unravelling the potential of different novel and advanced metallic materials suited for many applications including aerospace, automobile, biomedical, and electronic applications. She is currently a Principal Investigator (P.I.) at Nanyang Technological University (NTU), Singapore, working in the domain of additively manufactured metallic materials as part of her NTU Presidential postdoctoral fellowship. She completed her BEng at Andhra University, India, in 2014 (Gold Medallist) and her PhD at the National University of Singapore (NUS) in 2018 for her work on 'Futuristic *in situ* and *ex situ* approaches for synthesis of superior magnesium nanocomposites'. She later went on to postdoctoral research at NUS on magnesium-based materials for aerospace/defence applications. Dr Sravya is extremely active in research with several visiting research roles at Helmholtz Zentrum Berlin, Germany, the Indian Institute of Science, and the Georgia Institute of Technology, USA, and has published over 24 journal publications, two book chapters, and 20 conference proceedings to date. She has also been a recipient of several early career researcher awards/honorariums such as the Asian Deans' Forum 2019 The Rising Stars, Sparrho's Early Career Researcher Hero, etc.

Dr Manoj Gupta

Dr Manoj Gupta was a former Head of the Materials Division of the Mechanical Engineering Department and Director designate of the Materials Science and Engineering Initiative at the National University of Singapore (NUS). He completed his BEng at VRCE-Nagpur in 1984 (first class with distinction), his MEng at IISC-Bangalore in 1987 (Gold Medallist), his PhD at the University of California, Irvine, USA (1992), and postdoctoral research at the University of Alberta, Canada (1992). In August 2017 he was highlighted among the top 1% of scientists of the world by The Universal Scientific Education and Research Network and among the top 2.5% of scientists as per ResearchGate. His credits include (i) a disintegrated melt deposition technique and (ii) a hybrid microwave sintering technique to synthesize alloys/micro-/nanocomposites. He has published over 560 peer-reviewed journal papers and owns two US patents and 1 trade secret. His current h-index is 67, his RG index is 47, and his citations are over 16 500. He has also co-authored six books, published by Wiley, Springer, and MRF—USA and has edited many others. He is Editor-in-chief/Editor of 12 international peer-reviewed journals. In 2018 he was announced as the World Academy Championship Winner in the area of biomedical sciences by the International Agency for Standards and Ratings. A multiple award winner, he actively collaborates/visits as an invited researcher, and a visiting and chair professor in Japan, France, Saudi Arabia, Qatar, China, USA, and India.

Chapter 1

Introduction to magnesium-based nanocomposites

This chapter presents an introduction to the unique potential of magnesium alloy nanocomposites. It is systematically divided into four sections. In the first section, the characteristics of elemental magnesium are introduced and discussed briefly. The second section presents an overview of the development of magnesium alloys and their properties in cast and wrought conditions. The suitability of magnesium alloys in engineering and biomedical applications are also highlighted. The benefits of reinforcement addition and the properties of magnesium-based metal matrix composites are summarized in the third section. Finally, the role of nano-lengthscale reinforcement in improving the mechanical properties of magnesium and magnesium alloys is introduced.

1.1 Introduction

Recent years have witnessed a renewed interest in the development of magnesium-based materials for engineering and biomedical applications. The replacement of the relatively heavier metallic alloys with magnesium offers significant mass savings to address the growing demands for fuel energy savings and emissions control [1, 2]. For example, in a V6 cylinder car the replacement of a cast iron engine block by magnesium reduced the weight from 86 kg to 30 kg, which in turn produces extensive fuel savings [1]. Similarly, the use of lightweight transmission casings made of magnesium alloys offer ~20%–25% weight savings in the VW Passat and Audi A4 and A6 [3, 4].

In addition to being lightweight, magnesium also exhibits a range of desirable properties such as excellent castability, machinability, damping capacity, heat dissipation, and electro-magnetic shielding characteristics [5]. However, its limitations, such as high reactivity, poor ductility and formability at room temperature, strength loss and creep at high temperatures, and poor corrosion resistance restrict

doi:10.1088/978-0-7503-3535-5ch1

the extended application of magnesium [5]. Most of these limitations are circumvented through the judicious addition of alloying elements and reinforcements [5]. Recently, the use of nano-lengthscale reinforcements was found to improve the properties of pure Mg and magnesium alloys and this book systematically summarizes the properties of magnesium-based nanocomposites containing reinforcements in the nano-lengthscale.

This chapter consists of four sections. While the first section introduces the characteristics of elemental magnesium, the second section systematically presents the effects of different alloying elements and the properties of magnesium alloys in cast and wrought conditions. The suitability of magnesium alloys in engineering and biomedical applications are also highlighted. In the third section, the role of reinforcement addition and the properties of magnesium-based metal matrix composites are summarized. Finally, the role of nano-lengthscale reinforcement in improving the mechanical properties of magnesium and magnesium alloys is introduced.

1.2 The characteristics of magnesium

Magnesium is a silvery white material that belongs to group 2 (alkaline earth metals) of the periodic table. It has a hexagonal close packed (HCP) crystal structure (lattice parameters: $a = 0.32$ nm, $c = 0.52$ nm) and a density of 1.74 g cc^{-1}, which is ~35% lighter than aluminium and ~75% lighter than iron [5–10]. It exhibits properties closer to that of aluminium, as shown in table 1.1.

Table 1.1. Properties of magnesium and aluminium [5–10].

Properties	Magnesium	Aluminium
Atomic number	12	13
Atomic weight	24.32	26.98
Crystal structure	HCP	FCC
Density	1.74	2.70
Melting point	650	660
Boiling point	1105	2520
Thermal expansion coefficient (μm mK^{-1})	25.2	23.6
Thermal conductance (cal cm^{-2} cm^{-1} $°C^{-1}$ s^{-1})	0.38	0.5
Electrical conductance ($\times 10^{-7}$ S m^{-1})	2.34	3.5
Specific heat (1.02 J g^{-1} K^{-1})	1.02	0.897
Latent heat of fusion (KJ mol^{-1})	8.954	10.79
Latent heat of vaporization (KJ mol^{-1})	127.4	293.4
Elastic modulus (GPa)	45	69
Yield strength (MPa)	15–140	35–150
Tensile strength (MPa)	60–220	75–170
Specific strength (kNm kg^{-1})	35–260	7–200
Specific stiffness (MNm kg^{-1})	21–29	25–38

Magnesium (Mg) is one of the most abundant elements in the Earth's crust (2.3% by weight) [11] and sea water (1300 ppm) [12]. Naturally, it exists in combination with other elements and has an oxidation state +2. Although elemental magnesium was first identified by Joseph Black in 1755, it was first isolated in 1808 by Sir Humphry Davy through the electrolysis of a mixture of magnesia (MgO) and mercuric oxide (HgO) [6, 13].

The industrial production of Mg through the electrolysis of magnesium chloride can be dated to the early twentieth century when it was mainly used for weapons manufacturing during World Wars I and II [6, 14]. As the military applications lost prominence in due course, the civilian markets broadened through the development of wrought products, photoengraving technology, and surface treatment systems. Today, the commercial production of Mg from naturally occurring brines (saltwater) and ores such as dolomite ($MgCO_3{\cdot}CaCO_3$) and magnesite ($MgCO_3$) follows two major methods: (i) the electrolysis of magnesium chloride and (ii) the thermal reduction of magnesium oxide through the Pidgeon process [15, 16].

Although magnesium exhibits excellent castability, the lower ignition temperature of Mg poses a fire hazard when exposed to heat [17, 18]. Hence, the molten metal surface of Mg is normally protected with either an inert gas or sulphur-based fluxes [16, 19]. Similarly, the usage of coolants together with proper tool design and process parameters can help in eliminating the fire hazard (by preventing heat build-up) during machining and grinding processes. However, water-based coolants cannot be used as they increase the risk of fire by promoting oxidation and hydrogen evolution [20].

With respect to the mechanical properties, the non-availability of five independent slip systems at room temperature poses a high resistance to the plastic deformation of HCP structured Mg (c/a value: 1.624), thereby affecting its room temperature ductility and formability [21, 22]. On the other hand, the activation of multiple deformation mechanisms including grain boundary sliding at high temperatures contributes to dynamic recrystallization and softening [23, 24].

The high reduction potential of Mg also makes it susceptible to galvanic corrosion when it is in electrical contact with other metals. Therefore, one of the promising ways to control the corrosion of Mg is to reduce the amount of heavy metal impurities such as Fe, Ni, and Cu, etc [25]. Protective alloying additions and chemically applied anodic coatings are also helpful in overcoming the reactivity issues of Mg [26].

1.3 The development of magnesium alloys

Magnesium-based alloys are developed through the addition of alloying elements such as aluminium (Al), zinc (Zn), zirconium (Zr), manganese (Mn), silicon (Si), and rare earth (RE) metals. Figure 1.1 shows a representative schematic of Mg alloy development and the effects of commonly used alloying elements [27].

Aluminium is the most commonly used alloying element in magnesium. The addition of Al enhances the castability and strength of Mg without impairing the weldability. However, the Mg–Al alloys suffer from creep and limited high

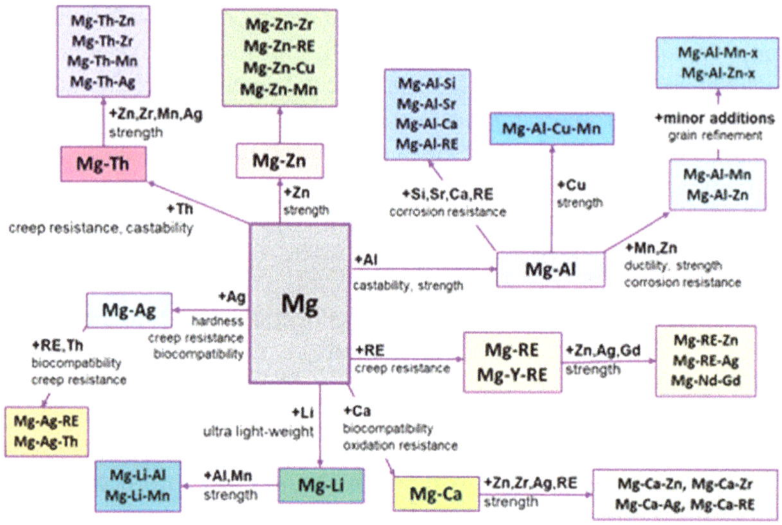

Figure 1.1. Development of magnesium alloys. Reprinted from [25], Copyright (2017), with permission from Elsevier.

temperature mechanical properties. When the Al addition exceeds 6%, the alloy becomes heat treatable. Magnesium alloys containing more than 1.5% Al are susceptible to stress corrosion and must be stress relieved after welding [10, 20].

Zinc improves the fluidity during casting. It is mostly used in combination with aluminium for strength enhancement. However, the amount of zinc addition is limited to less than 2% to prevent hot-shortness. It is also used in combination with zirconium and RE metals to produce precipitation hardened magnesium alloys with good strengths. Zinc combined with aluminium also helps to circumvent the harmful corrosive effects of iron and nickel impurities [10, 20].

Zirconium is a grain refiner in Mg as its lattice parameters (a = 0.323 nm, c = 0.514 nm) are very close to those of Mg (a = 0.320 nm, c = 0.520 nm). It is commonly used in combination with Zn and REs. Since it forms stable phases with Al, it is typically not used in Al containing alloys [10, 20].

Manganese helps in removing heavy metals such as iron from the melt. Hence, a small quantity of Mn improves the saltwater corrosion resistance of Mg–Al and Mg–Al–Zn alloys. No significant effects on the mechanical properties due to its addition have been reported so far [10, 20].

The addition of *silicon* increases the fluidity of the metal in the molten state and decreases the corrosion resistance of Mg alloys containing iron. It is widely used in the development of die cast Mg alloys [10, 20].

Copper improves the temperature and high temperature strengths of Mg. However, the addition of Cu in excess of 0.05% adversely affects the corrosion resistance of magnesium [10, 20].

Strontium is used in combination with other alloying elements to improve the creep performance of Mg [10, 20].

Thorium increases the creep strength of Mg up to 370 °C. The commonly used alloys contain 2%–3% of thorium in combination with Zn, Zr, or Mn. It also improves the weldability of magnesium alloys containing zinc [10, 20].

Silver improves the mechanical properties of magnesium alloys by improving the response to age hardening. Because of its antibacterial properties and biocompatibility in the human body, the addition of silver is recommended for biomedical alloys [10, 20].

Tin addition improves the ductility of the Mg alloy. It also reduces the cracking tendency of Mg alloys during a hot working process [10, 20].

RE elements, in general, improve the high temperature creep and corrosion resistance of magnesium. REs narrow down the metal freezing range of Mg alloys, and weld cracking defects and porosities in cast materials are eliminated. The addition of cerium in small amounts (only 0.2% Ce) improves the ductility of magnesium [10, 20, 28]. Yttrium has a relatively high solid solubility in magnesium (12.4%). It improves the creep resistance of Mg alloys up to 300 °C when added together with other RE elements. For example, the addition of cerium or neodymium improves the creep resistance, thermal stability, corrosion resistance, and deformation behaviour (better ductility and work hardening) of wrought Mg–Y alloys [10, 20, 28, 29].

Lithium with density (0.54 g cc^{-1}) is an attractive alloying element for developing lightweight magnesium alloys [10, 20]. It also has a relatively high solid solubility in magnesium. When added at more than 11.5 wt.% (~31 at.%), the alloys change their crystal structure from HCP to BCC which improves the formability of magnesium alloys tremendously [30, 31]. In general, Li addition enhances the ductility. However, it negatively affects the strength and flammability of Mg [32].

The addition of *calcium* is generally restricted to 0.5%–1% to reduce the oxidation of magnesium in the molten condition as well as during subsequent heat treatment. It also improves the rollability of magnesium sheets. An excess of calcium would lead to cracking during the welding process [20]. Calcium, a major component of human bone, is also used in the development of biomedical Mg alloys [33, 34].

The addition of *beryllium* in small quantities (<30 ppm) reduces the surface melt oxidation significantly during the casting, melting, and welding processes. Beryllium addition results in grain coarsening in sand casting [10, 35].

Iron and *nickel* are generally regarded as harmful alloying elements for Mg from a corrosion perspective, as they tend to deteriorate the corrosion resistance of Mg alloys when present in even small quantities (>0.005%) [10, 20].

Mg alloys are named as per the ASTM designation which uses a standard four-part system to highlight the major alloying additions and the temper condition [10, 20]. For example, in AZ91E-T6 the first part, 'AZ', indicates that the major alloying additions are Al and Zn. The second part, '91', indicates the composition of the major alloying additions (Al = 9%, Zn = 1%). The third part, 'E', indicates that this is the fifth alloy standardized with approximately 9% Al and 1% Zn as the principal alloying additions. The fourth part, 'T6', denotes that the alloy is solution treated and artificially aged. The conventions used to designate the alloying constituents and temper conditions are shown in table 1.2.

Table 1.2. The ASTM designation of Mg alloys [10].

Part 1		Part 2	Part 3	Part 4	
Two letters indicating the two principal alloying additions. Arranged in order of decreasing amount.		Two numbers (whole numbers, rounded off) indicating the amount of principal alloying additions in part 1.	Distinguishes between different alloys with the same percentages of the two principal alloying elements. A one letter alphabet (except I and O) is assigned in order as compositions become standard.	Indicates the temper condition. Consists of a letter followed by a number (separated from the third part of the designation by a hyphen).	
				F	As fabricated
				O	Annealed and recrystallized (wrought products)
				H	Strain hardened
Aluminium	A			H1	Strain hardened only
Copper	C			H2	Strain hardened and partially annealed
Rare earth	E			H3	Strain hardened and stabilized
Thorium	H			W	Solution treated (unstable temper)
Strontium	J			T	Thermally treated to produce stable tempers (other than O, H, F)
Zirconium	K			T1	Naturally aged
Lithium	L			T2	Annealed (cast products)
Manganese	M			T3	Solution heat treated and cold worked
Silver	Q			T4	Solution heat treated and naturally aged
Silicon	S			T5	Artificially aged
Tin	T			T6	Solution heat treated and artificially aged
Gadolinium	V			T7	Solution heat treated and stabilized
Yttrium	W			T8	Solution heat treated, cold worked, and artificially aged
Calcium	X			T9	Solution heat treated, artificially aged, and cold worked
Zinc	Z			T10	Artificially aged and cold worked

1.3.1 Cast magnesium alloys

The low melting point and thermal conductivity of Mg have promoted casting methods for high-volume production. The widely used casting methods include high pressure die casting, sand casting, permanent mould casting, and investment casting. Other relevant technologies are squeeze casting, thixo-casting, and thixo-moudling [36].

Mg–Al alloys form a major group of magnesium castings used in noncritical applications. Since these alloys exhibit micro-shrinkage upon sand casting, high pressure die casting is one of the preferred fabrication methods. While Al and Zn additions along with heat treatment improves the strength of AZ (Mg–Al–Zn–Mn) alloys, the ductility, creep, and high temperature response are often compromised. In this regard, the AM (Mg–Al–Mn) alloys exhibit higher toughness and elongation while retaining comparable corrosion resistance [10]. The AS (Mg–Al–Si–Mn) and AE (Mg–Al–RE) series alloys exhibit good creep resistance up to 150 °C [37]. The silicon and RE additions are, in particular, found to be beneficial for low Al containing Mg alloys, as in the case of AS21 and AE42 alloys which offer better mechanical strength and creep resistance compared to other alloys of the same series. For high temperature and creep requirements, AJ (Mg–Al–Sr) series alloys are also being developed as strontium addition promotes the formation of thermally stable Al4Sr intermetallics [29]. Tables 1.3 and 1.4 show the composition and properties of cast Mg alloys developed through Al addition.

Aluminium free Mg alloy castings containing a small amount of zirconium as a grain refiner together with zinc (ZK), rare earth (ZE), yttrium (WE), and silver (QE) additions are also used for parts operating at temperatures up to 300 °C for extended periods of time [37, 38]. While ZK (Mg–Zn–Zr) and ZE (Mg–Zn–RE) alloys offer better strength, they suffer from hot-shortness resulting from larger amounts of Zn addition [39]. QE series alloys also exhibit improved room temperature and elevated temperature mechanical properties due to the significant grain refinement obtained by Zr addition and heat treatment [24]. While the yttrium containing alloys such as WE43 and WE54 exhibit comparable corrosion resistance to AZ91D, they have superior high temperature properties [40]. Some of the latest developments include alloying additions such as Gd, La, Sm, and Yb to improve the high temperature performance and corrosion resistance of cast Mg alloys [41].

1.3.2 Wrought magnesium alloys

Because of the fine microstructure, wrought Mg alloys exhibit better mechanical properties compared to the cast alloys [10, 21, 39]. Although the HCP crystal structure of magnesium makes it essential to carry out wrought alloy processing (extrusion, forging, rolling) at temperatures above 300 °C (to activate multiple slip systems), most commercial alloys are available in different wrought forms such as rolled plates, forgings, and extruded bars [21, 39, 43].

Wrought Mg alloys with high strength and formability are developed using RE additions that are effective in texture modification and grain refinement [44, 45]. Recent research works also highlight the heat treatment methods to develop ultra-high strength Mg–RE alloys containing very fine nanostructured precipitates and

Table 1.3. General characteristics and properties of cast Mg alloys [5, 10, 14, 20, 29, 39, 41].

Alloy	Composition	Process[a]	Density ($g\ cc^{-1}$)	Heat capacity ($kJ\ kg^{-1}\ K^{-1}$)	CTE ($\mu\ K^{-1}$)	Thermal conduct. ($W\ mK^{-1}$)	General characteristics
AE42	Al: 4%, RE: 2.4%, Mn: 0.25%	DC	1.79	1020	16.1	84	Creep resistant up to 150 °C. Good room temperature mechanical properties.
AE44	Al: 4%, RE: 4%	HPDC	1.82	1020	25.2	85	Good strength, ductility, creep resistance and corrosion resistance.
AJ62	Al: 6%, Sr: 2.4%	HPDC	1.8	1150	27.2	77	Good castability, high temperature strength, creep and corrosion resistance.
AM100A	Al:10%, Mn: 0.1%	SPC	1.83	1050	25	73	Used in pressure tight sand and permanent mould castings, good mechanical properties.
AM20	Al: 2%, Zn: 0.2%, Mn: 0.2%, Si: 0.05%	DC	1.75	1020	26	94	High ductility and impact strength.
AM50A	Al: 5%, Mn: 0.35%	DC	1.77	1020	26	65	Good castability, strength, ductility, and corrosion resistance.
AM60A, B	Al: 6%, Mn: 0.3%	DC	1.8	1000	25.6	67	High ductility comparable to AZ91, excellent saltwater corrosion resistance.
AS21	Al: 2.2%, Si: 1%, Mn: 0.1%	DC	1.78	1020	26.1	84	Creep resistant up to 150 °C. Good room temperature mechanical properties.
AS41A and B	Al: 4.25%, Mn: 0.35%	DC	1.78	1010	26.1	68	Higher strength and castability than AS21.
AZ63A	Al: 6%, Zn: 3%, Si: 0.3, Mn: 0.1%	SPC	1.83	—	26.1	77	Used in sand castings. Good strength, ductility, and saltwater corrosion resistance.

AZ81A	Al: 7.6%, Zn: 0.7%, Mn: 0.13%	SPC	1.8	800	27.2	51.1	Used in sand and permanent mould castings. Good strength and excellent ductility.
AZ91A, C, C, D, and E	Al: 9%, Zn: 0.7%, Mn: 0.3%	A, B, D—DC C, E—SPC	1.81	1050	26	72	Commonly used cast alloy. Useful mechanical properties up to 175 °C. Excellent saltwater corrosion resistance.
EQ21A	Nd: 2.5%, Ag: 1.5%, Zr: 0.7%	SPC	1.81	1000	26.7	113	Cast alloy solution treated for high yield strength at elevated temperatures.
EZ33A	Nd: 3.5%, Zr: 0.7%	SPC	1.8	1040	26.4	99.5	Sand cast, good castability, pressure tight, damping and creep resistance up to 250 °C.
HK31	Th: 3%, Zr: 0.7%	DC	1.8	1000	26.8	92	Sand cast alloy for high temperature applications. Excellent weldability.
QE22A	Ag: 2.5%, RE: 2%, Zr: 0.6%	SPC	1.81	1000	26.7	113	Heat treatable, high temperature alloy, useful properties up to 250 °C, weldable.
WE43A	Y: 4.75%, Nd: 3%, Zr: 0.7%, Zn: 0.2%	SPC	1.8	965	26.7	51.3	High strength casting alloy. Good corrosion resistance and weldability.
WE54A	Y:5.25%, RE: 1.75%, Nd: 1.75	SPC	1.8	—	24.6	52	Exceptional strength at room and high temperature.
ZC63A	Zn: 6.5%, Cu: 3%, Mn: 0.75%	SPC	1.87	960	26	122	Good castability. High strength.
ZE41A	Zn: 4.2%, RE: 1.2%, Zr: 0.7%	SPC	1.84	—	15.1	124	Good castability, pressure tight, weldable. Used in low temperature applications.
ZE63A	Zn: 5.7%, RE: 2.5%, Zr: 0.7%	SPC	1.87	960	—	109	Weldable, pressure tight, specially designed for thin walled castings.
ZK51A	Zn: 4.5%, Zr: 0.7%	SPC	1.83	1000	—	110	Used in sand castings. High yield strength and ductility.

[a] Process: DC—die casting; HPDC—High pressure die casting; SC—sand casting; SPC—sand and permanent mould casting.

Table 1.4. Mechanical properties of cast Mg alloys [5, 10, 14, 20, 29, 39, 41, 42].

Alloy	Process[a]	Temper[b]	Tensile strength (MPa)	Yield strength (MPa)	Elongation (%)	Remarks
AE42	DC	F	230	145	6	
AE44	DC	F	245	142	10	
AJ62	DC	F	240	143	3	
AM100A	SPC	F	150	83	2	Aging: 232 °C/5 h; solution: 432 °C/
		T6	275	110	4	16–24 h; age hardening: 232 °C/5 h
AM20	DC	F	210	90	20	
AM50	DC	F	230	125	8	
AM60	DC	F	220	130	6	
AS21	DC	F	220	120	7	
AS41	DC	F	240	140	8	
AZ63	SPC	F	200	97	6	Aging: 260 °C/4 h; solution: 391 °C/
		T4	275	97	12	10–14 h; age hardening: 218 °C/5 h
		T5	200	105	4	
		T6	275	130	5	
AZ81	SPC	T4	275	85	15	Solution: 418 °C/16–24 h
AZ91	A, B, D—DC	F	185	150	3	Solution: 418 °C/16–24 h
		T4	250	80	4.5	Age hardening: 168 °C/16 h
		T6	230	130	2.3	
	C, E—SPC	F	165	95		
		T4	275	85		
		T6	275	195		
EQ21	SPC	T6	234	172	2	Solution: 530 °C/4–8 h; age hardening: 200 °C/16 h
EZ33	SPC	T5	160	105	3	Aging: 175 °C/16 h

QE22	SPC	T6	275	205	4	Solution: 530 °C/4–8 h; age hardening: 204 °C/8 h
WE43	SPC	T6	235	190	4	Solution: 535 °C/4–8 h; age hardening: 250 °C/16 h
WE54	SPC	T6	270	190	4	Solution: 535 °C/4–8 h; age hardening: 250 °C/16 h
ZC63	SPC	T6	210	125	4	Solution: 445 °C/4–8 h; age hardening: 200 °C/16 h
ZE41	SC	T5	205	140	3.5	Aging: 329 °C/2 h
ZE63	SPC	T6	295	190	7	Solution: 491 °C/10–72 h; age hardening: 141 °C/48 h
ZK51	SC	T5	276	165	3	Aging: 177 °C/12 h

[a] Process: DC—die casting; HPDC—High pressure die casting; SC—sand casting; SPC—sand and permanent mould casting.
[b] Temper conditions: F—as fabricated; T4—solution heat treated and naturally aged; T5—artificially aged; T6—solution heat treated and artificially aged.

ordered phases. However, these alloys involve a large amount of RE additions which are expensive.

Li addition is also beneficial in improving the formability of Mg. Increasing Li content reduces the axial ratio (c/a) of HCP-Mg and hence increases the activation of non-basal prismatic planes [31]. While Li addition in excess of 5.5 wt.% transforms the Mg–Li alloys into a duplex (HCP + BCC) structure, the alloys change entirely into BCC structure when the Li amount is greater than 11.5 wt.% [30–32]. Although Li additions are beneficial for mechanical properties, Mg–Li alloys suffer from poor corrosion behaviour and flammability concerns [30]. Some of the other alloying additions such as Ca, Sr, Sn, Ag, and Y are also being explored for wrought Mg alloy development. In this regard, Sr addition has a remarkable similarity to REs in producing a two-component texture for formability improvement in Mg–Zn alloys [44, 46, 47]. The composition and properties of commercial wrought Mg alloys are listed in table 1.5.

1.3.3 Applications of magnesium alloys

In automobiles, Mg alloys are currently used in seat frames, wheels, housings, transmissions cases, engine blocks, steering wheels, columns, etc [1, 14, 29, 49, 50]. Volkswagen was the first to use Mg alloys in the powertrain of its Beetle model cars in 1936. Porsche also developed a casting technique for the use of Mg in engine blocks in 1968. Although the use of Mg provided good mileage in the automobile and aerospace sectors, there were significant issues related to high temperature creep, flammability, room temperature formability, and corrosion. As a result, researchers started to focus on Mg alloy development in the 1990s, which led to increasing use of magnesium in the automotive industry. Currently, the use of Mg in automobiles is largely focused on (i) engine and transmission parts, (ii) interior parts, (iii) chassis, and (iv) body components (figure 1.2 and table 1.6) [1, 13, 14, 29, 49–52].

The aerospace industry also employs magnesium alloys to derive weight savings in the manufacture of satellite components and rocket propellants. Some of the examples include the B-36 aircraft bomber which contained 19 000 lbs of magnesium and the Titan I rocket that used 1100 lbs of magnesium sheets [53]. Over the years, Mg usage has expanded in both civil and military aircrafts as thrust reverser (for Boeing 737, 747, 757, 767), gearbox (Rolls-Royce), engines, and helicopter transmission casings, etc (figure 1.3). Military aircrafts, such as the Eurofighter Typhoon, Tornado, and F16, also benefit from the lightweight characteristics of magnesium alloys for transmission casings. Magnesium alloys are also used in consumer electronics and sports applications.

In recent years, Mg and its alloys have been actively considered for biomedical applications as the properties of Mg, such as its density, elastic modulus and tensile strength, are close to those of human cortical bone. As a biodegradable material, Mg-based temporary implants (figure 1.4) do not require corrective surgery for their removal and help to reduce the risks of multiple surgeries and associated heath care costs.

Table 1.5. The properties of wrought Mg alloys [5, 10, 20, 29, 39, 43, 48].

Alloy name	Density ($g\ cc^{-1}$)	CTE ($\mu\ K^{-1}$)	Specific heat capacity ($J\ kg^{-1}\ K^{-1}$)	Thermal conductivity ($W\ m^{-1}\ K^{-1}$)	Form	Tensile strength (MPa)	Yield strength (MPa)	Ductility (%)	Comp. yield strength (MPa)	Hardness (HB500)
AZ31B (Al = 2.5%–3.5%; Zn = 0.6%–1.4%; Mn = 0.2% min)	1.77	26	1000	96	Extrusion	260 (F)	200 (F)	15 (F)	95 (F)	49 (F)
					Plates and sheets	255 (T4) 290 (T6)	150 (T4) 220 (T6)	21 (T4) 15 (T6)	110 (T4) 180 (T6)	73 (T6)
					Forging	260 (F)	195 (F)	9 (F)	85 (F)	—
AZ61A (Al = 2.5%–3.5%; Zn = 0.6%–1.4%; Mn = 0.2% min)	1.77	26	1000	96	Extrusion	310 (F)	230 (F)	16 (F)	130 (F)	60 (F)
					Forging	195 (F)	180 (F)	12 (F)	115 (F)	—
AZ80A (Al = 7.8%–9.2%; Zn = 0.2%–0.8%; Mn = 0.12% min)	1.8	26	1050	76	Extrusion	340 (F) 380 (T5)	250 (F) 275 (T5)	11 (F) 7 (T5)	140 (F) 240 (T5)	82 (T5)
	1.8	26	1050	76	Forging	315 (F) 345 (T5) 345 (T6)	215 (F) 235 (T5) 250 (T6)	8 (F) 6 (T5) 5 (T6)	170 (F) 195 (T5) 185 (T6)	—
ZC71 (Zn = 6%–7%; Cu = 1%–1.5%; Mn = 0.5%–1%)	1.83	26	—	122	Extrusion	360 (F)	340 (F)	5 (F)	—	80 (F)
ZK60A (Zn = 4.8%–6.2%; Zr = 0.45% min)	1.83	26	1000	120	Extrusion	340 (F) 365 (T5)	250 (F) 305 (T5)	14 (F) 11 (T5)	185 (F) 250 (T5)	88 (T6)
	1.83	26	1000	120	Forging	305 (T5) 325 (T6)	20 (T5) 270 (T6)	16 (T5) 11 (T6)	195 (T5) 170 (T6)	—
HM31 (Th = 2.5%–3.5%; Mn = 1.2% min)	1.8	26	1000	104	Extrusion	290 (F) 300 (T5)	230 (F) 270 (T5)	10 (F) 10 (T5)	—	—
HK31A (Th = 2.5%–3.5%; Mn = 1.2% min)	1.8	26	1000	104	Plates and sheets	255 (T6)	200 (T6)	9 (T6)	160 (T6)	68 (T6)

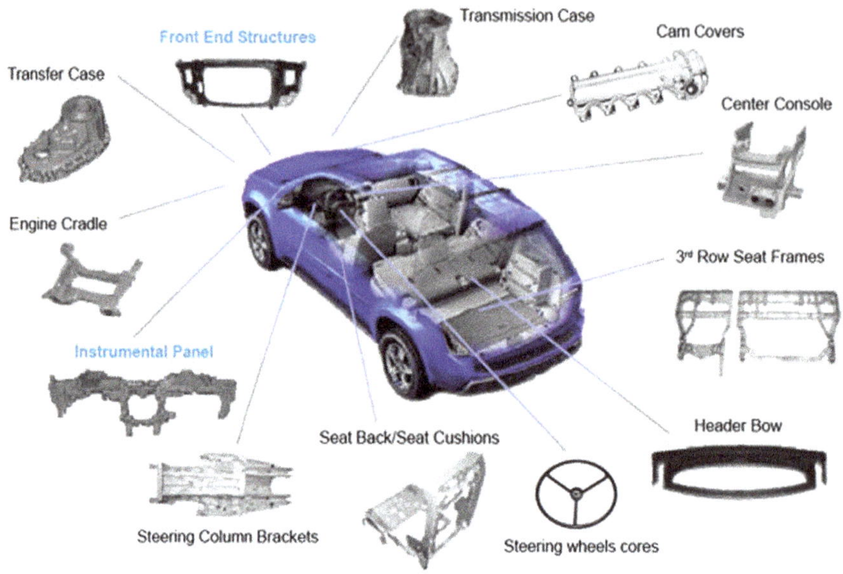

Figure 1.2. The use of magnesium-based materials in automobiles. Adapted from [49].

1.4 Magnesium-based metal matrix composites

Despite the development of new alloys, most of the commercially available Mg alloys exhibit low thermal stability and are stable only up to ~150 °C. Hence, Mg-based metal matrix composites (Mg-MMC) have been developed with ceramic reinforcements. The properties of some of the common particulate reinforcements used in MG-MMCs are listed in table 1.7.

The end properties of Mg-MMCs are controlled by variables such as the processing types, morphology, volume fraction, and distribution of the reinforcements in addition to the properties of the matrix materials, defects, and matrix/reinforcement interface. An overview of the theoretical predictions applicable for the properties of metal matrix composites are summarized below.

The relative proportions of matrix and reinforcement constituents are expressed as volume and weight fractions. While the volume fraction is used in the computation of properties, it is easy to weigh the relative proportions of the matrix and reinforcement using an analytical balance. Hence, it is essential to understand the conversion between weight and volume fractions.

Considering $V_c = V_m + V_r$, the weight and volume fractions can be related using the rule of mixtures [21, 22] as follows:

$$W_m = \frac{\rho_m}{\rho_c} V_m \text{ and } W_R = \frac{\rho_r}{\rho_c} V_r, \tag{1.1}$$

where W and V represent the weight and volume fractions of the matrix (subscript 'm') and reinforcement (subscript, 'r') phases, and ρ_m, ρ_r, and ρ_c represent the densities of the matrix, reinforcement, and composite, respectively.

Table 1.6. Examples of Mg alloy applications in automobiles [13, 29, 50, 51].

Component	Car models
	Engine and transmission parts
Engine block	BMW (N52), Ford, Porsche (911)
Intake manifold	BMW (V8 motor), GM (V8 North Star motor), Chrysler
Lower crankcase	Chrysler (Jeep), Alfa Romeo (GTV), GM (Oldsmobile), McLaren Motors (F1–V12)
Transmission case	AutoZAZ-Daewoo (Tavria, Slavuta, Daewoo-Sens), Volvo Motors (LCP), Porshe AG (911 series), Volkswagen (Volkswagen Passat), Audi (A4, A6), Mercedes-Benz, Ford (Bronco, Aerostar 1994)
Gearbox and controls housing	AutoZAZ-Daewoo (Tavria, Slavuta, Daewoo-Sens)
Cylinder block	GM (Pontiac Gran AM, Corvette)
Cylinder head	Dodge (Dodge Raw), Honda Motors (City Turbo), Alfa Romeo (GTV), AutoZAZ-Daewoo (Tavria, Slavuta, Daewoo-Sens), Honda, BMW, Ford, Isusu, Volvo Motors (LCP), Chrysler
Camshaft drive chain case (AZ91D)	Porsche AG (911 series), Chrysler (Jeep 1993, Viper)
Clutch case (AZ91B, AZ91D, AZ91HP)	Alfa Romeo (GTV), AutoZAZ-Daewoo (Tavria, Slavuta, Daewoo-Sens), Volvo Motors (LCP), Ford (Ranger), GM (Corvette)
Oil pan and pump housing	Ford (Ranger), Chrysler (Jeep), Chrysler (Viper), McLaren Motors (F1–V12)
Engine cradle	Ford, GM (Corvette)
Brackets (brake pedal, clutch, and accelerator)	Ford, GM, Chrysler
	Interior parts
Steering wheel frame	Ford (Thunderbird, Cougar, Taurus, Sable, Ranger), Chrysler (Chrysler Plymouth, LH Midsize 1993), Toyota, BMW (MINI), Lexus (Lexus LS430), GM (Oldsmobile, Pontiac, Buick)
Steering link bracing	GM (LH Midsize)
Instrument panel	GM, Chrysler (Jeep), Ford, Audi (A8), Toyota (Toyota Century)
Seat frame	GM (Impact), Mercedes-Benz (Mercedes Roadster 300/400/500 SL), Lexus (Lexus LS430)
Brake and clutch pedal	GM Oldsmobile, Pontiac, Buick
Door inner	Aston Martin (DB9 2004)
Console bracket	Ford
Air bag retainer	Chrysler
	Chassis components
Wheel rims	Toyota (Toyota 2000GT, Toyota Supra), Alfa Romeo (GTV), Porshe AG (944 and 911)
Lift gate	Chrysler (Pacifica 2017), Lincoln (MKT Luxury Crossover 2010), Mercedes (E-Class T-Model 2009), Aston Martin (Vanquish S 2017)
Valve cover	GM (Corvette)
ABS mounting bracket	Ford, Chrysler

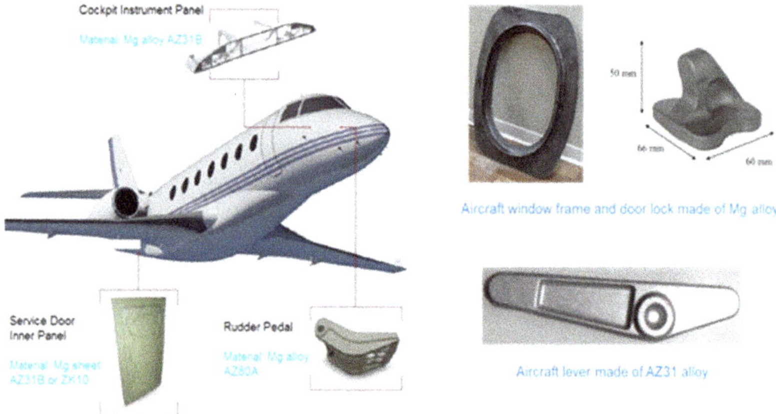

Figure 1.3. Magnesium-based materials in aerospace applications. Adapted from [49].

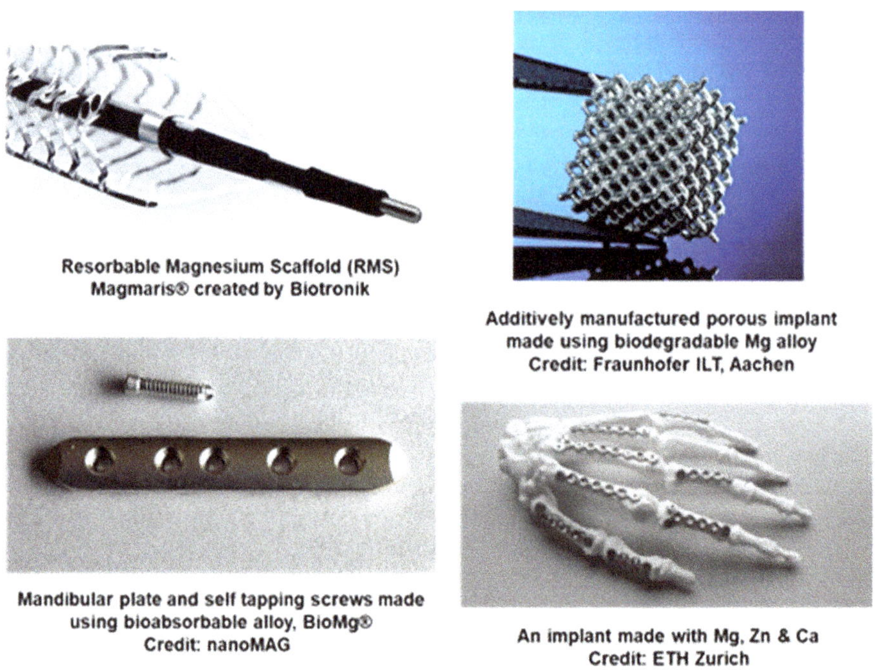

Figure 1.4. The use of magnesium-based materials in biomedical applications. Adapted from [54–57].

The density of the composite can be calculated using the rule of mixtures [21, 22] as follows:

$$\rho_c = \rho_m V_m + \rho_r V_r \text{ (or) } \rho_C = \frac{1}{\frac{W_m}{\rho_m} + \frac{W_r}{\rho_r}}. \tag{1.2}$$

Table 1.7. Properties of the commonly used reinforcement materials [5, 58].

Metal	Crystal structure	Density (g cc^{-1})	Melting point (°C)	Thermal conductivity (W mK^{-1})	Thermal expansion coefficient (μm mK^{-1})	Modulus (GPa)
Al_2O_3	Hex.	3.9	2050	25	7.92	410
AlN	Hex.	3.25	2300	10	4.84	350
B_4C	Rhom.	2.52	2450	29	6.08	450
BN	Hex.	2.2	3000	25	3.8	90
C (graphite)	Hex.	2.18	3642 (sublimation)	2300	−1.44	690
MgO	Cub.	3.58	2852	30	11.61	317
MoS_2	Ordered.	5.06	2375	3	7.6	280
NbC	Cub.	7.6	3608	14	6.84	338
SiC	Hex.	3.21	2300	59	5.4	480
Si_3N_4	α-trigonal/β-Hex./γ-cub.	3.29	1900	29	1.44	310
SiO_2		2.66	1713	12	1.08	73
TaC	Cub./Hex.	13.9	3880	21	6.46	366
TiB_2	Hex.	4.5	2900	27	8.28	414
TiC	Cub.	4.93	3140	29	7.6	270
TiN	Cub.	5.24	2950	29	9.4	600
VC	Cub.	5.77	2810		7.16	434
WC	Hex	15.7	2800	110	5.09	690
ZrB_2	Hex.	6.09	3246	127	8.28	503
ZrC	Cub.	6.73	3540	21	6.66	360
ZrO_2	Monoclinic/tetragonal	5.89	2945	3	12.01	200

Based on the rule of mixtures, the thermal expansion coefficient [21, 22] of the composite can be computed as follows:

$$\alpha_c = \alpha_m V_m + \alpha_r V_r, \tag{1.3}$$

where α_c, α_m, and α_r refer to the CTE values of the composite, matrix, and reinforcement, respectively.

The thermal expansion coefficient of particle reinforced MMCs can also be calculated based on the Turner model [59] as below:

$$\alpha_c = \frac{(\alpha_m V_m K_m + \alpha_r V_r K_r)}{(V_m K_m + V_r K_r)}, \tag{1.4}$$

where K_m and K_r are the bulk moduli of the matrix and reinforcement, respectively.

Kerner also proposed a model to predict the thermal expansion of a composite as below [60]:

$$\alpha_c = \alpha_r V_r + \alpha_m \alpha_m + (\alpha_r - \alpha_m) V_r \frac{V_m(K_r - K_m)}{V_r K_r + V_m V_m + \frac{3K_r K_m}{4G_m}}. \tag{1.5}$$

The elastic modulus of the composite depends on the type and volume fraction of the reinforcement. It generally improves with the addition of the reinforcement [21, 22]:

$$E_C = E_m V_m + E_r V_r, \tag{1.6}$$

where E_m and E_r represent the elastic moduli of the matrix and the reinforcement, respectively.

For particle (with aspect ratio s) reinforced composites, the Halpin–Tsai model [61] can be used to calculate the elastic modulus as follows:

$$E_c = \frac{E_m(1 + 2sqV_r)}{1 - qV_r};\ q = \frac{\left(\frac{E_r}{E_m} - 1\right)}{\left(\frac{E_r}{E_m} + 2s\right)}. \tag{1.7}$$

In general, metal matrix composites display better yield strength (σ_c) than unreinforced matrix material (σ_m) due to the strengthening contribution from reinforcements ($\Delta\sigma_i$) by (i) load transfer, (ii) thermal residual stresses, (iii) enhanced dislocation density, (iv) grain refinement, and (v) Orowan strengthening mechanisms, as described below [62, 63].

Load transfer

The strengthening effect due to the transfer of load from a relatively soft matrix material to the hard and strong reinforcements is given by [64, 65]

$$\Delta\sigma_{LT} = V_r \sigma_m \frac{(L + t)A}{4L}, \tag{1.8a}$$

where σ_m is the yield strength of the matrix material, L is the length of the reinforcement material, and t represents its thickness. Therefore, for small aspect reinforcements, the strengthening effect due to load transfer can be expressed as $\Delta\sigma_{LT} = 0.5 V_r \sigma_m$.

Thermal residual stresses

Thermal cycling at high temperatures often results in residual stresses that strengthen the matrix material as per the following equation [66, 67]:

$$\Delta\sigma_{Ts} = \frac{E_r E_m}{E_r V_r + E_m V_m} V_r \Delta\alpha \Delta T \tag{1.8b}$$

where $\Delta\alpha$ represents the difference in CTE values and ΔT is the difference in processing temperature.

Enhanced dislocation density

The difference in CTE values and Young's modulus between the matrix and reinforcement materials increases the dislocation density which strengthens the matrix material [21, 22, 66]:

$$\Delta\sigma_{\mathrm{CTE}} = \frac{A\Delta\alpha\Delta T V_{\mathrm{r}}}{bdV_{\mathrm{m}}} \tag{1.8c}$$

$$\Delta\sigma_{\mathrm{EM}} = \frac{6V_{\mathrm{r}}}{\pi d^3}\varepsilon, \tag{1.8d}$$

where A is a geometric constant, b represents the Burgers vector, and d is the size of the reinforcement particles.

Grain refinement strengthening

In general, the grain boundaries in polycrystalline metals obstruct the dislocation motion and increase the stress required for continuing the deformation process [21, 22, 68]. In this regard, as reinforcement additions are known to refine the grain size of MMCs resulting in a larger volume of grain boundaries, the grain boundary strengthening contribution can be significant for MMCs and the same can be related to its average grain size as per the Hall–Petch equation

$$\Delta\sigma_{\mathrm{HP}} = \frac{K_y}{\sqrt{d_{\mathrm{m}}}}, \tag{1.8e}$$

where K_y is the strengthening coefficient which is a characteristic constant of each material.

Orowan strengthening

Orowan strengthening specifically applies to very fine particle reinforcements that are closely spaced and effective in impeding the dislocation movement [21, 22]. Since the interparticle distance and the particle size are critical for Orowan strengthening, its contribution is significant for sub-micron and nano-lengthscale reinforcements [69]:

$$\Delta\sigma_{\mathrm{OR}} = \frac{0.13bG}{d\left(\sqrt[3]{\frac{1}{2}V_{\mathrm{r}}} - 1\right)}\ln\left(\frac{d}{2b}\right), \tag{1.8f}$$

where G is the matrix shear modulus.

Therefore, the overall yield strength of the MMC (σ_{c}) can be calculated as follows:

$$\sigma_{\mathrm{c}} = \sigma_{\mathrm{m}} + \sqrt{\sum \Delta\sigma_{\mathrm{i}}^2}. \tag{1.9}$$

There are numerous reports in the available literature indicating the efficacy of particulate reinforcements in improving the mechanical properties of Mg. Table 1.8 lists the mechanical properties of a few Mg-MMC reinforced with ceramic particles.

1.5 Magnesium-based nanocomposites: the role of nano-lengthscale reinforcements

Although traditional micron-sized reinforcements in the form of fibres and particles offer significant improvements to the thermal stability, elastic modulus, stiffness, and mechanical strength of Mg composites, the ductility and fracture toughness of

Table 1.8. The mechanical properties of Mg-MMCs reinforced with micron-sized reinforcements.

Matrix	Reinforcement	Reinforcement size (μm)	Vol. (%)	Processing method*	Composite formulation	Yield strength (MPa)	Tensile strength (MPa)	Elongation (%)	Remarks*	Reference
Mg	Al_2O_3	1	1.1	PM + HE	$1.1Al_2O_3/Mg$	172	227	16.8		[70]
		0.3	1.1		$1.1Al_2O_3/Mg$	182	237	12.1		
Mg	Al_2O_3	0.3	0	MWS-PM + HE	Mg	117	168	2.8		[71]
			5		$5.0Al_2O_3/Mg$	160	215	3.9		
			2.5		$2.5Al_2O_3/Mg$	130	167	4		
Mg	Al_2O_3	0.3	0.7	DMD + HE	$0.7Al_2O_3/Mg$	214	261	12.5		[72]
			1.1		$1.1Al_2O_3/Mg$	200	251	8.6		
			2.5		$2.5Al_2O_3/Mg$	222	281	4.5		
Mg	SiC	0.5–25	0	SPS + HE	Mg	115				[73]
			5		5SiC/Mg	130				
			10		10SiC/Mg	140				
			15		15SiC/Mg	120				
Mg	SiC	25	0	DC	Mg	110	199	10.2		[74]
			4.3		4.3SiC/Mg	112	191	6.8		
			8.7		8.7SiC/Mg	112	177	4.7		
			16.8		16.8SiC/Mg	114	154	1.8		
Mg	SiC	40	30	SC	30SiC/Mg	229	258	2		[75]
Mg	SiC	38	6.6	DMD + HE	6.6SiC/Mg	119	197	3.9	Ext. $T = 350$ °C	[76]
						139	201	3.7	Ext. $T = 250$ °C	
						148	204	3.6	Ext. $T = 150$ °C	
						148	201	3.6	Ext. $T = 100$ °C	
						130	191	12.8	Ext. $T = 100$ °C + HT 5 h	
						126	190	10.4	Ext. $T = 100$ °C + HT 10 h	

Mg	SiC	0.6	0	DMD + HE	Mg	153	207	9.2		[77]
			2.7		2.7SiC/Mg	182	219	2.1		
			5.8		5.8SiC/Mg	171	221	1.5		
			9		9SiC/Mg	155	207	1.4		
			2.7		2.7SiC/Mg	200	233	2.9	HT 5 h	
			9		9SiC/Mg	168	213	3.6	HT 5 h	
Mg	SiC	35	0	DMD + HE	Mg	153	219	3.1		[78]
			33.6		33.6SiC/Mg	158	205	12.1		
AZ31	SiC	0.5–25	5	SPS + HE	5SiC/AZ31	160	—	—		[73]
			10		10SiC/AZ31	165	—	—		
			15		15SiC/AZ31	150	—	—		
Mg–9Al	SiC		15	LPS	15SiC/Mg–9Al	339	364	1.2	Sintering: 470 °C/1 h	[79]
		20	15		15SiC/Mg–9Al	271	355	3.2	Sintering: 470 °C/10 h	
ZK51	SiC	37	0	PM	ZK51	229	315	8.5		[80]
AZ91	SiC	0.2	0	SC	AZ91	216	332	15.5		[81]
			2		2SiCp/AZ91	285	377	4.2		
AZ91	SiC	1	0	SC(U)	AZ91	70	130	2		[83]
			3		3SiC/AZ91	100	155	1.8		
			5		5SiC/AZ91	105	140	0.7		
AZ91	SiC	0.2	0	SC	AZ91	225	255	—		[84]
			0.5		0.5SiC/AZ91	260	300	—		
			1		1.0SiC/AZ91	275	330	—		
			1.5		1.5SiC/AZ91	280	360	—		
			2		2.0SiC/AZ91	280	365	—		
			3		3.0SiC/AZ91	265	350	—		
			5		5.0SiC/AZ91	260	330	—		

(Continued)

Table 1.8. *(Continued)*

Matrix	Reinforcement	Reinforcement size (μm)	Vol. (%)	Processing method*	Composite formulation	Yield strength (MPa)	Tensile strength (MPa)	Elongation (%)	Remarks*	Reference
AZ91	SiC	0.2	2	SC + HT + HF + HE	2SiC0.2μ $A^{-1}Z^{-91}$	280	370	—	HT T4 For. T = 420 °C Ext. T = 370 °C Ext. R 16:1	[85]
			5		5SiC0.2μ $A^{-1}Z^{-91}$	250	320	—		
			10		10SiC0.2μ $A^{-1}Z^{-91}$	250	310	—		
		5	2		2SiC5μ/AZ91	260	325	—		
			5		5SiC5μ/AZ91	270	330	—		
			10		10SiC5μ/AZ91	270	315	—		
		10	2		2SiC10μ/AZ91	—	325	—		
			5		5SiC10μ/AZ91	300	385	—		
			10		10SiC10μ/AZ91	280	350	—		
AZ91	SiC	12.8	15	SC(V) + HT	AZ91	84	225	7.2	HT T4	[82]
			15		15SiC/AZ91	178	218	1.1		
AZ91	SiC	10	0	SC + T4 + HE	AZ91	225	350	8	Ext. T = 250 °C Ext. R 12	[86]
			10		10SiC/AZ91	225	290	2		
			0		AZ91	170	290	20	Ext. T = 350 °C Ext. R 12	
			10		10SiC/AZ91	250	320	5		
			0		AZ91	170	290	—	Ext. T = 350 °C Ext. R 5	
			10		10SiC/AZ91	230	300	—		

AZ91	SiC	10	10	SSM-PDC + HT	10SiC/AZ91	116	139	0.8	Before forging	[87]
			10	(T4) + HF	10SiC/AZ91	—	222	0.5	Forg. $T = 320$ °C	
			10		10SiC/AZ91	—	239	0.5	Forg. $T = 370$ °C	
			10		10SiC/AZ91	238	251	0.8	Forg. $T = 420$ °C	
			10		10SiC/AZ91	226	242	0.8	Forg. $T = 470$ °C	
			10		10SiC/AZ91	215	230	0.8	Forg. $T = 520$ °C	
ZK51	(TA) Ti alloy	37	0	PM	ZK51	229	315	8.5		[80]
			10		10TA/ZK51	295	386	5.6		
AZ91	TiC	*In situ*	0	BM + DC	AZ91 alloy	95	198	2.5		[88]
			3.1		3.1TiC/AZ91	115	235	1		
AZ91	TiC	*In situ*	0	SHS	AZ91	—	160	9		[89]
			10		10TiC/AZ91	—	214	4		

BM: ball milling; DC: die casting; DMD: disintegrated melt deposition;
HE: hot extrusion; Ext. T: extrusion temperature; Ext. R: extrusion ratio; HF: hot forging; Forg. T: forging temperature;
HT: heat treatment;
LPS: liquid phase sintering;
PM: powder metallurgy; MWS-PM: microwave sintering assisted powder metallurgy;
SC: stir casting; SC(U): ultrasonic stir casting; SC(V): vacuum stir casting;
SPS: spark plasma sintering;
SSS-PDC: semi-solid stirring and pressure die casting;
SHS: self-propagating high temperature synthesis

these composites with micron-sized reinforcements are detrimentally affected. However, when the reinforcement size is reduced to the nano-scale, the larger surface-to-volume ratio of the nano-reinforcement is noticeably beneficial for an improved mechanical response without any adverse effects on the ductility and fracture toughness. In general, the strength enhancement due to nano-scale reinforcements can be due to the following mechanisms [62, 90, 91].

Load bearing effect. In the event of a strong cohesion between the matrix and reinforcing phases, the hard and strong reinforcements can actively contribute towards load-sharing [62].

Enhanced dislocation density. The difference in CTE values between the matrix and reinforcements can induce stress and plastic deformation at the matrix/reinforcement interface which increase the dislocation density and contributes towards matrix strengthening [62].

Orowan strengthening. This refers to the strengthening of the matrix by closely spaced reinforcement particles that restrict dislocation movement. Since, the interparticle spacing is usually smaller for nano-sized particle reinforced MMC, Orowan strengthening plays a significant role in the strengthening of nanocomposites [90].

Grain refinement. During solidification, the nanoparticle reinforcements act as heterogeneous nucleation sites for grain nucleation and refinement [91, 92]. In particular, in the case of wrought Mg nanocomposites, many nanoscale reinforcements also contribute towards texture weakening, hence activating non-basal slip systems.

Considering the above mentioned strengthening mechanisms, Zhang and Chen [90] proposed a simplified model to predict the yield strength of nanocomposites:

$$\sigma_c = (1 + 0.5V_r)\left(\sigma_m + A + B + \frac{AB}{\sigma_m}\right), \tag{1.10}$$

where $A = 1.25G_m b\sqrt{\frac{12\Delta\alpha\Delta T V_r}{bdV_m}}$ and $B = \frac{0.13G_m b}{d\sqrt[3]{\frac{1}{2V_r}} - 1}\ln\left(\frac{d}{2b}\right)$.

The ductility enhancement in Mg nanocomposites can be attributed to the activation of non-basal slip and the delay in the deformation process brought forth by a change in the stress state (from plane strain to plane stress) due to the uniformly distributed nanoscale reinforcements which also control (i) the matrix grain size and (ii) the size, sharpness, and quantity of second phases. In particular, in the case of wrought Mg nanocomposites, the nanoscale reinforcements also contribute towards texture weakening to activate non-basal slip systems.

Subsequent chapters summarize the processing methods used to fabricate Mg-based metal matrix composites reinforced with nano-scale reinforcements and the properties of Mg-based nanocomposites in terms of:

- Tensile and compressive properties at room temperature.
- Temperature and strain rate effects on the tensile and compressive properties.
- Fatigue behaviour.
- Friction and wear properties.

- Oxidation properties and ignition properties.
- Creep response.
- Machinability.
- Wet corrosion and biocompatibility.

References

[1] Tharumarajah A and Koltun P 2007 Is there an environmental advantage of using magnesium components for light-weighting cars? *J. Clean. Prod.* **15** 1007–13

[2] Mattes W G 2007 The BMW approach to Tier 2 Bin 5 *13th DEER Conference* pp 1–29

[3] Kardys G 2017 Magnesium car parts: a far reach for manufacturers? *Engineering 360* https://insights.globalspec.com/article/7243/magnesium-car-parts-a-far-reach-for-manufacturers-part-1

[4] Barclay S 2007 Lightweight magnesium alloys for the automotive industry *Automotive Industries* https://ai-online.com/2007/03/lightweight-magnesium-alloys-for-the-automotive-industry/

[5] Gupta M and Sharon N M L 2010 *Magnesium, Magnesium Alloys, and Magnesium Composites* (New York: Wiley)

[6] Emley E F 1966 *Principles of Magnesium Technology* (Oxford: Pergamon)

[7] King J F 2006 Technology of magnesium and magnesium alloys *Magnesium Technology: Metallurgy, Design Data, Applications* (Berlin: Springer)

[8] Kainer K U 2003 *Magnesium—Alloys and Technologies* (Weinheim: Wiley)

[9] Watarai H 2004 Trend of research and development for magnesium alloys—reducing the weight of structural materials in motor vehicles *Sci. Technol. Trends Quarter.* **7** 84–97

[10] Moosbrugger C 2017 Engineering properties of magnesium alloys (Materials Park, OH: ASM International)

[11] Royal Society of Chemisty Minerals, elements and the Earth's crust, Teachers resources *Royal Society of Chemisty* https://rsc.org/Education/Teachers/Resources/jesei/minerals/students.htm

[12] Lenntech Magnesium and water: reaction mechanisms, environmental impact and health effects *Lenntech* https://lenntech.com/periodic/water/magnesium/magnesium-and-water.htm

[13] Mathaudhu S N, Luo A A, Neelameggham N R, Nyberg E A and Sillekens W H 2014 *Essential Readings in Magnesium Technology* (New York: Wiley)

[14] Mordike B L and Ebert T 2001 Magnesium: properties—applications—potential *Mater. Sci. Eng.* A **302** 37–45

[15] Wulandari W, Brooks G A, Rhamdhani M A and Monaghan B J 2008 Magnesium: current and alternative production routes *Chemeca 2010 Engineering Edge (Adelaide, Australia, 26–29 September 2010)*

[16] Neelameggham N R 2013 Primary production of magnesium *Fundamentals of Magnesium Alloy MetallurgyMetals and Surface Engineering* (Cambridge: Woodhead)

[17] Han D, Zhang J, Huang J, Lian Y and He G 2020 A review on ignition mechanisms and characteristics of magnesium alloys *J. Magnesium Alloys* **8** 329–44

[18] Tekumalla S and Gupta M 2017 An insight into ignition factors and mechanisms of magnesium based materials: a review *Mater. Des.* **113** 84–98

[19] Subramanian J, Seetharaman S and Gupta M 2015 Processing and properties of aluminum and magnesium based composites containing amorphous reinforcement: a review *Metals* **5** 743–62

[20] Avedesian M M and Baker H 1999 *ASM Specialty Handbook: Magnesium and Magnesium Alloys* (Materials Park, OH: ASM International)

[21] Dieter G E and Bacon D 1988 *Mechanical Metallurgy* SI metric edition (New York: McGraw-Hill) pp 81–220

[22] Callister W and Rethwisch D 2007 Materials science and engineering: an introduction *Mater. Sci. Eng.* **94** 142–404

[23] Zhong T, Rao K P, Prasad Y V R K, Zhao F and Gupta M 2014 Hot deformation mechanisms, microstructure and texture evolution in extruded AZ31—nano-alumina composite *Mater. Sci. Eng.* A **589** 41–9

[24] von Buch F and Mordike B L 2003 High-temperature properties of magnesium alloys *Magnesium—Alloys and Technology* (Weinheim: Wiley) pp 106–29

[25] Esmaily M *et al* 2017 Fundamentals and advances in magnesium alloy corrosion *Prog. Mater. Sci.* **89** 92–193

[26] Atrens A *et al* 2015 Review of recent developments in the field of magnesium corrosion *Adv. Eng. Mater.* **17** 400–53

[27] Gupta M and Seetharaman S 2017 Magnesium based nanocomposites for cleaner transport *Nanotechnology for Energy Sustainability* (Weinheim: Wiley)

[28] Rokhlin L L 1998 Advanced magnesium alloys with rare-earth metal additions *Advanced Light Alloys and Composites* (Dordrecht: Springer) pp 443–8

[29] Kulekci M K 2008 Magnesium and its alloys applications in automotive industry *Int. J. Adv. Manuf. Technol.* **39** 851–65

[30] Wu R *et al* 2015 Recent progress in magnesium–lithium alloys *Int. Mater. Rev.* **60** 65–100

[31] Białobrzeski A, Saja K and Hubner K 2007 Ultralight magnesium–lithium alloys *Arch. Foundry Eng.* **11** 21–4

[32] Muga C O and Zhang Z W 2016 Strengthening mechanisms of magnesium–lithium based alloys and composites *Adv. Mater. Sci. Eng.* **2016** 1078187

[33] Li Z, Gu X, Lou S and Zheng Y 2008 The development of binary Mg–Ca alloys for use as biodegradable materials within bone *Biomaterials* **29** 1329–44

[34] Salahshoor M and Guo Y 2012 Biodegradable orthopedic magnesium–calcium (MgCa) alloys, processing, and corrosion performance *Materials* **5** 135–55

[35] Cao P, Qian M and St John D H 2004 Grain coarsening of magnesium alloys by beryllium *Scr. Mater.* **51** 647–51

[36] Pan F, Yang M and Chen X 2016 A review on casting magnesium alloys: modification of commercial alloys and development of new alloys *J. Mater. Sci. Technol.* **51** 1211–21

[37] Luo A A 2004 Recent magnesium alloy development for elevated temperature applications *Int. Mater. Rev.* **49** 13–30

[38] Luo A and Pekguleryuz M O 1994 Cast magnesium alloys for elevated temperature applications *J. Mater. Sci.* **29** 5259–271

[39] Kainer K U and von Buch F 2003 The current state of technology and potential for further development of magnesium applications *Magnesium—Alloys and Technology* (Weinheim: Wiley) pp 1–22

[40] Aghion E, Bronfin B, Von Buch F, Schumann S and Friedrich H 2003 A comparative study of new magnesium alloys developed for elevated temperature applications in automotive industry *SAE Technical Papers* 2003-01-0191

[41] Song J, She J, Chen D and Pan F 2020 Latest research advances on magnesium and magnesium alloys worldwide *J. Magnesium Alloys* **8** 1–41

[42] Naryshkin A A 1959 Heat treatment and mechanical properties of magnesium alloy parts *Met. Sci. Heat Treat. Met.* **1** 51–3

[43] Becker J and Fischer G 2003 Manufacturing and potential of extruded and forged magnesium products *Magnesium—Alloys and Technology* (Weinheim: Wiley) pp 90–105

[44] You S, Huang Y, Kainer K U and Hort N 2017 Recent research and developments on wrought magnesium alloys *J. Magnesium Alloys* **5** 239–53

[45] Bettles C and Barnett M 2012 *Advances in Wrought Magnesium Alloys: Fundamentals of Processing, Properties and Applications* (Cambridge: Woodhead)

[46] Sadeghi A and Pekguleryuz M 2011 Recrystallization and texture evolution of Mg–3%Al–1%Zn–(0.4–0.8)%Sr alloys during extrusion *Mater. Sci. Eng.* A **528** 1678–85

[47] Sadeghi A, Hoseini M and Pekguleryuz M 2011 Effect of Sr addition on texture evolution of Mg–3Al–1Zn (AZ31) alloy during extrusion *Mater. Sci. Eng.* A **528** 3096–104

[48] Yang Z, Li J P, Zhang J X, Lorimer G W and Robson J 2008 Review on research and development of magnesium alloys *Acta Metall. Sin. (English Lett.)* **313** 313–28

[49] Dziubińska A, Gontarz A, Dziubiński M and Barszcz M 2016 The forming of magnesium alloy forgings for aircraft and automotive applications *Adv. Sci. Technol. Res. J.* **10** 158–68

[50] Kumar D S, Sasanka C T, Ravindra K and Suman K N S 2015 Magnesium and its alloys in automotive applications—a review *Am. J. Mater. Sci. Technol.* **4** 12–30

[51] Cizek L, Pawlica L, Greger M, Pavelcová J and Tański T 2002 Mechanical properties of the model casting magnesium alloy AZ 91 *11th International Scientific Conference: Achievements in Mechanical and Materials Engineering (Polish Academy of Sciences)* pp 47–50

[52] Mathaudhu S N and Nyberg E A 2014 Magnesium alloys in US military applications: past, current and future solutions *Essential Readings in Magnesium Technology* (New York: Wiley)

[53] Furuya H, Kogiso N, Matunaga S and Senda K 2000 Applications of magnesium alloys for aerospace structure systems *Mater. Sci. Forum* **350–351** 341–8

[54] Biotronik. n.d. Resorbable Magnesium Scaffold (RMS) Magmaris, www.biotronik.com/en-de/products/coronary/magmaris

[55] Fraunhofer Institute of Laser Technology 2018 SLM adapted to work with soft alloys and materials *Optics.org* https://optics.org/news/7/3/8

[56] NanoMAG Craniomaxillofacial applications for bioabsorable magnesium *Nano Mag* https://nanomag.us/applications.html

[57] Caggiano A 2019 Magnesium alloys for implants *Materials World Magazine* https://iom3.org/materials-world-magazine/news/2019/nov/26/magnesium-alloys-implants

[58] Kainer K U 2006 Basics of metal matrix composites *Combination of Materials for Light Metal Matrix Composites* (Weinheim: Wiley)

[59] Turner P S 1946 Thermal-expansion stresses in reinforced plastics *J. Res. Natl Bur. Stand.* **37** 239–50

[60] Kerner E H 1956 The elastic and thermo-elastic properties of composite media *Proc. Phys. Soc. Sect.* B **69** 808

[61] Halpin J C 1969 Effects of environmental factors on composite materials *Technical Report* AFML-TR-67-423 (Air Force Materials Lab Wright-Patterson AFB OH)

[62] Clyne T W and Withers P J 1995 *An Introduction to Metal Matrix Composites* (Cambridge: Cambridge University Press)

[63] Arsenault R J 1988 Relationship between strengthening mechanisms and fracture toughness of discontinuous SiC/Al composites *J. Compos. Technol. Res.* **10** 140–5

[64] Clyne T W 1989 A simple development of the shear lag theory appropriate for composites with a relatively small modulus mismatch *Mater. Sci. Eng.* A **122** 183–92
[65] Fukuda H and Chou T W 1981 An advanced shear-lag model applicable to discontinuous fiber composites *J. Compos. Mater.* **15** 79–91
[66] Arsenault R J and Shi N 1986 Dislocation generation due to differences between the coefficients of thermal expansion *Mater. Sci. Eng.* **81** 175–87
[67] Kumar S, Dieringa H and Kainer K U 2005 Effect of particulate content on the thermal cycling behaviour of the magnesium alloy based hybrid composites *Composite* A **36** 321–5
[68] Hansen N 2004 Hall–Petch relation and boundary strengthening *Scr. Mater.* **51** 801–6
[69] Lloyd D J 1994 Particle reinforced aluminium and magnesium matrix composites *Int. Mater. Rev.* **39** 1–23
[70] Hassan S F and Gupta M 2006 Effect of length scale of Al_2O_3 particulates on microstructural and tensile properties of elemental Mg *Mater. Sci. Eng.* A **425** 22–7
[71] Wong W L E, Karthik S and Gupta M 2005 Development of high performance Mg–Al_2O_3 composites containing Al_2O_3 in submicron length scale using microwave assisted rapid sintering *Mater. Sci. Technol.* **21** 1063–70
[72] Hassan S F and Gupta M 2008 Effect of submicron size Al_2O_3 particulates on microstructural and tensile properties of elemental Mg *J. Alloys Compd.* **457** 244–50
[73] Muhammad W N A W, Sajuri Z, Mutoh Y and Miyashita Y 2011 Microstructure and mechanical properties of magnesium composites prepared by spark plasma sintering technology *J. Alloys Compd.* **509** 6021–9
[74] Lim S C V, Gupta M and Lu L 2001 Processing, microstructure, and properties of Mg–SiC composites synthesised using fluxless casting process *Mater. Sci. Technol.* **17** 823–32
[75] Saravanan R A and Surappa M K 2000 Fabrication and characterisation of pure magnesium-30 vol.% SiCP particle composite *Mater. Sci. Eng.* A **276** 108–16
[76] Lim S C V and Gupta M 2001 Enhancing the microstructural and mechanical response of a Mg/SiC formulation by the method of reducing extrusion temperature *Mater. Res. Bull.* **36** 2627–36
[77] Reddy S U, Srikanth N, Gupta M and Sinha S K 2004 Enhancing the properties of magnesium using SiC particulates in sub-micron length scale *Adv. Eng. Mater.* **6** 957–64
[78] Soon L P and Gupta M 2002 Synthesis and recyclability of a magnesium based composite using an innovative disintegrated melt deposition technique *Mater. Sci. Technol.* **276** 92–8
[79] Xie W *et al* 2007 Influence of sintering routes to the mechanical properties of magnesium alloy and its composites produced by PM technique *J. Alloys Compd.* **431** 162–6
[80] Xi Y L, Chai D L, Zhang W X and Zhou J E 2006 Titanium alloy reinforced magnesium matrix composite with improved mechanical properties *Scr. Mater.* **54** 19–23
[81] Nie K B, Deng K K, Xu F J, Wang X J and Wu K 2015 Development of microstructure in submicron particles reinforced magnesium matrix composite processed by room temperature deformation *Mater. Chem. Phys.* **149-150** 21–6
[82] Gui M C, Han J M and Li P Y 2004 Microstructure and mechanical properties of Mg–Al_9Zn/SiCp composite produced by vacuum stir casting process *Mater. Sci. Technol.* **20** 765–71
[83] Nie K B *et al* 2011 Microstructure and tensile properties of micro-SiC particles reinforced magnesium matrix composites produced by semisolid stirring assisted ultrasonic vibration *Mater. Sci. Eng.* A **528** 8709–14

[84] Deng K K, Wu K, Wu Y W, Nie K B and Zheng M Y 2010 Effect of submicron size SiC particulates on microstructure and mechanical properties of AZ91 magnesium matrix composites *J. Alloys Compd.* **504** 542–7
[85] Deng K K *et al* 2012 Effect of particle size on microstructure and mechanical properties of SiCp/AZ91 magnesium matrix composite *Mater. Sci. Eng.* A **543** 158–63
[86] Wang X J *et al* 2011 Influences of extrusion parameters on microstructure and mechanical properties of particulate reinforced magnesium matrix composites *Mater. Sci. Eng.* A **530** 446–51
[87] Deng K K *et al* 2010 Microstructure evolution and mechanical properties of a particulate reinforced magnesium matrix composites forged at elevated temperatures *Mater. Sci. Eng.* A **527** 1630–5
[88] Zhang X, Liao L, Ma N and Wang H 2006 Mechanical properties and damping capacity of magnesium matrix composites *Composites* A **37** 2011–6
[89] Jiang Q C, Li X L and Wang H Y 2003 Fabrication of TiC particulate reinforced magnesium matrix composites *Scr. Mater.* **48** 713–7
[90] Zhang Z and Chen D L 2006 Consideration of Orowan strengthening effect in particulate-reinforced metal matrix nanocomposites: a model for predicting their yield strength *Scr. Mater.* **54** 1321–6
[91] Sanaty-Zadeh A 2012 Comparison between current models for the strength of particulate-reinforced metal matrix nanocomposites with emphasis on consideration of Hall–Petch effect *Mater. Sci. Eng.* A **531** 112–8
[92] Mirza F and Chen D 2015 A unified model for the prediction of yield strength in particulate-reinforced metal matrix nanocomposites *Materials* **8** 5138–53

Chapter 2

Processing of magnesium-based (nano) composites

This chapter presents an overview of the processing methods that are applicable for magnesium-based composites including nanocomposites. Broadly, the processing methods are classified under the categories of liquid state processing, solid-state processing, and semisolid processing. Based on the nature of reinforcement addition, the processing methods can also be classified as either *ex situ* or *in situ* methods. This chapter will help researchers and industries to properly identify a suitable manufacturing method by considering the advantages and limitations of each method.

2.1 Introduction

With the latest advancements in nanotechnology, making composites with nano-lengthscale reinforcements has emerged as a promising approach to engineer desired properties in metallic alloys. In the nanocomposite approach, the interaction between the dislocations and the nano-lengthscale reinforcements plays a determining role for the end properties of nanocomposites. Therefore, the homogeneous dispersion of nanoreinforcements through suitable processing methods is the key to achieving targeted end properties.

Like the conventional processing of metal alloys, the processing methods for Mg-based metal matrix nanocomposites can be grouped into three major categories, depending on the physical state of the metallic matrix:

- Solid-state processing.
- Liquid state processing.
- Semisolid processing.

Similarly, based on the nature of reinforcement addition, the processing methods can also be classified as either *ex situ* (externally added) or *in situ* (internally formed due to a chemical reaction) processing methods. These categories and the different methods are described in subsequent sections.

2.2 Liquid state processing

Liquid state processing of Mg nanocomposites involves the preparation of a composite slurry by dispersing the nanoscale reinforcements in molten Mg. The prepared composite slurry is then solidified into the required shape. Here, the reinforcement dispersion can be performed in multiple ways such as (i) melt infiltration, (ii) stir casting or compo-casting, and (iii) melt deposition.

2.2.1 Melt infiltration

Melt infiltration involves either the spontaneous or the forced infiltration of a liquid metal alloy into a porous preform containing reinforcements [1]. While spontaneous infiltration (also known as pressureless infiltration) without any external pressure or force is used to process a number of MMCs based on Al–Si, Al–Mg, and Al–Zn alloys with better flowability, the poor wettability between the matrix and reinforcing phase slows down the fabrication process, which results in undesirable reaction products at the interface. In forced infiltration, an external mechanical force or pressure is used to overcome the issues associated with poor wetting and adhesion characteristics.

2.2.1.1 Mechanical infiltration

In mechanical infiltration (figure 2.1), the molten composite slurry is force injected into the mould using an injection barrel [2, 3]. The low cost and high precision capability are the key advantages of this method.

2.2.1.2 Gas pressure infiltration

In this process, the reinforcement preforms are infiltrated using a pressurized inert gas as shown in figure 2.2. Generally, the gas pressure is applied in combination with a vacuum at the other end of the preform to avoid air entrapment and help easy penetration at lower pressures [5]. Since reaction times are relatively shorter in this method, magnesium-based reactive materials can be processed without difficulty compared to other methods.

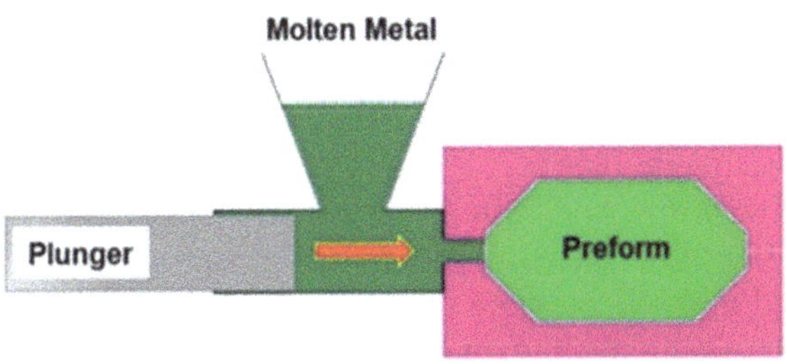

Figure 2.1. Pressure die infiltration process. Reprinted from [4], Copyright 2019, with permission from Elsevier.

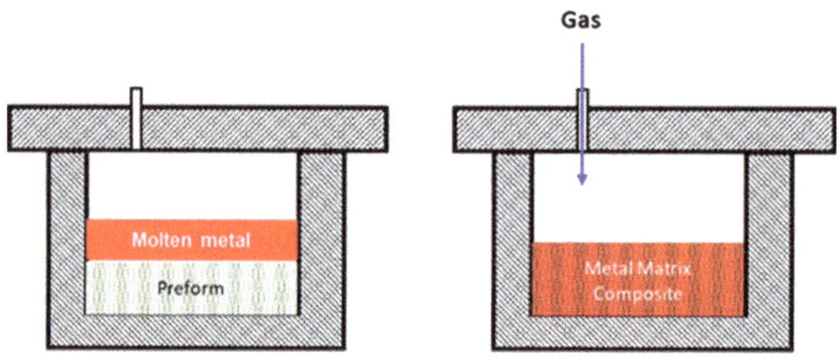

Figure 2.2. Gas pressure infiltration. Reprinted from [4], Copyright 2019, with permission from Elsevier.

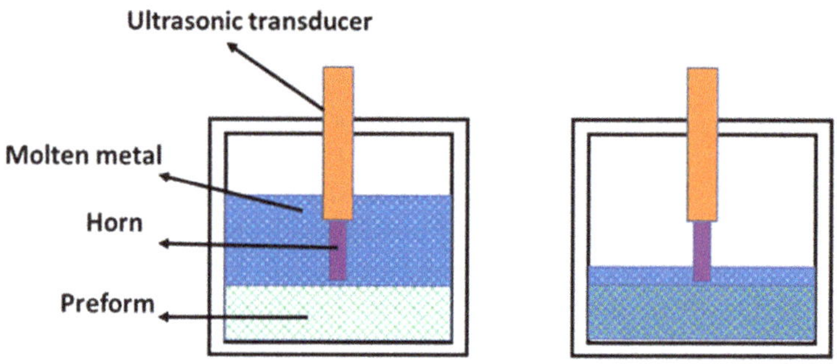

Figure 2.3. Ultrasonic infiltration [6, 8].

2.2.1.3 Ultrasonic infiltration

In this method, the infiltration is achieved using shock waves that are created due to the collapse of acoustic cavitation bubbles (figure 2.3) [6]. In addition to ultrasonic power, parameters such as the horn hole dimension and fabrication speed also affect the quality of the fabricated parts. While past studies showed better results for an optimum hole diameter of 5 mm, an increase in the fabrication speed adversely affected the infiltration ratio [6, 7].

2.2.2 Squeeze casting

Squeeze casting involves forcing the molten metal into a preform with a pressure until the solidification is complete and the mode of pressure application can be either direct or indirect, as shown in figure 2.4. While the pressure is directly applied onto the melt in direct squeeze casting, it is forced through a gate system in indirect squeeze casting that helps to eliminate the oxide residue and offers better control over the melt volume [4]. In general, the molten metal is first injected into a die with a low pressure and then the pressure is gradually increased during the solidification stage. This would prevent any damage to the preforms and eliminates casting defects such as porosity and shrinkage cavities. Since squeeze casting involves fabricated preforms, it can be effectively employed for the infiltration of both fibre and particle-

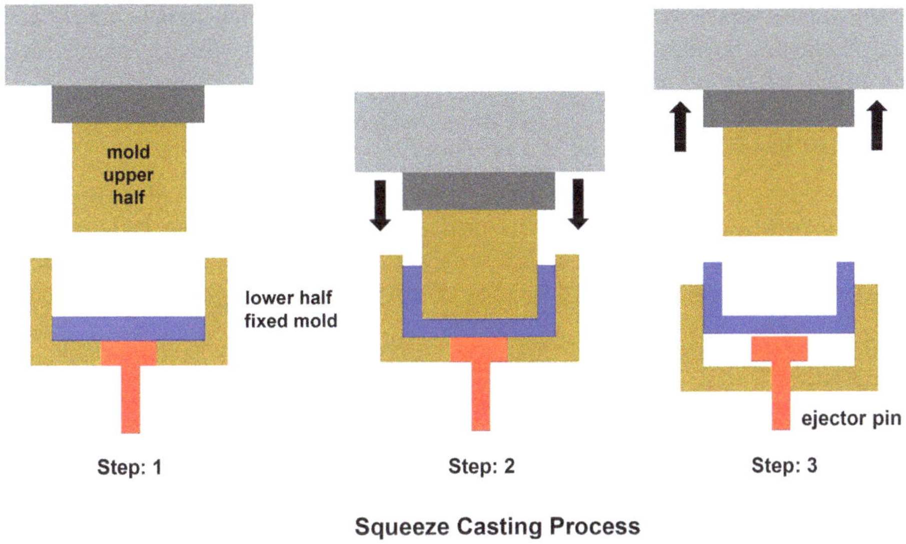

Figure 2.4. Squeeze casting. Reprinted from [4], Copyright 2019, with permission from Elsevier.

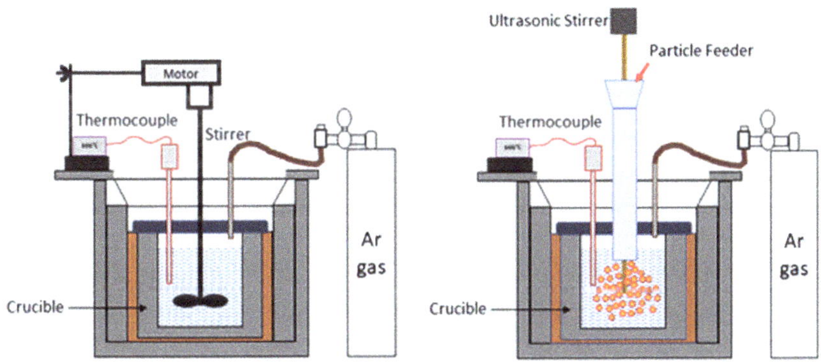

Figure 2.5. Schematic of (a) mechanical stir casting and (b) ultrasonic stir casting. Reprinted from [12], Copyright 2020, with permission from Elsevier.

based reinforcement. One of the major advantages of squeeze casting is that it involves a shorter infiltration time and hence is highly effective for processing reactive metals such as Mg [9, 10].

2.2.3 Stir casting

Stir casting or compo-casting is one of the most common methods used to produce MMCs [1, 11]. Figure 2.5 illustrates the schematic of stir casting methods based on mechanical (shear force) stirring and ultrasonic (acoustic) cavitation to effectively disperse the reinforcement particles in the molten matrix. While the stirring conditions, including the mechanism, stirring speed, duration of stirring, stirrer geometry, and melt temperature, play an important role in the distribution of reinforcement particles, the interfacial bonding between the matrix material and reinforcements is improved by the use of wetting agents which also avoids any unwanted reaction at high temperatures.

2.2.4 Melt deposition

While deposition techniques including immersion plating, electroplating, chemical vapour deposition, and physical vapour deposition are used extensively to produce fibre reinforced MMCs, spray, and disintegrated melt deposition (DMD) methods are increasingly being used in the fabrication of nanocomposites [13].

2.2.4.1 Spray deposition

Figure 2.6 shows a schematic of the spray deposition process in which particle reinforcements are injected into an atomized stream of molten metal to create a layer of MMC on the substrate [14]. Here, the morphology of reinforcement particles plays a major role in facilitating their effective integration into the molten spray. Since spray deposited materials exhibit high porosity, they are often densified by suitable post-processing methods.

2.2.4.2 Disintegrated melt deposition

DMD is one of the most promising methods used to fabricate Mg-based traditional and nanocomposites [15, 16]. Figure 2.7 shows a schematic of the DMD process which inherits the benefits of the stir casting and spray processing methods. Similar to stir casting, this method also involves the preparation of a composite slurry by the vortex mixing of reinforcement particles in a molten matrix. The prepared composite slurry is deposited onto a metallic substrate after disintegration by jets of inert gases to incorporate the benefits of spray processing. Unlike spray deposition, DMD employs a lower impinging velocity to achieve a bulk composite. Hence, it offers the features of (i) the fine grain structure and low segregation of

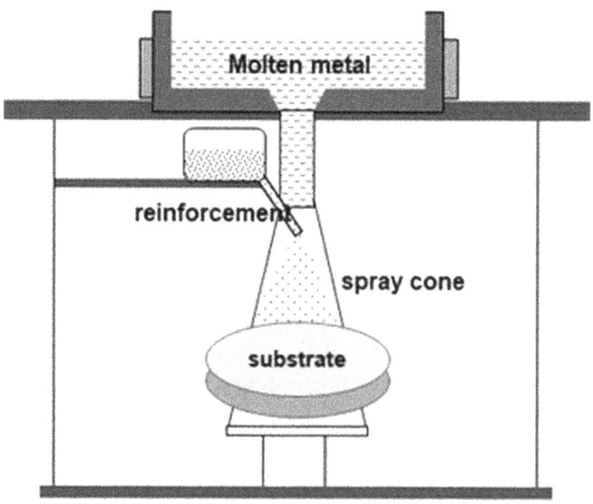

Figure 2.6. Spray deposition. Reprinted from [4], Copyright 2019, with permission from Elsevier.

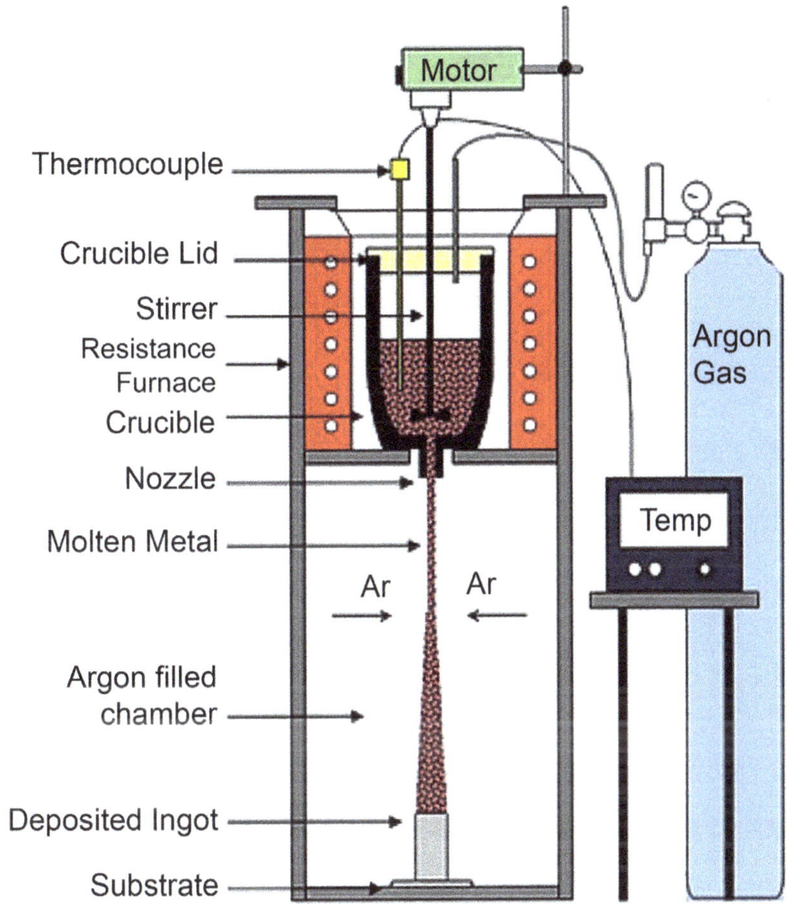

Figure 2.7. Disintegrated melt deposition. Reprinted from [16], Copyright 2019, with permission from Elsevier.

reinforcements of the spray process and (ii) the simplicity and cost effectiveness of the conventional stir cast foundry process.

2.3 Solid-state processing

Solid-state processing methods applicable for magnesium-based nanocomposites can be classified into the following types:

- Powder metallurgy methods based on powder consolidation and sintering.
- Melting or binding based powder additive manufacturing methods.
- Solid-state welding using frictional force.

2.3.1 Powder metallurgy

In general, powder metallurgy methods involve the following steps: (i) the preparation of a composite powder, (ii) the consolidation or joining, and (iii) hot deformation. The composite powder mixture can be prepared by either directly

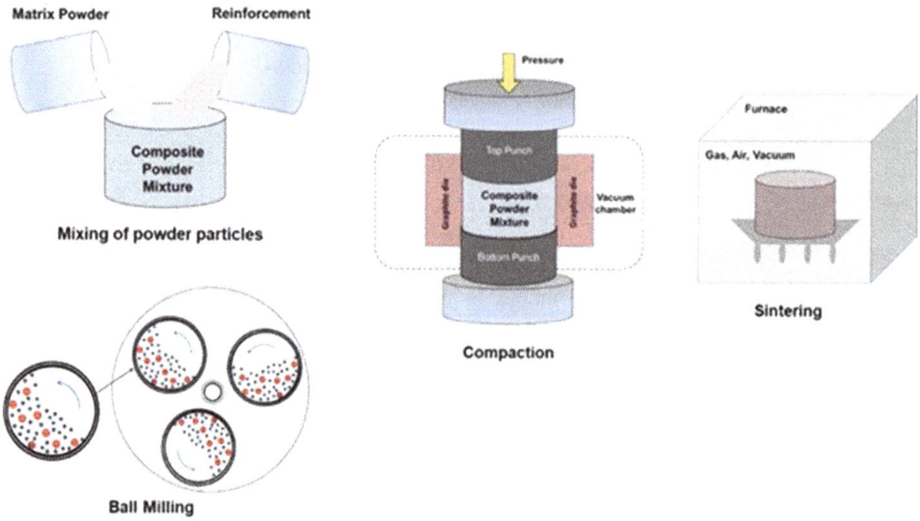

Figure 2.8. Schematic of the powder metallurgy route [16, 20]. Reprinted from [12], Copyright 2020, with permission from Elsevier.

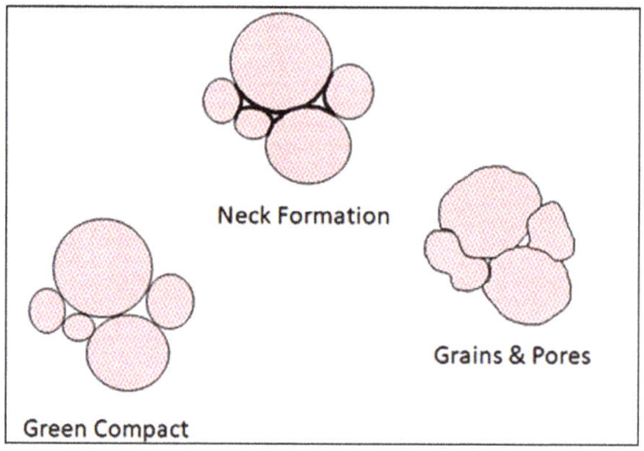

Figure 2.9. Different stages in sintering. Stage 1: Particle bonding in green compact. Stage 2: Growth of contact points into 'necks'. Stage 3: Final microstructure showing grains and pores. Reprinted from [12], Copyright 2020, with permission from Elsevier.

mixing (blending) or ball milling the required amounts of matrix and reinforcement powder materials (figure 2.8). Due to the nature of the high dislocation densities generated during ball milling and the homogenous distribution of reinforcing constituents, the ball milling assisted powder metallurgy method can be effectively applied to develop a range of equilibrium/non-equilibrium alloys and composites with excellent sets of properties. The prepared powder mixture is then compacted and heat treated/sintered for better densification and strengthening (figure 2.8). During sintering, the green compact is heated to a temperature closer to the solidus line of the matrix alloy for a certain period in order to allow atomic diffusion and inter-particle bonding [18, 19]. In most cases, the sintering of the green powder compact facilitates grain recrystallization and the removal of residual binder (figure 2.9).

2.3.1.1 Microwave assisted rapid sintering

Although microwave processing of ceramic materials is well established, the consolidation of metallic powders using microwaves has emerged as a recent advancement [19, 21, 22]. The microwave heating of metallic materials is fundamentally different to conventional heating, as shown in figure 2.10. While conventional heating is based on the transfer of heat from the surface to the interior of the material by conduction, convection, and radiation, microwave heating involves the interaction between the electric and magnetic fields which heats the core of the material. Since microwaves exhibit an inverse temperature distribution, the microwave heating can be substantially quick (~80%) when compared to conventional heating. However, the rapid heating often results in microstructural variations across the thickness which can be minimized by using susceptors.

2.3.2 Additive manufacturing

2.3.2.1 Powder additive manufacturing

Nanoparticle reinforced metal matrix composites can be fabricated using additive manufacturing methods such as binder/binderless jetting and laser melting [24]. While the jetting processes involve the selective application of a glue (in binder

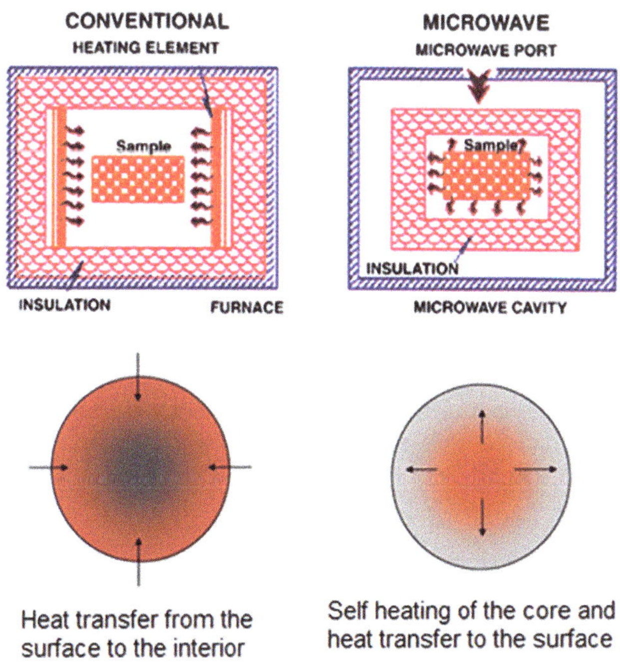

Figure 2.10. Schematic of microwave sintering and the heat transfer principle. Adapted from [23] reproduced with permission of MDPI.

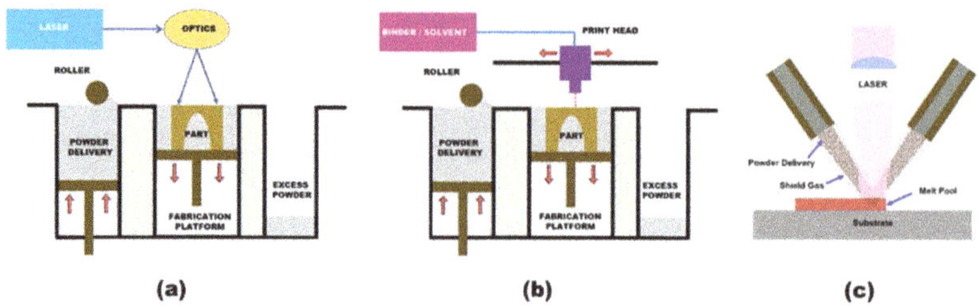

Figure 2.11. Schematic of the (a) SLM, (b) binder/binderless jetting, and (c) laser deposition methods [31].

jetting) or single-phase solvent material (in binderless jetting) to join powdered raw materials, the laser melting methods use one or more lasers to selectively melt or deposit powdered raw materials in a layer-by-layer fashion (figure 2.11) [25, 26]. One of the major advantages of the jetting process is that it does not involve the melting of powder materials and hence is safe for processing Mg-based materials. However, parts fabricated using binder jetting methods always require post-processing steps such as infiltration, sintering, and isostatic pressing, etc, as the products fabricated using this method lack sufficient strength [26].

Based on the powder feed mechanism, laser melting can be classified into two categories: (i) laser powder bed fusion or selective laser melting (SLM) in general and (ii) laser deposition [25]. While SLM involves the spreading of powdered raw material over previous layers, laser deposition uses one or more powder injection nozzles for powder delivery. The advantages of SLM over laser deposition are that the SLM produced parts have greater feature accuracy and resolution. Similarly, laser deposition methods offer faster deposition rates, feature regeneration, and multi-material capabilities.

While composite powder preparation plays an important role in determining the end properties of the bulk composites fabricated using AM methods, some recent reports also highlight the *in situ* formation of hard reinforcement phases in SLM processed materials [27–30].

2.3.2.2 Paste extrusion deposition

In this process, a syringe is used to extrude a paste of powered material onto a base plate which moves relatively to form the desired 3D contours (figure 2.12) [32]. The deposited part is then dried without any heat treatment. This process is particularly valuable for making magnesium implants as the drugs required for the healing process can be added to the paste. However, the part strength is often lower compared to other fusion-based AM processes.

2.3.3 Friction stir processing

Friction stir processing (FSP) is a solid-state welding-based AM method used to fabricate Mg-based composite materials through surface modification (figure 2.13) [33, 34]. The effective dispersion of reinforcements depends on the frictional heating at the interface between a rotating tool and the workpiece (the matrix material).

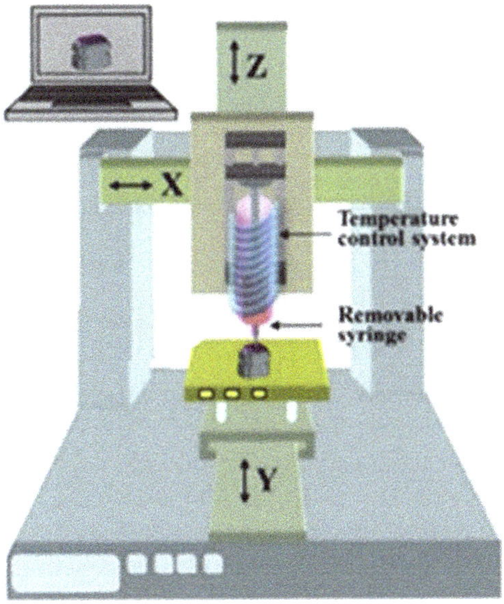

Figure 2.12. Set-up for paste extrusion deposition. Reprinted from [32], Copyright 2014, with permission from Elsevier.

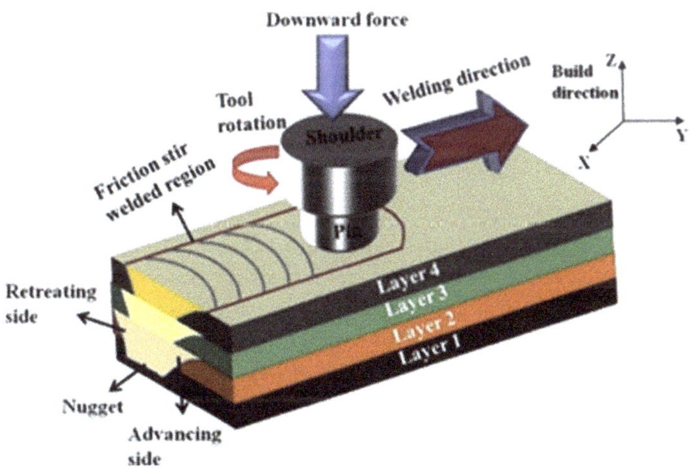

Figure 2.13. Solid-state welding. Reprinted from reference [34], Copyright 2015, with permission from Elsevier.

2.4 Semisolid processing

Semisolid or thixo-processing is carried out between the liquidus and solidus temperatures to combine the advantages of casting and forging [35]. This process works on the basis of the thixotropic fluid property of metals which allows fluid flow upon shearing. It involves the controlled cooling of a molten slurry to a temperature suitable for injection into hardened steels dies. Similar to vacuum die casting,

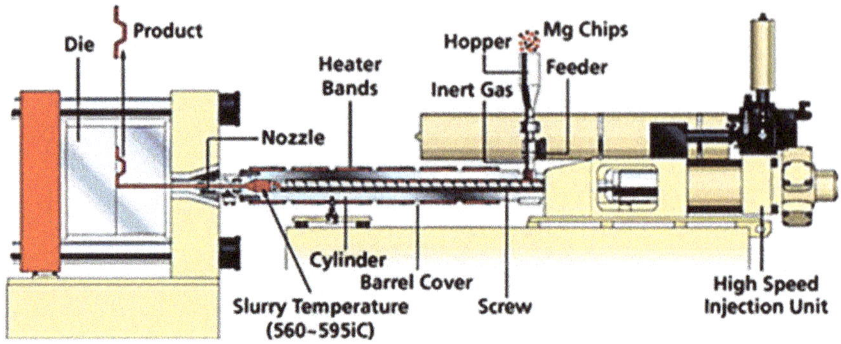

Figure 2.14. Schematic of a thixo-moudling machine. Reprinted from [36], Copyright 2013, with permission from Elsevier.

semisolid processing yields less porosity and improved fatigue strength. As the molten metal can be obtained in several ways, semisolid processing can be tailored and optimized for MMNC. Figure 2.14 shows the schematic of a thixo-moudling machine.

2.5 *In situ* synthesis

In situ synthesis refers to the synthesis of nanoparticles during processing as a result of a sequence of chemical reactions. In the *in situ* synthesis of magnesium nanocomposites, for example, reactions between the matrix and reinforcement occur during processing, leading to the evolution of uniformly distributed nanoparticles, contrary to conventional casting or powder metallurgy where the homogeneous dispersion of nanoparticles in a magnesium matrix is a huge concern as the nanoparticles are incorporated *ex situ*.

Recently, a novel way of synthesizing and processing the nanocomposites was suggested through the *in situ* evolution of the nanoparticles in the matrix [37]. In this approach, the thermodynamic reactions between the matrix and reinforcements and the kinetics of the reactions play a crucial role in the *in situ* evolution. The nanoparticles added to the thermodynamically favourable matrix react with the matrix at processing temperatures such as 750 °C to release new nanoparticles in the matrix, as in the case of the Mg-1.8Y/1.53ZnO nanocomposite. While this work is still in its nascent stages, it is suggested that the control of kinetics and thermodynamics could lead to control of the interface, therefore affecting the microstructure and properties. Thus, *in situ* fabrication of magnesium matrix nanocomposites can be performed by optimizing the combination of alloying and nano-reinforcement additions that satisfy the thermodynamic criterion to obtain maximized benefit in terms of the properties. Further, this route also offers the advantages of uniform distribution of the reinforcement and breaking nanoparticle agglomerates, which are typically the disadvantages of using other processing routes. The following methods can be employed for *in situ* synthesis of nanocomposites.

Table 2.1. Benefits and limitations of processing methods used to fabricate magnesium-based (nano) composites. Reproduced with permission from [38], © Wiley-VCH.

Method	Benefits	Limitations
	Solid–solid processing	
Blending	Capable of using different types and volume fractions of reinforcement. Less interfacial reaction.	Complex parts cannot be fabricated. Ineffective dispersion of reinforcement. Difficulty in removal of binder materials.
Mechanical alloying/ball milling	Effective dispersion of reinforcement. High strength equilibrium and non-equilibrium alloy can be developed. Strengthening due to high dislocation density.	Danger in handling of highly reactive powders. Not suitable for bulk production. Inability to produce complex parts.
Microwave sintering	Rapid process. Energy efficient technology. Interfacial reactions can be controlled.	Reinforcement segregation at larger volume fraction.
	Liquid–solid processing	
Stir casting	Able to produce parts in large quantities. Economical process. Capable of using larger reinforcement quantities (up to 30%).	Reinforcement clustering. Undesirable brittle interfacial reaction products. Increase in the viscosity of molten metal slurry. Damage to reinforcement due to high pressure. Casting defects such as porosity.
Ultrasonic stirring	Improved reinforcement dispersion.	Undesirable interfacial reaction. Scalability.
Disintegrated melt deposition	Combined benefits of stir casting and spray processing. Flexibility of using different types and volume fractions of reinforcement. Effective reinforcement distribution and lesser reinforcement segregation issues. Finer grain structure due to faster cooling rates. No need of flux or greenhouse gases. Scalable.	Incorporation of nanoparticles beyond 5% by weight.
	Semisolid processing (or) thixo-moudling	
Semisolid processing (or) thixo-moudling	Combined benefits of solid and liquid state processing. Completely automated and controlled. Better process efficiency, low porosity, and tighter tolerances.	Higher die development and raw material costs.
	***In situ* synthesis**	
In situ synthesis	Better coherency and bonding between the matrix and reinforcing phases.	Process kinetics need to be controlled carefully to obtain *in situ* reinforcements of preferred chemistry and size distribution.

Reactive alloying element. Elements such as Y, Ca, Th, Sm, etc, which are more reactive than Mg at processing/elevated temperatures can be employed such that they react with reinforcements in the matrix.

Ceramic reinforcements. Ceramic reinforcements such as CeO_2, La_2O_3, ZnO, Al_2O_3, and so on, which dissolve in the matrix and release ions that react with the alloying elements in the matrix to form stable *in situ* nanoparticles.

When a more reactive element such as Y (which forms a stable oxide) is present in the Mg matrix, introducing particles such as ZnO can trigger a sequence of thermodynamically feasible interfacial reactions at elevated temperatures. The novel *in situ* reactions taking place during the processing aid in the uniform and homogeneous dispersion of nanoparticles in the magnesium matrix, which otherwise is not possible in traditional Mg nanocomposites fabricated using conventional routes. By controlling the kinetics of the thermodynamically feasible *in situ* reactions, the industrial viability of Mg nanocomposites can be achieved using standard routes of processing. This work opens a large window for the research community working in the domain of Mg-based materials. Controlling the kinetics, and identifying and optimizing the alloying and reinforcement combinations as well as tailoring the interfaces will be an interesting direction of research and aid in light-weighting the future transportation sector as well as the electronic, biomedical, and sports sectors.

2.6 Summary

Magnesium-based nanocomposites can be fabricated using a variety of processing methods that are broadly classified into the (i) solid-state, (ii) liquid state, and (iii) semisolid processing methods. As the homogeneous distribution of reinforcement particulates is very much essential to achieve the expected property outcomes, each processing method must be carefully assessed for its benefits and limitations. From the published literature, it is clear that the liquid state processing techniques including manual or ultrasonic based stir casting and DMD methods, and solid-state techniques such as powder metallurgy (based on blending, mechanical alloying, and microwave sintering) and FSP are widely used to fabricate Mg-based nanocomposites. Some of the recent literature also explores the feasibility of the semisolid or thixo-moudling process to fabricate defect-free Mg nanocomposites. Table 2.1 lists the individual benefits and limitations of these methods.

References

[1] Kainer K U 2006 *Basics of Metal Matrix Composites* (Weinheim: Wiley)

[2] Cook A J and Werner P S 1991 Pressure infiltration casting of metal matrix composites *Mater. Sci. Eng.* A **144** 189–206

[3] Altinkok N, Demir A and Ozsert I 2003 Processing of Al_2O_3/SiC ceramic cake preforms and their liquid Al metal infiltration *Composites* A **34** 577–82

[4] Garg P, Jamwal A, Kumar D, Sadasivuni K K, Hussain C M and Gupta P 2019 Advance research progresses in aluminium matrix composites: manufacturing and applications *J. Mater. Res. Technol.* **8** 4924–939

[5] Daoud A 2004 Wear performance of 2014 Al alloy reinforced with continuous carbon fibers manufactured by gas pressure infiltration *Mater. Lett.* **58** 3206–13
[6] Pan J, Yang D M, Wan H and Yin X F 1995 Application of ultrasonic infiltration in metal matrix composites *Key Eng. Mater.* **104–107** 275–82
[7] Jiang S, Huang L J, An Q, Geng L, Wang X J and Wang S 2018 Study on titanium–magnesium composites with bicontinuous structure fabricated by powder metallurgy and ultrasonic infiltration *J. Mech. Behav. Biomed. Mater.* **81** 10–5
[8] Dieringa H 2018 Processing of magnesium-based metal matrix nanocomposites by ultrasound-assisted particle dispersion: a review *Metals* **8** 431
[9] Jayalakshmi S, Kailas S V, Seshan S and Fleury E 2006 Properties of squeeze cast Mg-6Zn-3Cu alloy and its saffil alumina short fibre reinforced composites *J. Mater. Sci.* **41** 3743–52
[10] Uozumi H, Kobayashi K, Nakanishi K, Matsunaga T, Shinozaki K, Sakamoto H, Tsukada T, Masuda C and Yoshida M 2008 Fabrication process of carbon nanotube/light metal matrix composites by squeeze casting *Mater. Sci. Eng.* A **495** 282–7
[11] Tzamtzis S, Barekar N S, Hari Babu N, Patel J, Dhindaw B K and Fan Z 2009 Processing of advanced Al/SiC particulate metal matrix composites under intensive shearing—a novel Rheo-process *Composites* A **40** 144–51
[12] Seetharaman S and Gupta M E 2020 *Fundamentals of Metal Matrix Composites* Reference Module in Materials Science and Materials Engineering (Elsevier)
[13] Harrigan W C 1998 Commercial processing of metal matrix composites *Mater. Sci. Eng.* A **244** 75–9
[14] Wood J 1997 Spray atomisation and deposition *Surf. Eng.* **13** 28–30
[15] Gupta M, Lai M O and Soo C Y 1995 Processing-microstructure-mechanical properties of an Al-Cu/SiC metal matrix composite synthesized using disintegrated melt deposition technique *Mater. Res. Bull.* **30** 1525–34
[16] Gupta M and Wong W L E 2015 Magnesium-based nanocomposites: lightweight materials of the future *Mater. Charact.* **105** 30–46
[17] Murugan S, Nguyen Q B and Gupta M 2019 Synthesis of magnesium based nanocomposites *Magnesium—The Wonder Element for Engineering/Biomedical Applications* (London: IntechOpen)
[18] Slotwinski J A, Garboczi E J, Stutzman P E, Ferraris C F, Watson S S and Peltz M A 2014 Characterization of metal powders used for additive manufacturing *J. Res. Natl Inst. Stand. Technol* **119** 460–93
[19] Padmavathi C, Upadhyaya A and Agrawal D 2011 Effect of microwave and conventional heating on sintering behavior and properties of Al–Mg–Si–Cu alloy *Mater. Chem. Phys.* **130** 449–57
[20] Suryanarayana C 2001 Mechanical alloying and milling *Prog. Mater Sci.* **46** 1–184
[21] Gupta M and Eugene W W L 2011 *Microwaves and Metals* (New York: Wiley)
[22] Wong W L E and Gupta M 2007 Development of Mg/Cu nanocomposites using microwave assisted rapid sintering *Compos. Sci. Technol.* **67** 1541–52
[23] Penchal Reddy M, Shakoor R A, Mohamed A M A and Gupta M 2016 Microwave rapid sintering of Al–metal matrix composites: a review on the effect of reinforcements, microstructure and mechanical properties *Metals* **6** 143
[24] Li X P, Ji G, Chen Z, Addad A, Wu Y, Wang H W, Vleugels J, Van Humbeeck J and Kruth J P 2017 Selective laser melting of nano-TiB_2 decorated $AlSi_{10}Mg$ alloy with high fracture strength and ductility *Acta Mater.* **129** 183–93

[25] Herderick E 2011 Additive manufacturing of metals: a review *Mater. Sci. Technol. Conf. Exhib.* **2** 1413–25

[26] Ziaee M and Crane N B 2019 Binder jetting: a review of process, materials, and methods *Addit. Manuf.* **28** 781–801

[27] AlMangour B, Grzesiak D and Yang J M 2017 Selective laser melting of TiB_2/H_{13} steel nanocomposites: influence of hot isostatic pressing post-treatment *J. Mater. Process. Technol.* **244** 344–53

[28] AlMangour B, Grzesiak D and Yang J M 2017 Selective laser melting of TiB_2/316L stainless steel composites: the roles of powder preparation and hot isostatic pressing post-treatment *Powder Technol.* **309** 37–48

[29] AlMangour B, Grzesiak D and Yang J M 2016 Nanocrystalline TiC-reinforced H13 steel matrix nanocomposites fabricated by selective laser melting *Mater. Des.* **96** 150–61

[30] Famodimu O H 2016 Additive manufacturing of aluminium–metal matrix composite developed through mechanical alloying *PhD Thesis* University of Wolverhampton http://hdl.handle.net/2436/620337

[31] Seetharaman S, Krishnan M, Goh Chung Wen F, Khan N A and Ng Ka Lai G 2016 Research updates on the additive manufacturing of nickel based superalloys, solid free. fabr *Proc. 27th Annu. Int. Solid Free. Fabr. Symp.—An Addit. Manuf. Conf.* pp 469–86 http://sffsymposium.engr.utexas.edu/sites/default/files/2016/034-Seetharaman.pdf

[32] Farag M M and Yun H S 2014 Effect of gelatin addition on fabrication of magnesium phosphate-based scaffolds prepared by additive manufacturing system *Mater. Lett.* **132** 111–5

[33] Mishra R S and Ma Z Y 2005 Friction stir welding and processing *Mater. Sci. Eng.* R **50** 1–78

[34] Palanivel S, Nelaturu P, Glass B and Mishra R S 2015 Friction stir additive manufacturing for high structural performance through microstructural control in an Mg based WE43 alloy *Mater. Des.* **65** 934–52

[35] Chen L, Zhao Y, Hou H, Zhang T, Liang J, Li M and Li J 2019 Development of AZ91D magnesium alloy-graphene nanoplatelets composites using thixomolding process *J. Alloys Compd.* **778** 359–74

[36] Luo A A 2013 Magnesium casting technology for structural applications *J. Magnes. Alloy* **1** 2–22

[37] Tekumalla S, Shabadi R, Yang C, Seetharaman S and Gupta M 2017 Strengthening due to the *in-situ* evolution of β_1' Mg-Zn rich phase in a ZnO nanoparticles introduced Mg-Y alloy *Scr. Mater.* **133** 29–32

[38] Gupta M and Seetharaman S 2017 Magnesium based nanocomposites for cleaner transport *Nanotechnology for Energy Sustainability* (Weinheim: Wiley)

IOP Publishing

Magnesium-Based Nanocomposites
Advances and applications
Sankaranarayanan Seetharaman, Sravya Tekumalla and Manoj Gupta

Chapter 3

Tensile and compressive properties of magnesium nanocomposites at room temperature

This chapter summarizes the influence of various nano-scale reinforcements on the room temperature quasi-static mechanical properties of Mg nanocomposites under tensile and compressive loads. Since several different types of nano-scale reinforcements have been used to date, this chapter is systematically divided into multiple sections and subsections based on the type of reinforcement, namely individual nano-scale oxides, carbides and carbon compounds, nitrides, borides, calcium phosphates, and elemental metallic particles. For example, the effects of Al_2O_3 nanoparticles on the tensile and compressive properties of Mg alloys are discussed in the first subsection under oxide nanoparticle reinforced Mg composites.

3.1 Introduction

Although magnesium alloys exhibit excellent weight saving attributes, they suffer from strength loss and creep at working temperatures above 150 °C. However, Mg-based metal matrix composites reinforced with hard and strong ceramic reinforcements in the form of particles or fibres exhibit satisfactory strengths at temperatures up to 250 °C [1–3]. However, the strength retention or increment due to reinforcements occurs at the expense of ductility and fracture toughness. In the last decade, the use of ceramic nanoparticles has been extensively researched to develop Mg-based nanocomposites with enhanced mechanical properties [4]. Interestingly, the increase in strength due to the addition of nano-scale reinforcement is often accompanied by increased ductility [4]. This chapter systematically reviews the state-of-the-art in the processing and mechanical properties of magnesium-based nanocomposites. Since the nano-scale particle reinforcements offer strengthening benefits through grain refinement, the matrix grain characteristics of different Mg nanocomposites are also summarized for each case.

doi:10.1088/978-0-7503-3535-5ch3

3.2 Oxide nanoparticle reinforced magnesium composites

3.2.1 The effect of alumina nanoparticles

Hassan [5] investigated the effects of nano-Al_2O_3 addition on the properties of pure Mg. In this study, the Mg/Al_2O_3 nanocomposites were processed using (i) disintegrated melt deposition (DMD) and (ii) powder metallurgy (PM) methods followed by hot extrusion. While the DMD method involved stirring of a molten composite slurry and its deposition onto a steel mould after disintegration by argon gas, the PM method involved blending, compaction, and sintering of powdered materials. It should be noted that the sintering of Mg/Al_2O_3 nanocomposites in this study was performed in a regular heat treatment furnace. The microstructural and mechanical properties of the DMD and PM processed Mg/Al_2O_3 nanocomposites were investigated in hot extruded condition and the results (as listed in table 3.1) indicated an overall improvement in both strength and ductility of Mg due to finer grain size and the effective dispersion of nano-sized Al_2O_3 particles (figure 3.1).

Mg/Al_2O_3 nanocomposites were also fabricated using the energy efficient microwave sintering based PM method [7]. This study reported an increase in the yield and tensile strength values for all the nanocomposites due to the incorporation of nano-Al_2O_3 particles, while the failure strain was found to improve up to 0.6 vol.% addition of nano-Al_2O_3 particulates (table 3.2).

AZ31B magnesium alloy reinforced with three different volume fractions of nano-Al_2O_3 particulates were developed using the DMD method followed by hot extrusion [8]. In this study, the results of mechanical property measurements (table 3.3) showed an increase in the ductility at the expense of yield and ultimate strengths under tensile loads. On the other hand, the compressive property measurements showed improvement in strengths largely without affecting the compressive fracture strain. Similar studies also report improvement in mechanical properties

Table 3.1. Grain size and tensile properties of PM and DMD processed Mg/Al_2O_3 nanocomposites developed by Hassan and Gupta [5].

Materials	Average grain size (μm)	Tensile yield strength (MPa)	Ultimate tensile strength (MPa)	Ductility (%)
PM processed				
Mg	60 ± 10	132 ± 7	193 ± 2	4.2 ± 0.1
Mg/0.22Al_2O_3	61 ± 18	169 ± 4	232 ± 4	6.5 ± 2.0
Mg/0.66Al_2O_3	63 ± 16	191 ± 2	247 ± 2	8.8 ± 1.6
Mg/1.11Al_2O_3	31 ± 13	194 ± 5	250 ± 3	6.9 ± 1.0
DMD processed				
Mg	49 ± 8	97 ± 2	173 ± 1	7.4 ± 0.2
Mg/0.22Al_2O_3	11 ± 3	146 ± 5	207 ± 11	8.0 ± 2.3
Mg/0.66Al_2O_3	14 ± 4	170 ± 4	229 ± 2	12.4 ± 2.1
Mg/1.11Al_2O_3	14 ± 2	175 ± 3	246 ± 3	14.0 ± 2.4

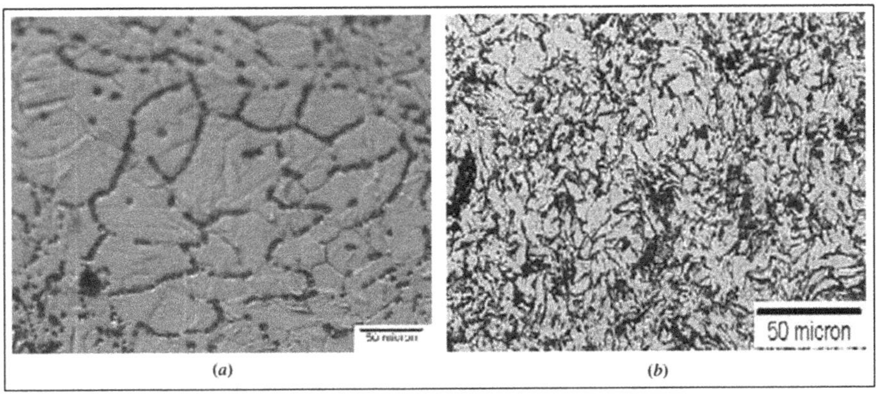

Figure 3.1. Microstructure of DMD processed Mg nanocomposite reinforced with Al_2O_3 nanoparticles. Reproduced from [6], with permission © Springer Nature.

Table 3.2. Grain size and tensile properties of Mg/Al_2O_3 nanocomposites fabricated using the microwave sintering assisted PM method [7].

Materials	Grain size	Yield strength (MPa)	Ultimate strength (MPa)	Ductility (%)
Mg	27 ± 7	116 ± 11	168 ± 10	6.1 ± 2.0
Mg/0.3Al_2O_3	—	119 ± 7	175 ± 8	7.5 ± 0.2
Mg/0.6Al_2O_3	24 ± 4	130 ± 5	180 ± 7	7.4 ± 0.3
Mg/1.0Al_2O_3	15 ± 3	154 ± 5	213 ± 12	6.3 ± 0.4
Mg/1.5Al_2O_3	18 ± 3	148 ± 10	209 ± 7	5.6 ± 0.3

due to the effectiveness of finely dispersed nano-Al_2O_3 particulates (figure 3.2) in the matrices of magnesium alloys [9, 10].

Nano-Al_2O_3 addition (0.5–1.5 vol.%) using DMD was also performed to improve the tensile properties of $MgY_{1.06}Zn_{0.76}Al_{0.42}$ (at.%) alloy [11]. In this study, the microstructure of the alloy (after homogenization heat treatment at 450 °C for 2 h) was composed of very fine Mg–Y–Zn–Al precipitates which together with the nano-Al_2O_3 particles promoted grain refinement by particle stimulated nucleation and dynamic recrystallization. The average grain size values for the alloy and composite formulations reported in this study are listed in table 3.4. The authors also highlighted the fact that the addition of nano-Al_2O_3 particles inhibited the formation of 14H LPSO phases during solidification. From the results of mechanical properties under tensile loads (table 3.4), the fine-grained microstructure and the weak basal texture in the materials was found to improve the strength and ductility of Al_2O_3 reinforced $MgY_{1.06}Zn_{0.76}Al_{0.42}$ alloy nanocomposites. The results of mechanical testing also highlighted the key role of the volume fraction of nano-Al_2O_3 particles. While the highest strength was reported for 0.5 vol.% nano-Al_2O_3 addition, the ductility was found to increase with increasing amounts of nano-Al_2O_3 particles.

Table 3.3. The effect of nano-Al_2O_3 addition on the properties of DMD processed Mg alloys [8–10].

		Tensile properties			Compression properties		
Materials	Average grain size (μm)	Yield strength (MPa)	Ultimate strength (MPa)	Failure strain (%)	Yield strength (MPa)	Ultimate strength (MPa)	Failure strain (%)
AZ31B	5.8 ± 0.4	201 ± 7	270 ± 6	5.6 ± 1.4	133 ± 4	444 ± 10	12.6 ± 1.0
AZ31B/ 0.66Al_2O_3	4.5 ± 0.3	149 ± 7	215 ± 15	14.6 ± 1.1	172 ± 17	468 ± 10	13.4 ± 1.8
AZ31B/ 1.11Al_2O_3	3.9 ± 0.1	148 ± 11	214 ± 16	25.5 ± 2.2	174 ± 2	478 ± 11	13.5 ± 0.8
AZ31B/ 1.50Al_2O_3	3.1 ± 0.03	144 ± 9	214 ± 16	29.5 ± 1.9	176 ± 7	486 ± 5	14.3 ± 1.2
AZ81	6.3 ± 1.2	225 ± 14	336 ± 10	7.9 ± 0.5	157 ± 17	487 ± 14	17.0 ± 0.1
AZ81/ 1.5Al_2O_3	6.1 ± 1.5	212 ± 12	339 ± 11	13.1 ± 1.0	146 ± 11	507 ± 19	20.0 ± 4.1
ZK60A	11.1 ± 3.0	182 ± 4	271 ± 1	6.7 ± 1.0	93 ± 8	498 ± 16	23.2 ± 4.6
ZK60A/ 1.5Al_2O_3	10.7 ± 2.5	175 ± 2	305 ± 2	18.1 ± 0.9	89 ± 8	531 ± 10	33.9 ± 9.3
AZ31-AZ91	5.1 ± 0.7	207 ± 4	316 ± 6	8.0 ± 0.1	117 ± 15	495 ± 13	19.6 ± 1.9
AZ31-AZ91/ 1.5Al_2O_3	4.8 ± 1.0	232 ± 13	339 ± 10	15.9 ± 0.5	123 ± 16	512 ± 3	18.2 ± 0.7

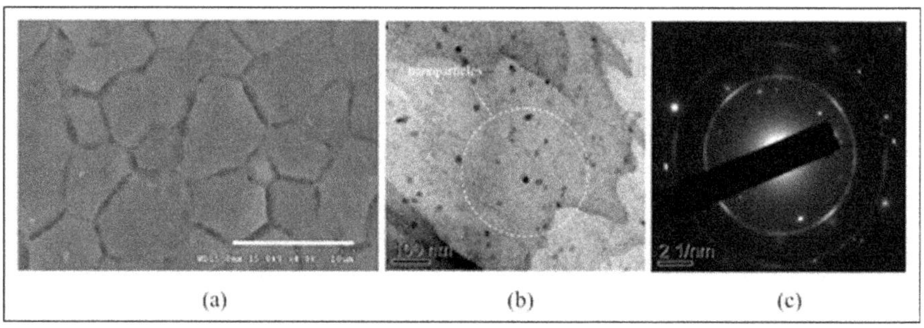

Figure 3.2. (a) Grain microstructure and the (b) distribution of nano-Al_2O_3 particles and fine intermetallics in AZ81/Al_2O_3 nanocomposite. (c) Representative SAED diffraction of Al_2O_3 nanoparticles. Reprinted from [9]. Copyright 2012, with permission from Elsevier.

3.2.2 The effect of bismuth oxide nanoparticles

Using the microwave sintering assisted PM method, Bi_2O_3 nanoparticle reinforced Mg composites were fabricated [12]. In this study, the hot extruded samples displayed an overall improvement in tensile and compressive strength properties while the failure strain was adversely affected under both types of loadings (table 3.5).

Table 3.4. The effects of nano-Al_2O_3 addition on the grain size and tensile properties of $MgY_{1.06}Zn_{0.76}Al_{0.42}$ alloy [11].

Materials	Grain size (μm)	Yield strength (MPa)	Ultimate strength (MPa)	Elongation (%)
$MgY_{1.06}Zn_{0.76}Al_{0.42}$	2.43 ± 1.10	376 ± 17	416 ± 5	10.8 ± 0.6
$MgY_{1.06}Zn_{0.76}Al_{0.42}/0.5Al_2O_3$	5.95 ± 3.16	310 ± 11	355 ± 11	15.6 ± 1.8
$MgY_{1.06}Zn_{0.76}Al_{0.42}/1.0Al_2O_3$	4.88 ± 2.35	260 ± 7	293 ± 1	16.3 ± 0.7
$MgY_{1.06}Zn_{0.76}Al_{0.42}/1.5Al_2O_3$	3.74 ± 1.89	239 ± 7	292 ± 2	17.4 ± 1.3

Table 3.5. Properties of microwave sintered Mg/Bi_2O_3 nanocomposites [12].

		Tensile properties			Compressive properties		
Materials	Grain size (μm)	Yield strength (MPa)	Ultimate strength (MPa)	Failure strain (%)	Ultimate strength (MPa)	Yield strength (MPa)	Failure strain (%)
Mg	20 ± 3	109 ± 09	161 ± 16	8.9 ± 1.1	93 ± 02	246 ± 23	19.4 ± 0.8
$Mg/0.22Bi_2O_3$	10 ± 2	135 ± 12	207 ± 13	7.5 ± 0.3	101 ± 09	341 ± 08	13.9 ± 0.7
$Mg/0.66Bi_2O_3$	11 ± 3	135 ± 13	201 ± 15	6.0 ± 0.5	102 ± 11	387 ± 17	12.0 ± 1.5
$Mg/1.11Bi_2O_3$	10 ± 2	139 ± 13	205 ± 18	4.9 ± 0.6	108 ± 03	411 ± 06	13.6 ± 1.4

3.2.3 The effect of calcium oxide nanoparticles

Mg/CaO nanocomposites containing different amounts of CaO nanoparticles (0–2 wt.%) were synthesized using the DMD method [13]. Microstructural analyses in this study revealed grain boundary pinning and grain refinement due to the presence of Mg_2Ca phases along with CaO nanoparticles. Because of the significant refinement (~77%) in grain size, a tremendous increment in strength was reported upon the progressive additions of CaO nanoparticles (figure 3.3). However, the ductility did not improve beyond 0.5 wt.% CaO addition due to the agglomeration of CaO nanoparticles. With 1 wt.% CaO showing promise in maintaining the strength and ductility of magnesium, Tekumalla *et al* [14] added 1 wt.% CaO to a ductile Mg–1.8Y matrix to realize the advantages of both the matrix and reinforcements in terms of strength and ductility. This resulted in the formation of an *in situ* magnesium nanocomposite as a result of the sequence of reactions during casting and extrusion. The *in situ* formed Y_2O_3 nanoparticles exhibited coherency with the matrix and led to the development of a large amount of elastic and plastic strain fields around them. These nanoparticles and secondary phases (Mg_2Ca and Mg_2Y) were reported as the reason for the extraordinary high tensile strength (~343 MPa) of the nanocomposite (figure 3.4). Further, the high ductility of 30% was attributed to the weak texture provided by the Y and Ca in the matrix.

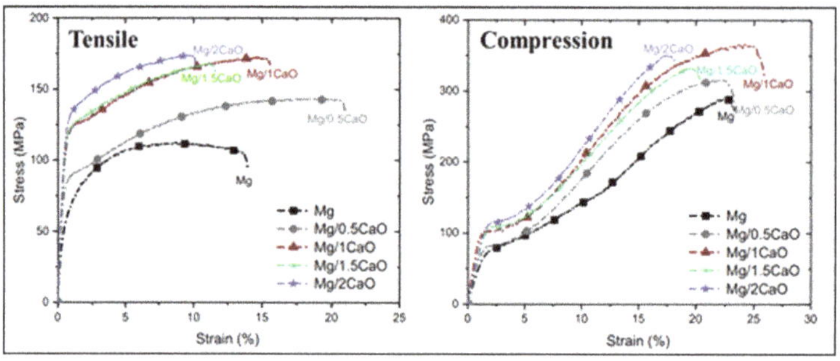

Figure 3.3. Tensile and compressive stress–strain curves of Mg/CaO nanocomposites. Reproduced with permission from [13] © Bentham Science.

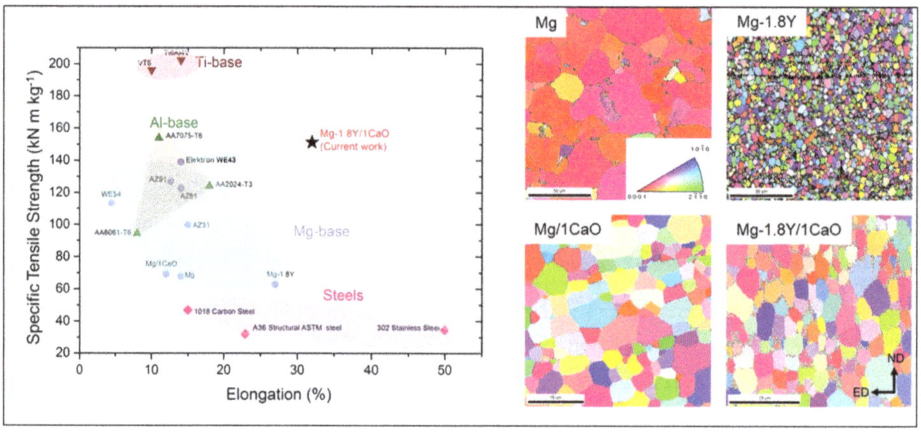

Figure 3.4. Mechanical properties and microstructure of nano-CaO added Mg–1.8Y. Reproduced from [14], with permission © Springer Nature.

3.2.4 Effect of iron oxide nanoparticles

Magnesium nanocomposites with 1–3 wt.% iron oxide (Fe_3O_4) nanoparticles were fabricated using an innovative cold deformation method called turning induced deformation (TID) [15]. Figure 3.5 shows the schematic of this process in which the chips collected from the turning (machining) process were cold compacted in a billet and then hot extruded into rods of preferred diameters. The authors named this process as 'turning induced deformation' as the chips or turnings resulting from a machining process involving extensive plastic deformation (as shown in figure 3.5). For comparison purposes, the authors also tested the properties of DMD processed Mg/Fe_3O_4 nanocomposites for mechanical properties and microstructure. From the results, they confirmed comparable mechanical properties for both DMD and TID processed Mg/Fe_3O_4 nanocomposites which suggest TID as an effective method for the recycling of Mg nanocomposites.

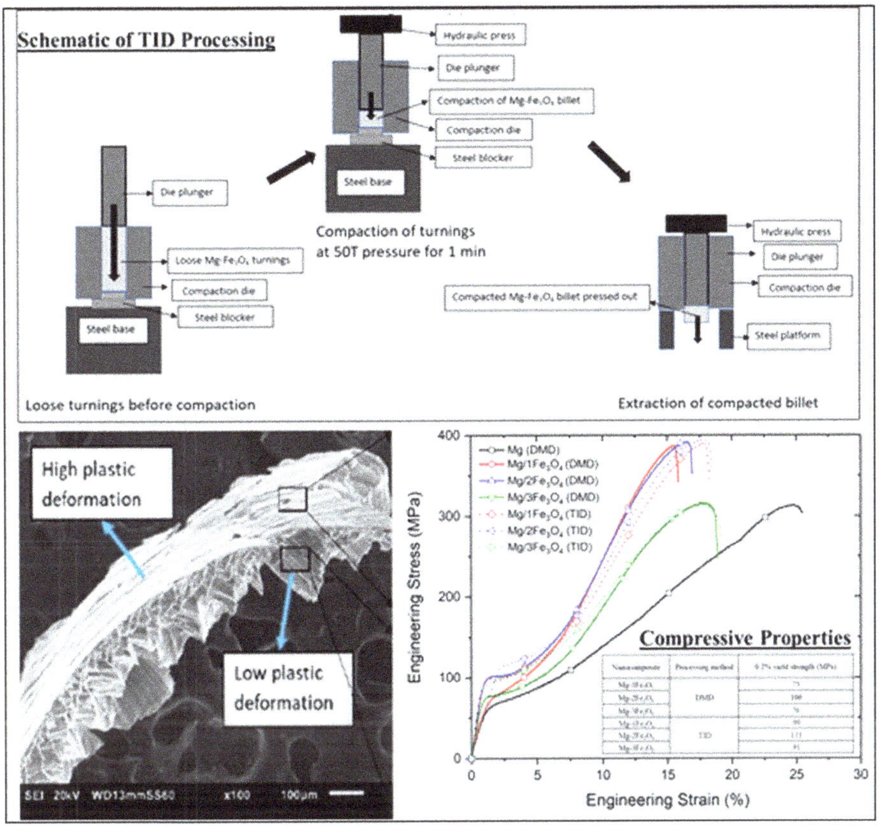

Figure 3.5. Turning induced deformation processing of Mg/Fe_3O_4 nanocomposites. Reproduced from [15], with permission of MDPI.

Table 3.6. Grain size and compressive properties of Mg/Sm_2O_3 nanocomposites [16].

Materials	Grain size (μm)	Yield strength (MPa)	Ultimate strength (MPa)	Failure strain (%)
Pure Mg	25.7 ± 2.6	74 ± 3	249 ± 6	17.4 ± 0.3
Mg/0.5 wt.%Sm_2O_3	21.4 ± 2.3	87 ± 1	285 ± 6	19.8 ± 0.6
Mg/1.0 wt.%Sm_2O_3	17.6 ± 1.8	118 ± 2	331 ± 7	20.1 ± 0.7
Mg/1.5 wt.%Sm_2O_3	12.9 ± 2.2	128 ± 5	395 ± 7	17.2 ± 0.5

3.2.5 The effect of samarium oxide nanoparticles

Mg nanocomposites containing different amounts (0–1.5 wt.%) of Sm_2O_3 nanoparticles were fabricated using a microwave sintering assisted PM method followed by hot extrusion [16]. In this study, the results of microstructural measurements showed a refinement in the matrix grain size due to well dispersed nano-Sm_2O_3 particles in the nanocomposites (table 3.6). Similarly, the mechanical properties under compression loads revealed continuous improvement in the room temperature compressive strengths with increasing amounts of Sm_2O_3 nanoparticles. With respect to failure strain, no significant effects were reported, although a slight improvement occurred for 0.5 and 1.0 wt.% Sm_2O_3 nanoparticle addition (table 3.6).

3.2.6 The effect of silica nanoparticles

Friction stir processing method (as shown in figure 3.6) was used to fabricate (5%–10%) SiO_2 nanoparticle reinforced AZ61 composites [17]. In this study, the results of microstructural studies confirmed the uniform distribution of amorphous SiO_2 nanoparticles in FSP processed specimens. While the size of SiO_2 particles measured around 20 nm after two direction–four pass FSP (2D4P), the grain size was refined from 8 μm to 0.8 μm (figure 3.7). Similarly, the room temperature strength of the AZ61 alloy was improved due to nano-SiO_2 reinforcements and FSP processing (figure 3.7). The authors also highlighted an improvement in the high strain rate super-plasticity of FSP processed AZ61/SiO_2 nanocomposite of over 400%.

Mg nanocomposites containing nano-SiO_2 particles were synthesized using a microwave sintering assisted PM method [18]. The results of tensile testing

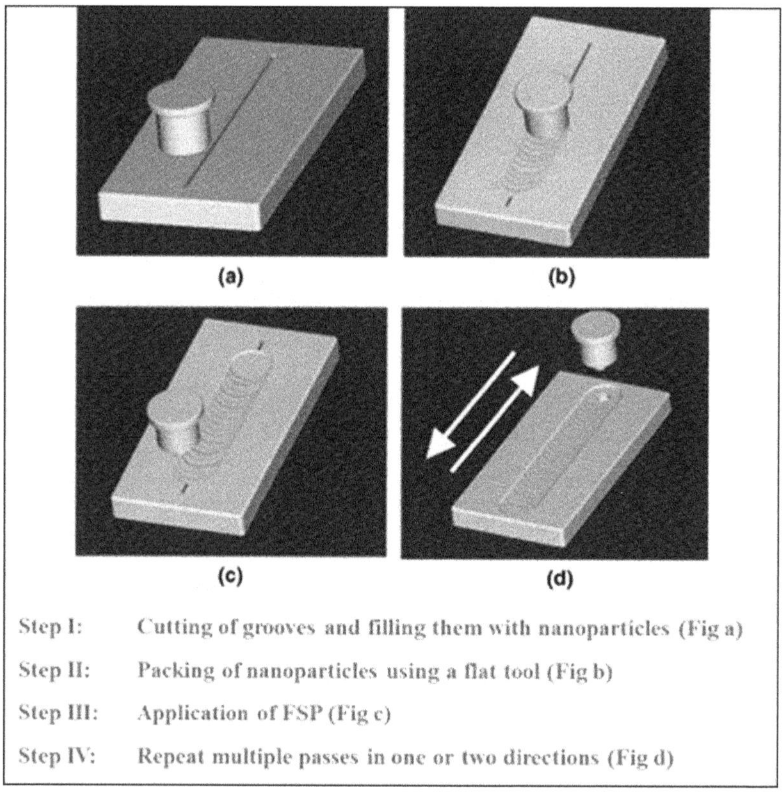

Figure 3.6. FSP procedure adopted to develop AZ61/SiO_2 nanocomposites. Data from [17].

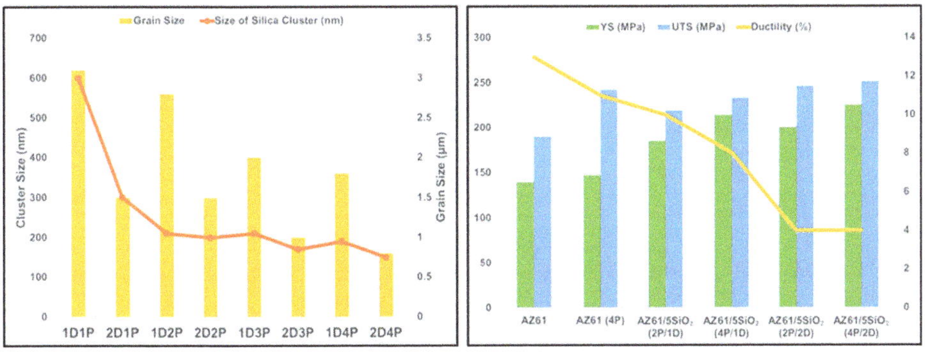

Figure 3.7. Microstructure and tensile properties of FSP processed AZ61/SiO_2 nanocomposites. Based on data from [17].

Table 3.7. The tensile properties of Mg/SiO_2 nanocomposites as reported by [18].

Material	Yield strength (MPa)	Ultimate strength (MPa)	Failure strain (%)
Pure Mg	87 ± 4	128 ± 6	8.7 ± 3.0
Mg/0.5SiO_2	85 ± 2	136 ± 1	10.1 ± 0.7
Mg/1.0SiO_2	94 ± 3	150 ± 2	14.3 ± 2.1
Mg/2.0SiO_2	114 ± 5	171 ± 8	9.5 ± 3.0

conducted on the hot extruded samples showed improvement in strength and ductility due to nano-SiO_2 addition (table 3.7). While the strength properties consistently improved with nano-SiO_2 content, superior ductility was reported for Mg/1SiO_2 which showed ~64% improvement when compared to unreinforced Mg. The theoretical calculations presented in this study identified Hall–Petch strengthening as the predominant strengthening mechanism.

3.2.7 The effect of titanium oxide nanoparticles

Mg-based nanocomposites containing 1.98% and 2.5% TiO_2 nanoparticles were produced using both the DMD and the microwave sintering assisted PM methods [19, 20]. The developed materials were tested for microstructural and mechanical properties in the as-extruded condition. While the microstructure of all the nanocomposites displayed fine grains compared to pure Mg, superior grain refinement was reported for the DMD processed nanocomposites. Similarly, the results of mechanical testing (figure 3.8) showed marginal improvement in the strength values of DMD processed Mg/2.5TiO_2 alongside ductility enhancement up to 50%. While the yield and ultimate strengths of PM processed Mg/TiO_2 nanocomposites were adversely affected under tensile loads, the tensile ductility increment for Mg/1.98TiO_2 was better than that of the DMD processed nanocomposite. The results of texture analysis also confirmed the dependence of both the processing method and the volume fraction of reinforcements.

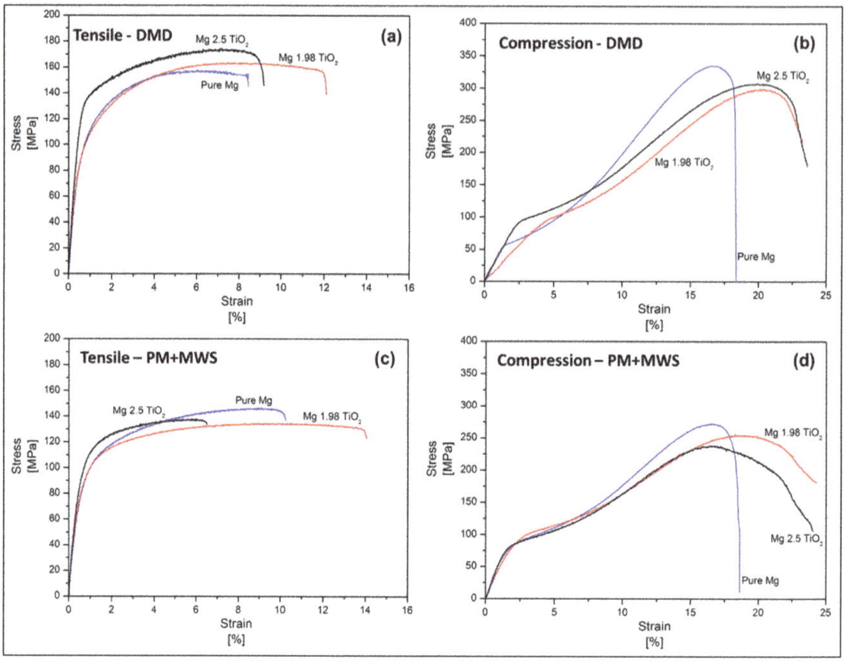

Figure 3.8. Tensile and compressive stress–strain curves of TiO_2 nanoparticle reinforced Mg nanocomposites. Reproduced from [19, 20], with permission © Springer Nature.

Table 3.8. Grain size and tensile properties of Mg/TiO_2 nanocomposites from [19–21].

Material	Synthesis	Reinforcement form	Grain size (μm)	Yield strength (MPa)	Ultimate strength (MPa)	Failure strain (%)
Pure Mg	DMD	—	45 ± 2.4	92 ± 5	156 ± 6	8.2 ± 0.2
	PM	—	32 ± 1.5	89 ± 4.5	142 ± 6	10 ± 0.3
Mg/1.98TiO_2	DMD	Nanoparticles	23 ± 5.5	102 ± 3	165.5 ± 3	11.5 ± 1
	PM		28 ± 1.5	88 ± 10	132 ± 8	14.5 ± 1
		Nanofibers	29 ± 2	80 ± 6	130 ± 9	13.5 ± 2
Mg/2.5TiO_2	DMD	Nanoparticles	21 ± 4	124 ± 8.8	170 ± 6	10 ± 1
	PM		25 ± 2.5	91.1 ± 5	134 ± 7	6 ± 1
		Nanofibers	26 ± 3	88 ± 5	136 ± 2	10 ± 1

The same authors also investigated the effects of reinforcement morphology by comparing the properties of Mg/TiO_2 nanocomposites reinforced with either TiO_2 nanoparticles or TiO_2 nanofibres. In this study, the microstructural studies revealed comparable grain refinement due to both the TiO_2 nanoparticles and the TiO_2 nanofibers (table 3.8). The results of tensile testing (table 3.8) showed an increase in

Table 3.9. Compressive properties of Mg/TiO_2 nanocomposites from [19–21].

Material	Synthesis	Reinforcement form	Yield strength (MPa)	Ultimate strength (MPa)	Fracture strain (%)
Pure Mg	DMD	—	57 ± 3	332 ± 10	18 ± 0
	PM		76 ± 2	275 ± 4	20 ± 1.5
$Mg/1.98TiO_2$	DMD	Nanoparticles	88.3 ± 1	297 ± 1	21.9 ± 1
	PM		89.9 ± 2	245 ± 8	25 ± 2
		Nanofibers	89.9 ± 7.4	300 ± 9	21 ± 0.5
$Mg/2.5TiO_2$	DMD	Nanoparticles	101 ± 9	305.5 ± 11	22 ± 2
	PM		81 ± 0.6	233 ± 6	24 ± 2
		Nanofibers	86 ± 1	294 ± 0.3	24 ± 1

the fracture strain values of both nanoparticle and nanofiber reinforced $Mg/1.98TiO_2$ nanocomposites, without any adverse effects on the strength values. Under compression loads, both nanoparticle and nanofiber reinforced Mg nanocomposites exhibited improved yield strength and fracture strain. However, their ultimate compressive strengths were adversely affected (table 3.9).

3.2.8 The effect of yttria nanoparticles

Magnesium nanocomposites containing nano-sized Y_2O_3 particles were prepared using PM and DMD methods [5]. Upon testing for tensile properties, the Mg/Y_2O_3 nanocomposites showed mechanical property enhancement due to significant grain refinement (table 3.10). While PM processed $Mg/0.22Y_2O_3$ exhibited the best ductility value, the DMD processed $Mg/0.66Y_2O_3$ had superior strength (table 3.10).

Tun and Gupta [22] used the microwave sintering assisted PM method to produce magnesium composites containing 0.2 and 0.7 vol.% nano-sized Y_2O_3 powder (particle size between 30 and 50 nm). The extruded samples developed in this study were investigated for their tensile properties and the results (table 3.11) indicated that the increasing amount of Y_2O_3 content increased the 0.2 yield strength (YS), ultimate tensile strength (UTS), and ductility.

The same authors investigated the influence of the microwave heating rate and extrusion ratio on the properties of $Mg/0.7Y_2O_3$ nanocomposite [23, 24]. Initially, the composites were sintered with two heating rates, 49 and 20 °C min^{-1}, and the sintered billets were then hot extruded into rods with an extrusion ratio of 25:1. Investigation of the mechanical properties revealed an increase in strength when sintered at high heating rates (table 3.11).

To study the effect of extrusion ratios, the nanocomposites sintered at 49 °C min^{-1} were extruded using three different extrusion ratios (12:1, 19:1, and 25:1) and the results of mechanical property measurements conducted on the extruded samples indicated an increase in strength values with increasing extrusion ratio and the same was attributed to the better distribution of Y_2O_3 nanoparticles and the smaller matrix grain size together with the reduced porosity (table 3.11 and figure 3.9).

Table 3.10. Grain size and tensile properties of Mg/Y_2O_3 nanocomposites fabricated by Hassan and Gupta [5].

Materials	Grain size (μm)	Tensile yield strength (MPa)	Ultimate tensile strength (MPa)	Tensile ductility (%)
		PM processed		
Mg	60 ± 10	132 ± 7	193 ± 2	4.2 ± 0.1
$Mg/0.22Y_2O_3$	25 ± 3	156 ± 1	211 ± 1	15.8 ± 0.7
$Mg/0.66Y_2O_3$	13 ± 2	151 ± 2	202 ± 2	12.0 ± 1.0
$Mg/1.11Y_2O_3$	12 ± 3	153 ± 3	195 ± 2	9.1 ± 0.2
		DMD processed		
Mg	49 ± 8	97 ± 2	173 ± 1	7.4 ± 0.2
$Mg/0.22Y_2O_3$	10 ± 1	218 ± 2	277 ± 5	12.7 ± 1.3
$Mg/0.66Y_2O_3$	6 ± 1	312 ± 4	318 ± 2	6.9 ± 1.6
$Mg/1.11Y_2O_3$	6 ± 2	—	205 ± 3	1.7 ± 0.5

Table 3.11. Properties of Mg/Y_2O_3 nanocomposites developed by Tun and Gupta [22–24].

Materials	Heating rate (°C min^{-1})	Extrusion rate	Tensile yield strength (MPa)	Ultimate tensile strength (MPa)	Tensile fracture strain (%)
Mg	49	25:1	134 ± 7	193 ± 2	7.5 ± 2.5
$Mg/0.2Y_2O_3$	49	25:1	144 ± 2	214 ± 4	8.0 ± 2.8
$Mg/0.7Y_2O_3$	49	25:1	157 ± 10	244 ± 1	8.6 ± 1.2
$Mg/0.7Y_2O_3$	20	25:1	111 ± 10	175 ± 8	9.2 ± 0.5
$Mg/0.7Y_2O_3$	49	19:1	105 ± 4	145 ± 4	9.0 ± 0.3
$Mg/0.7Y_2O_3$	49	12:1	90 ± 7	132 ± 2	6.5 ± 0.4

Kumar *et al* [25] investigated the effects of nano-Y_2O_3 addition on the microstructural and mechanical properties of Mg–3Al–2.5La alloy in comparison to Mg–3Al alloy. While the addition of nano-Y_2O_3 reinforcement refined the average grain size, the SEM micrographs of both the Mg–3Al–2.5La alloy and the Mg–3Al–2.5La/1.9Y_2O3 nanocomposite revealed very fine Al_2La, $Al_{2.12}La_{0.88}$, and $Al_{11}La_3$ intermetallic phases, unlike the dominant $Mg_{17}Al_{12}$ phases in Mg–3Al (figure 3.10). The results of mechanical testing showed improved strengths under a compression load, without any adverse effect on failure strain.

3.2.9 The effect of zirconia nanoparticles

Using blend-press-sinter based PM and liquid metallurgy based DMD methods followed by hot extrusion, ZrO_2 nanoparticles were incorporated to improve the

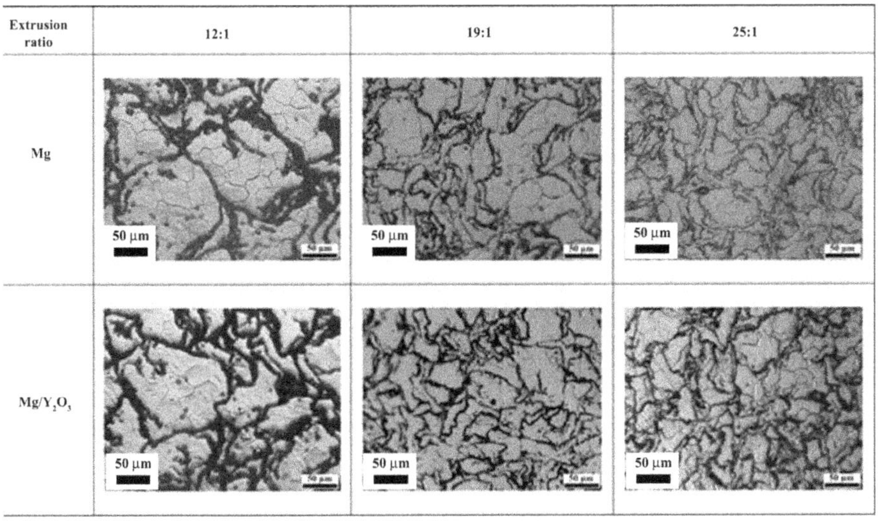

Figure 3.9. Grain morphology of pure magnesium and Mg/Y_2O_3 nanocomposites extruded at different extrusion ratios. Reproduced from [24], with permission © Springer Nature.

mechanical response of Mg [5]. The overall enhancement in mechanical properties obtained in this study was attributed to the fine grain size and other strengthening effects due to different nano-reinforcement addition (table 3.12). Similar improvements were also reported when nano-ZrO_2 particles were added to AZ31 alloy using a friction stir processing method [26].

3.2.10 The effect of zinc oxide nanoparticles

Tun *et al* [27] investigated the properties of Mg-based nanocomposites containing nano-ZnO particles developed using a similar microwave sintering method. The results showed similar tensile and compressive yield strengths and minimal tension–compression yield asymmetry (table 3.13) due to a heterogeneous grain size distribution and resultant weak basal texture (figure 3.11). With regards to the failure strain, significant improvement under tensile loads and a compromise in compressive failure strain were reported. The improvement in tensile fracture strain was attributed to non-basal slip activation and associated texture changes. Nano-ZnO particle reinforced Mg composites were also developed using the DMD method and the characterization of developed materials under tensile and compressive loads showed an overall enhancement in the mechanical properties of Mg due to grain refinement and texture weakening [28].

In a similar study, the addition of nano-ZnO to Mg–3Al–0.4Ce by DMD processing resulted in better yield and ultimate strengths under tensile loads (figure 3.12). While a maximum yield strength of 180 MPa and ultimate tensile strength of 294 MPa was achieved for the Mg–3Al–0.4Ce/2.5ZnO nanocomposite, the strain to fracture remained unaffected [29]. In another study [30], the kinetics

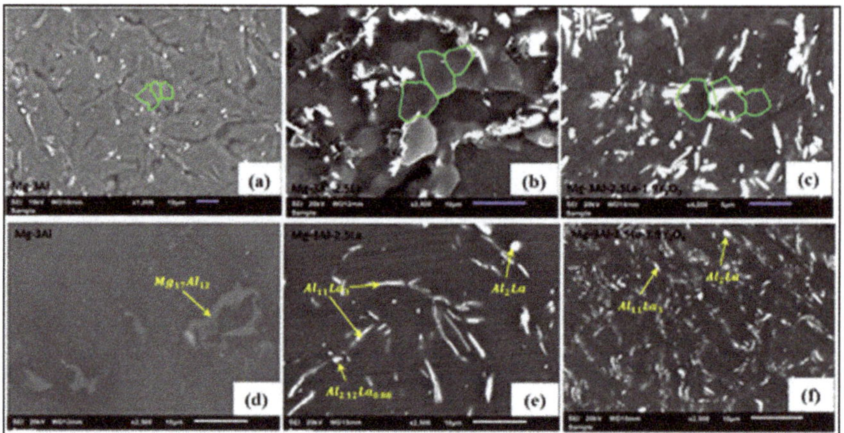

Mg-3Al | Mg-3Al-2.5La | Mg-3Al-2.5La/1.5Y_2O_3

Grain Size and Compressive Properties

Materials	Grain Size (μm)	Compressive Yield Strength (MPa)	Ultimate Compressive Strength (MPa)	Compressive Fracture Strain (%)
Mg-3Al	7.7 ± 1.5	154 ± 2	481 ± 7	24 ± 0.5
Mg–3Al–2.5La	6.6 ± 1.0	141 ± 4	456 ± 3	18 ± 1
Mg–3Al–2.5La/1.9Y_2O_3	3.6 ± 0.5	156 ± 5	520 ± 8	18 ± 0.70

Figure 3.10. Properties of Mg–3Al–2.5La/1.9Y_2O_3 nanocomposites. Reprinted from [25], with permission from MDPI.

during melt processing at 750 °C facilitated the *in situ* formation of the $\beta 1'$ phase and Y_2O_3 which in turn contributed to the grain size reduction and significant strengthening of Mg1.8Y/1.53ZnO nanocomposite. The tensile yield strength and the ultimate tensile strength of the nanocomposite were increased by 124% and 61%, respectively, compared to Mg–1.8Y alloy (figure 3.13). The authors attributed this significant enhancement in strength properties to the grain refinement and the *in situ* evolution of the nano-scale Mg–Zn rich $\beta 1'$ phase.

Using the method of DMD, nano-ZnO addition was also incorporated in Mg–4Zn–3Gd–1Ca alloy. The resultant Mg–4Zn–3Gd–1Ca/2ZnO nanocomposite exhibited exceptional yield and ultimate strength values of 355 MPa and 703 MPa under compressive loads (figure 3.14) [31]. The compressive failure strain was reported as 10.6% which is comparable to that of the conventional Mg alloys. The significant enhancement in compressive strengths was mainly attributed to the

Table 3.12. Grain size and tensile properties of Mg/ZrO_2 nanocomposites from [5, 26].

Materials	Grain size (μm)	Yield strength (MPa)	Ultimate strength (MPa)	Fracture strain (%)
		PM processed		
Mg	60 ± 10	132 ± 7	193 ± 2	4.2 ± 0.1
$Mg/0.22ZrO_2$	13 ± 3	140 ± 3	160 ± 8	6.4 ± 1.5
$Mg/0.66ZrO_2$	13 ± 2	163 ± 3	202 ± 6	11.4 ± 0.9
$Mg/1.11ZrO_2$	11 ± 3	146 ± 1	199 ± 5	10.8 ± 1.3
		DMD processed		
Mg	49 ± 8	97 ± 2	173 ± 1	7.4 ± 0.2
$Mg/0.22ZrO_2$	8 ± 2	186 ± 2	248 ± 4	4.7 ± 0.2
$Mg/0.66ZrO_2$	5 ± 2	221 ± 5	271 ± 6	4.8 ± 0.7
$Mg/1.11ZrO_2$	2 ± 1	216 ± 4	250 ± 6	3.0 ± 0.2
		FSP processed		
AZ31 billet	70	100	160	9
AZ31 (after 4 pass FSP)	6	120	204	18
$AZ31/10ZrO_2$ (after 4 pass FSP)	3	143	232	6
$AZ31/20ZrO_2$ (after 4 pass FSP)	2	167	255	6

Table 3.13. Properties of Mg/ZnO nanocomposites from [27, 28].

		Tensile properties			Compressive properties		
Materials	Grain size (μm)	Yield strength (MPa)	Ultimate strength (MPa)	Fracture strain (%)	Yield strength (MPa)	Ultimate strength (MPa)	Fracture strain (%)
			PM processed				
Mg	20.4 ± 3.4	121 ± 5	179 ± 6	11 ± 1	103 ± 4	263 ± 5	22 ± 2
Mg/0.5ZnO	8.9 ± 4.9	119 ± 9	203 ± 17	16 ± 2	111 ± 8	344 ± 15	14 ± 2
Mg/1.0ZnO	8.8 ± 5.4	125 ± 4	231 ± 13	17 ± 2	106 ± 11	372 ± 15	13 ± 1
Mg/1.5ZnO	8.8 ± 5.1	125 ± 4	229 ± 4	17 ± 2	109 ± 3	368 ± 21	14 ± 1
			DMD processed				
Pure Mg	37 ± 11	95 ± 8	133 ± 7	8.3 ± 2.9	66 ± 8	228 ± 7	22.2 ± 0.9
Mg/0.16ZnO	19 ± 9	119 ± 11	204 ± 9	15 ± 1.4	98 ± 5	284 ± 6	22.3 ± 0.5
Mg/0.48ZnO	22 ± 10	131 ± 6	210 ± 8	16.3 ± 1.4	110 ± 2	308 ± 9	22.5 ± 1.2
Mg/0.8ZnO	16 ± 7	147 ± 9	237 ± 8	11.88 ± 1.7	125 ± 8	345 ± 4	21.7 ± 2.1

grain refinement effects of nano-ZnO particles (figure 3.14). It is also worth noting that the compressive strengths reported in this study exceed most commercial magnesium alloys (i.e. WE43, WE54, ZK60, and ME21) and mild steels (i.e. S275 and S355).

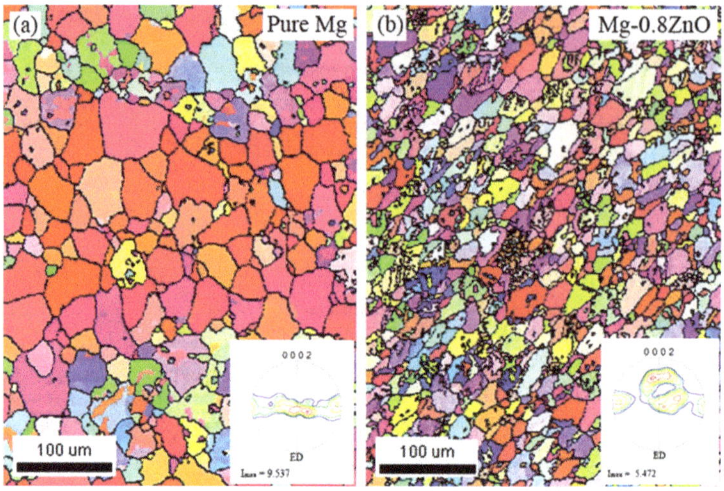

Figure 3.11. EBSD generated microstructure of DMD processed (a) pure Mg and (b) Mg/0.8ZnO nanocomposite showing grain refinement and texture weakening due to nano-ZnO addition. Reprinted from [28], Copyright 2014, with permission from Elsevier.

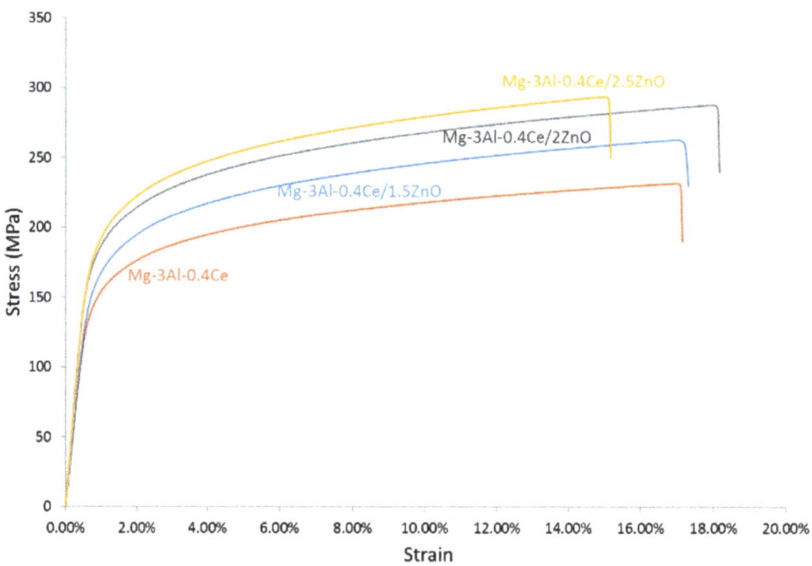

Figure 3.12. Mechanical properties of Mg–3Al–0.4Ce/ZnO nanocomposites. Reprinted from [29], with permission from MDPI.

3.3 Carbide and carbon-based reinforcement containing magnesium nanocomposites

3.3.1 The effect of boron carbide nanoparticles

Mg composites containing B_4C nanoparticles were developed using the PM and DMD methods followed by hot extrusion [32, 33]. While the incorporation of B_4C

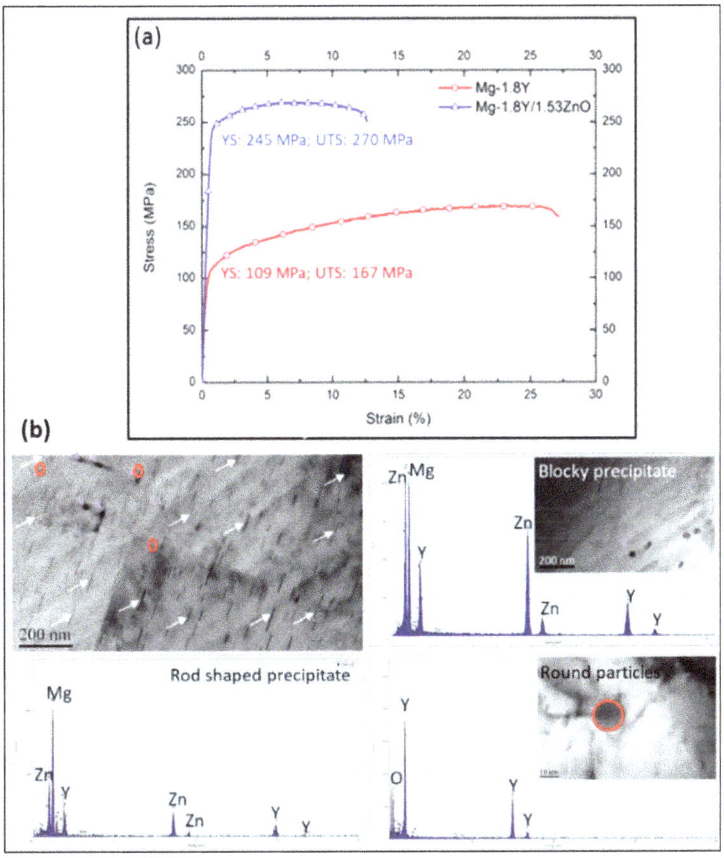

Figure 3.13. (a) Tensile properties of a Mg–1.8Y/1.53ZnO nanocomposite and (b) rod and block shaped phases of the Mg–Zn rich phase along with Y_2O_3 nanoparticles in an extruded Mg–1.8Y/1.53ZnO nanocomposite. Reprinted from [30], Copyright 2017, with permission from Elsevier.

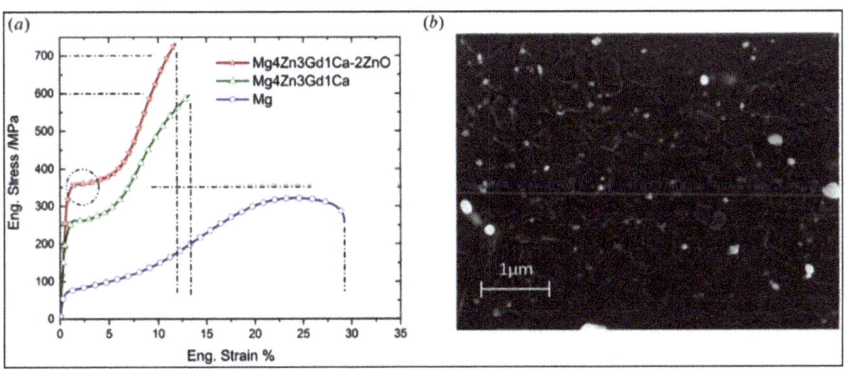

Figure 3.14. Compressive properties and microstructure of Mg–4Zn–3Gd–1Ca/2ZnO nanocomposite. Reproduced from [31], with permission © Springer Nature.

Table 3.14. Properties of B_4C reinforced Mg nanocomposites from [32–34].

		Tensile properties			Compression properties		
Materials	Grain size (μm)	Yield strength (MPa)	Ultimate strength (MPa)	Fracture strain (%)	Yield strength (MPa)	Ultimate strength (MPa)	Fracture strain (%)
			Microwave sintering assisted powder metallurgy method				
Mg	20 ± 3	109 ± 09	161 ± 16	8.9 ± 1.1	93 ± 02	246 ± 23	19.4 ± 0.8
Mg/0.22B_4C	11 ± 2	110 ± 15	159 ± 14	9.9 ± 0.6	97 ± 04	337 ± 14	10.0 ± 1.6
Mg/0.66B_4C	10 ± 3	120 ± 05	164 ± 06	10.0 ± 0.3	100 ± 05	335 ± 11	11.8 ± 1.8
Mg/1.11B_4C	10 ± 2	82 ± 11	119 ± 17	5.5 ± 1.2	105 ± 03	331 ± 10	13.3 ± 1.4
			DMD method				
Mg	28 ± 9	120 ± 9	169 ± 11	06.4 ± 0.7	—	—	—
Mg-0.35B_4C	21 ± 5	127 ± 6	202 ± 6	11.8 ± 1.6	—	—	—
Mg-1.04B_4C	18 ± 9	137 ± 5	215 ± 8	17.4 ± 2.0	—	—	—
Mg-1.74B_4C	11 ± 6	160 ± 2	240 ± 5	12.4 ± 1.7	—	—	—
AZ91-ZK60A	5 ± 1	225 ± 4	321 ± 4	16.1 ± 0.3	106 ± 5	508 ± 17	19.5 ± 1.7
AZ91-ZK60A/ 1.5B_4C	4.2	232 ± 2	345 ± 2	25.1 ± 1.2	140 ± 20	498 ± 2	12.1 ± 1.1

nanoparticles through the PM and DMD methods consistently improved the strength properties of Mg, the ductility values were found to improve only up to a critical volume fraction of nano-B_4C addition (i.e. 0.66 vol.% for PM and 1.04 vol.% for DMD) (table 3.14). Under compression loads, the strength property enhancement in PM processed Mg/B_4C nanocomposites occurred with a drop in compressive failure strain. While the grain refinement and the dispersion strengthening effects due to nano-B_4C addition improved the strength properties, the ductility enhancements upon nano-B_4C addition was mainly attributed to the grain refinement and the texture weakening effects that facilitate the activation of multiple basal/non-basal slip modes (figure 3.15). Similar improvement in strength properties and tensile ductility at the expense of compressive failure strain were also reported when 1.5% B_4C was added to a hybrid (AZ91-ZK60) alloy using the DMD method [34].

Nano-B_4C particles were added to Elektron 21 (E21) alloy using the method of DMD followed by hot extrusion [35]. Microstructural characterization of E21/B_4C nanocomposites exhibited grain refinement with the progressive addition of nano-sized boron carbide particles. The evaluation of mechanical properties under compression loads indicated significant improvement in the yield strength of the E21 alloy due to nano-B_4C addition (figure 3.16) which was attributed to the refined grain size and the uniform distribution of nano-B_4C particles.

3.3.2 The effect of silicon carbide nanoparticles

Using PM and mechanical alloying methods, Ferkel and Mordike [36] prepared sub-microcrystalline magnesium composites containing nano-SiC particles. The results

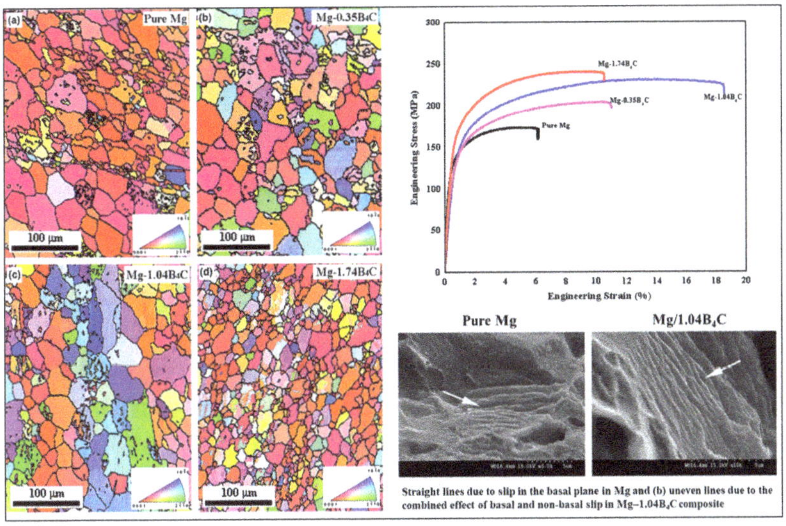

Figure 3.15. Microstructure and tensile properties of DMD processed Mg/B$_4$C nanocomposites. Reprinted from [32], Copyright 2014, with permission from Elsevier.

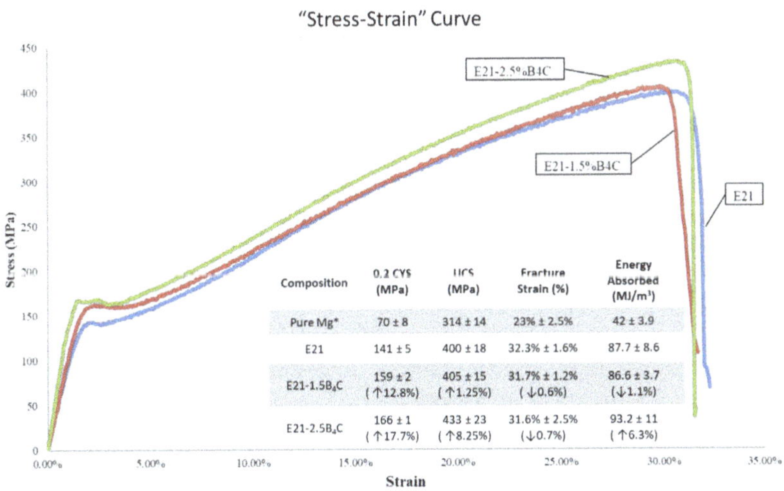

Figure 3.16. Properties of Elektron 21/B$_4$C nanocomposites. Reprinted from [35], with permission from MDPI.

of tensile tests (figure 3.17) performed on the extruded rods of ball milled composites revealed excellent strength (at the expense of ductility) due to smaller grain size (Hall–Petch strengthening). However, the composite prepared by blending and mixing showed a slight reduction in the strength alongside adverse effects on ductility.

Wong and Gupta [37] employed the microwave sintering method to produce nano-sized SiC particle reinforced Mg nanocomposites. In this study, magnesium composites containing different volume percentages of nano-sized SiC particles were fabricated using a microwave sintering aided PM method followed by hot extrusion.

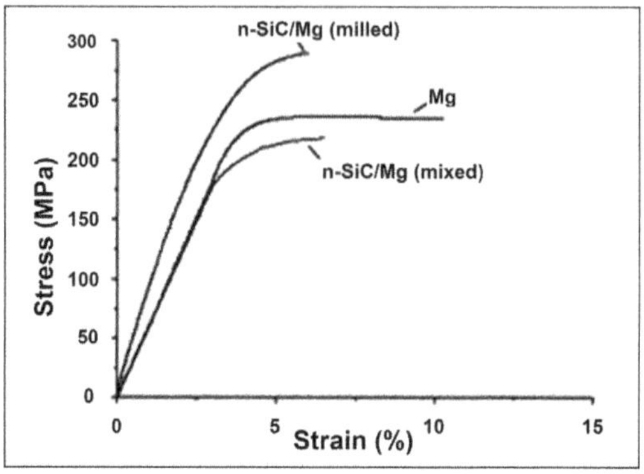

Figure 3.17. Properties of Mg–SiC nanocomposites developed by Ferkel and Mordike. Reprinted from [36], Copyright 2001, with permission from Elsevier.

Table 3.15. Tensile properties of Mg/SiC nanocomposites as reported by [37].

S. no.	Materials	Yield strength (MPa)	Ultimate strength (MPa)	Fracture strain (%)
		Direct extrusion without microwave sintering		
1	Mg	106 ± 7	160 ± 8	5.8 ± 1.8
2	Mg/0.35SiC	116 ± 12	169 ± 17	5.2 ± 1.4
3	Mg/0.5SiC	107 ± 10	161 ± 11	6.5 ± 0.2
4	Mg/1.0SiC	125 ± 2	181 ± 4	6.1 ± 0.9
		Microwave sintered		
1	Mg	125 ± 15	172 ± 12	5.8 ± 0.9
2	Mg/0.35SiC	132 ± 14	194 ± 11	6.3 ± 1.8
3	Mg/0.5SiC	144 ± 12	194 ± 10	7.0 ± 2.0
4	Mg/1.0SiC	157 ± 22	203 ± 22	7.6 ± 1.5

The properties of extruded Mg/SiC nanocomposites were evaluated with particular emphasis on the effectiveness of microwave sintering. For this purpose, the properties of the composites were also investigated under extrusion conditions without microwave sintering. The results (table 3.15) revealed: (a) simultaneous improvement in microhardness, 0.2% yield strength (0.2%YS), UTS, and ductility due to nano-SiC addition, and (b) superior overall mechanical response of microwave sintered composites when compared to their respective un-sintered counterparts. Amongst the microwave sintered composites, Mg/1SiC showed the best overall improvement with an increase of ~10% in microhardness of the matrix, ~26% increase in 0.2% YS, ~18% in UTS, and ~31% increase in ductility.

Nie *et al* [38] investigated the influence of extrusion temperature and process parameters on the microstructure and tensile properties of Mg/SiC nanocomposites fabricated using an ultrasonic vibration assisted semisolid stirring method. The developed SiCp/AZ91 nanocomposites were first homogenized at 415 °C for 24 h (T4 treatment) before extrusion at different temperatures (250 °C, 300 °C, and 350 °C) with an extrusion ratio and speed of 12:1 and 15 mm s^{-1}, respectively. Microstructural analysis showed gradual refinement in the average grain size of the SiC/AZ91 nanocomposite with increasing amounts of SiC nanoparticles. Further, an increase in extrusion temperature from 250 °C to 350 °C also refined the average grain size of the SiC/AZ91 nanocomposite. Reduced stirring time also reduced the average grain size due to the improved distribution of SiC nanoparticles. The results of mechanical property testing (figure 3.18) also showed gradual improvement in the yield and ultimate tensile strength of the nanocomposites with increasing extrusion temperatures and reduced stirring times.

Cao *et al* [39, 40] also used the ultrasonic cavitation method to disperse nano-SiC particles in pure Mg and Mg–6Zn alloy. In these studies, the addition of nano-SiC particles reduced the grain size (figure 3.19) and improved the strengths without significant adverse effects on the ductility (table 3.16). However, De Cicco *et al* [41] reported ductility enhancement alongside strengthening in the case of Mg–4Zn, Mg–6Zn, and Mg–8Zn alloys due to the ultrasonically dispersed nano-SiC particles

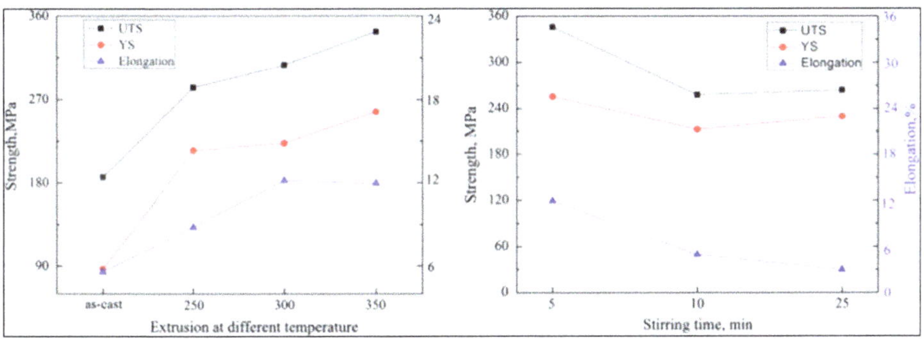

Figure 3.18. The effect of extrusion temperature and stirring time on the mechanical properties of AZ91/SiC nanocomposites as reported by Nie *et al*. Reprinted from [38], Copyright 2012, with permission from Elsevier.

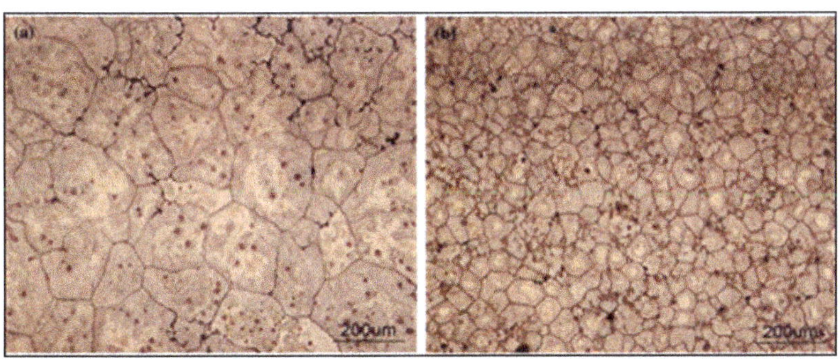

Figure 3.19. Microstructure of (a) Mg–6Zn and (b) Mg–6Zn/1.5%SiC nanocomposites fabricated by ultrasonic cavitation-based solidification processing. Reproduced Cao *et al* [40], with permission © Springer Nature.

Table 3.16. Tensile properties of nano-SiC added pure Mg, Mg–Zn, and AZ91 alloys. From [39–43].

Material	Yield strength (MPa)	Ultimate strength (MPa)	Fracture strain (%)	References
Mg	20	89.6	14.0	Cao *et al* [39]
Mg/0.5 wt.%SiC	28.3	120.7	15.5	
Mg/1.0 wt.%SiC	30.3	124.1	14.2	
Mg/2.0 wt.%SiC	35.9	131.0	12.6	
Mg/4.0 wt.%SiC	47.6	106.9	5.5	
Mg–4Zn	44 ± 2	112 ± 14	5 ± 1	De Cicco *et al* [41]
Mg–4Zn/1.5 wt.%SiC	67 ± 4	199 ± 6	10 ± 1	
Mg–6Zn	51 ± 4	136 ± 19	5 ± 1	
Mg–6Zn/1.5 wt.%SiC	79 ± 5	194 ± 15	7 ± 1	
Mg–8Zn	81	152	3	
Mg–8Zn/3.0 wt.%SiC	111	222	7	
Mg–6Zn (T5)	125	200	3.5	Cao *et al* [40]
Mg–6Zn/1.5 wt.%SiC (T5)	165	230	3.0	
AZ91	100	175	3	Wang *et al* [42]
AZ91/0.1 wt.%SiC	120	190	5	
AZ91/0.3 wt.%SiC	120	200	5.8	
AZ91/0.5 wt.%SiC	120	210	6.2	
AZ91	65	133	2.2	Shiying *et al* [43]
AZ91D/0.5 wt.%SiC	140	180	2.1	
AZ91D/1.0 wt.%SiC	141	191	2.4	
AZ91D/2.0 wt.%SiC	117	137	1.3	

(table 3.16). Similar results were also reported by Wang *et al* [42] when nano-SiC particles were added to the AZ91 alloy. Shiying *et al* [43] used (0.5–2.0 wt.%) SiC nanoparticles to reduce the average grain size of AZ91D alloy. With respect to tensile properties, an improvement in tensile and yield strength was reported. However, the ductility was affected when more than 1 wt.% nano-SiC particles were added (table 3.16).

3.3.3 The effect of titanium carbide nanoparticles

Paramsothy *et al* [44] used the DMD method to fabricate ZK60A/TiC nanocomposites. In this study, although the addition of nano-scale TiC particles reduced the volume fraction of intermetallic phases and improved the strengths and ductility of ZK60A alloy, the yield strength was found to decrease under compression loads (figure 3.20). However, the ultimate strength and fracture strain values remained significantly improved. The decrease in compressive yield strength (and the increased

Material	Grain Size (μm)	Tensile Properties			Compressive Properties		
		Yield Strength (MPa)	Ultimate Strength (MPa)	Failure Strain (%)	Yield Strength (MPa)	Ultimate Strength (MPa)	Failure Strain (%)
ZK60	8.9 ± 2.0	163 ± 3	268 ± 3	6.6 ± 0.6	128 ± 11	522 ± 11	19.6 ± 0.9
ZK60A/1.5TiC	7.0 ± 1.6	184 ± 2	309 ± 3	11.6 ± 1.4	106 ± 7	577 ± 6	25.3 ± 1.6

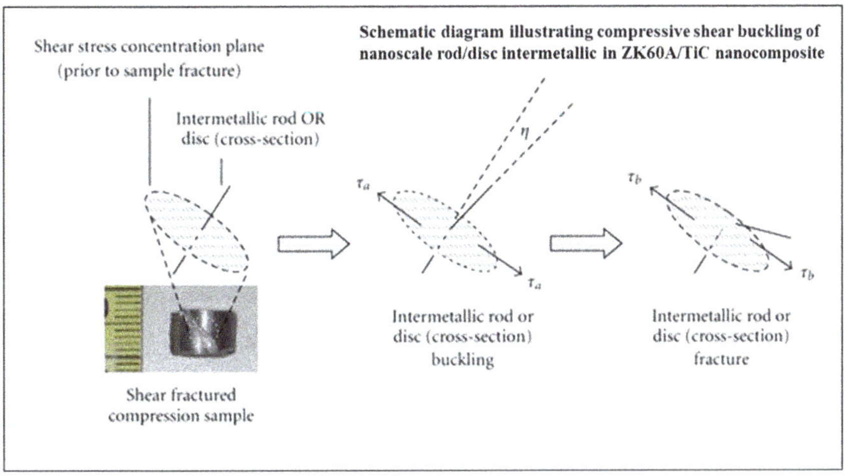

Figure 3.20. Compressive properties and the schematic of compressive shear buckling of nano-rod/disc intermetallics in ZK60A/TiC nanocomposite. Reprinted from [44], with permission of Hindawi.

yield asymmetry) was attributed to the compressive shear buckling of nano-scale rod and disc shaped Mg–Zn intermetallics (figure 3.20).

3.3.4 The effect of carbon nanotubes

There is a growing interest in carbonaceous nanomaterials, such as graphene and carbon nanotubes (CNTs), as reinforcements in MMCs. In general, CNTs are grouped into single-walled (SWCNTs) and multi-walled (MWCNTs). While SWCNTs are made by wrapping a sheet of carbon atoms into a seamless cylinder with a diameter of 1–2 nm, MWCNTs consist of multiple concentric single walls with diameters of 4–20 nm (schematic shown in figure 3.21).

Although graphene and CNTs have unique physical and mechanical properties for use as reinforcements in MMNC, there are limitations due to their agglomeration tendency and poor wetting behaviour. Over the last decade, several processing methods including PM, mechanical alloying, thermomechanical processing, metal casting, and additive manufacturing have been employed to produce graphene or CNT reinforced MMNCs.

CNT [46] reinforced Mg composites with enhanced strength and ductility were developed using the microwave sintering based PM method. The improved mechanical properties of Mg/CNT nanocomposites (as listed in table 3.17) were

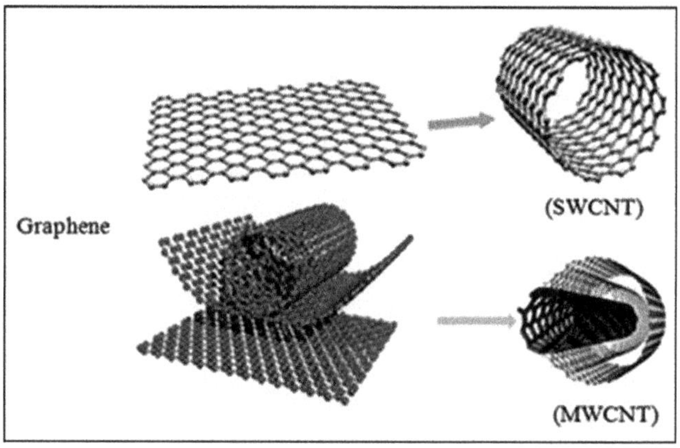

Figure 3.21. Graphene, SWCNT, and MWCNT structures. Reprinted from [45], with permission of Fronteirs.

Table 3.17. Tensile properties of CNT reinforced pure Mg and AZ31 alloy nanocomposites from [46, 48].

Materials	Yield strength (MPa)	Ultimate strength (MPa)	Ductility (%)
Mg	127 ± 5	205 ± 4	9 ± 2
Mg/0.06 wt.%CNT	133 ± 2	203 ± 1	12 ± 1
Mg/0.18 wt.%CNT	138 ± 4	206 ± 7	11 ± 1
Mg/0.30 wt.%CNT	146 ± 5	210 ± 6	8 ± 1
AZ31	188 ± 5.2	280 ± 4.5	14.4 ± 1.2
AZ31/0.5 wt.%CNTs	215 ± 3.5	291 ± 3.8	13.3 ± 1.8
AZ31/1.0 wt.%CNTs	235 ± 4.8	293 ± 5.5	10.9 ± 1.1
AZ31/2.0 wt.%CNTs	241 ± 4.3	297 ± 3.6	9.2 ± 2.0
AZ31/4.0 wt.%CNTs	228 ± 6.5	283 ± 5.8	6.9 ± 2.8

attributed to the strong adherence between the Mg matrix and the CNT reinforcement (figure 3.22). In a similar study, PM processed Mg/CNT nanocomposites were also tested for compressive properties along the plane normal and in-plane orientations [47]. In this study, the increase in CNT amounts improved the yield strength until a critical CNT weight percentage and then decreased. When the CNT percentage was less than or equal to 0.5%, twinning operated predominantly in the in-plane samples. However, twinning did not operate in all plane normal samples and the in-plane samples with 1% or higher CNT content. Overall, the yield strengths were improved by 35%–129% compared to that of pure Mg. Comparison with a theoretical model incorporating the Orowan and thermal expansion mismatch strengthening confirmed their effective contribution in improving the yield strength of Mg/CNT nanocomposites. Zhou *et al* [48] also used a PM method to fabricate CNT reinforced AZ31 alloy nanocomposites with improved strengths and a compromise in ductility. The results of microstructural analysis

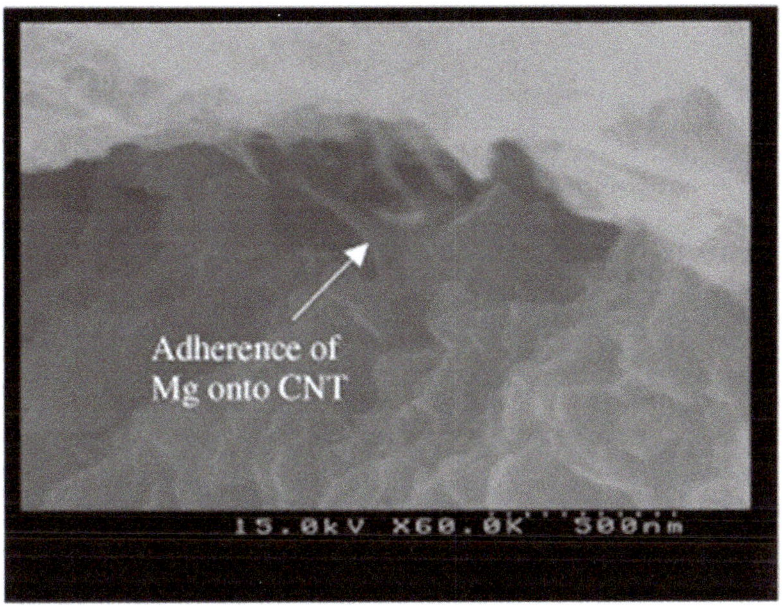

Figure 3.22. Microstructure of Mg/CNT nanocomposite showing good adherence between the Mg matrix and CNT. From [46], © IOP Publishing. Reproduced with permission. All rights reserved.

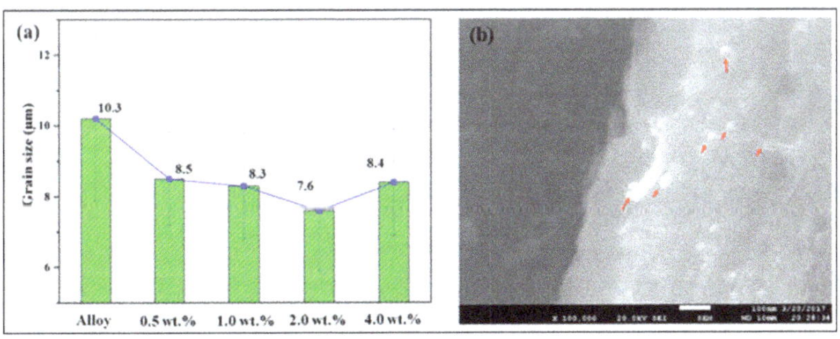

Figure 3.23. (a) Matrix grain size and (b) Mg–CNT interfacial characteristics of AZ31/CNT nanocomposites. Reprinted from [48], with permission from MDPI.

(figure 3.23) conducted on the extruded rods of nanocomposites showed a slightly smaller grain size due to the uniform distribution of CNTs.

Improvements in tensile failure strain due to CNTs were reported for DMD processed AZ81 and ZK60A alloys (table 3.18). However, the strength properties remained largely unaffected [49, 50]. Under compression loads, while CNT addition improved the failure strain of ZK60A, no significant changes were reported for the AZ81 alloy.

Hou *et al* [51] used a PM method to synthesize MWCNT reinforced Mg–9Al nanocomposites. Based on the results of microstructural analysis performed in this study, the authors confirmed the efficient dispersion of CNTs without any structural

Table 3.18. Properties of CNT reinforced AZ81 and ZK60 Mg nanocomposites from [49, 50].

		Tensile properties			Compressive properties		
Materials	Grain size (μm)	Yield strength (MPa)	Ultimate strength (MPa)	Fracture strain (%)	Yield strength (MPa)	Ultimate strength (MPa)	Fracture strain (%)
AZ81	6 ± 1	225 ± 14	336 ± 10	7.9 ± 0.5	157 ± 17	487 ± 14	17.0 ± 0.1
AZ81/1.5CNT	6 ± 1	209 ± 9	328 ± 4	13.7 ± 2.2	129 ± 19	488 ± 13	16.0 ± 1.8
ZK60A	9 ± 2	163 ± 3	268 ± 3	6.6 ± 0.6	128 ± 11	522 ± 11	19.6 ± 0.9
ZK60A/1.0CNT	3 ± 1	180 ± 6	295 ± 8	15.0 ± 0.7	110 ± 7	547 ± 3	33.2 ± 6.2

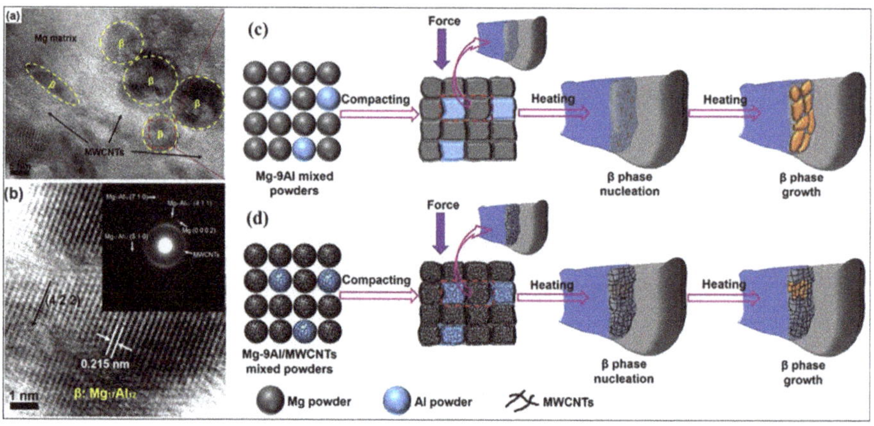

Figure 3.24. HRTEM microstructure of (a) Mg–9Al/0.4CNT composite, (b) enlarged view of a selected region with the SAED pattern in the inset. Schematic of $Mg_{17}Al_{12}$ formation in (c) Mg–9Al alloy and Mg–9Al/CNT nanocomposites. Reprinted from [51], Copyright 2019, with permission from Elsevier.

damage and a defect free matrix–CNT interface which together refined the size of $Mg_{17}Al_{12}$ intermetallics from the micron- to nano-lengthscale (figure 3.24). With respect to mechanical properties, a significant enhancement (~150%) in tensile ductility was achieved due to CNT dispersion when compared to the Mg–9Al alloy, although the strength values showed only a moderate improvement (table 3.19).

In a similar study [52], a nano-scale thick Ni coating was applied on CNTs to enhance the interfacial characteristics in Mg/CNT nanocomposites processed using the microwave sintering technique. From the microstructure, it is clear that the nickel coating on the CNTs facilitated Mg_2Ni formation at the interface (figure 3.25) which enhanced the bonding between the Mg matrix and CNT reinforcement for better load transfer. While the results of microstructural measurements showed grain refinement and the improved dispersion of CNT reinforcements, mechanical

Table 3.19. Tensile properties of Mg–9Al/CNT nanocomposites as reported by [51].

Materials	Grain size (μm)	Yield strength (MPa)	Ultimate strength (MPa)	Elongation (%)
Mg–9Al	12.9 ± 2.1	235 ± 3	301 ± 5	6 ± 2
Mg–9Al/0.2 wt.%CNT	10.9 ± 1.5	242 ± 3	346 ± 4	14 ± 1
Mg–9Al/0.4 wt.%CNT	11.1 ± 1.7	248 ± 5	355 ± 7	15 ± 3
Mg–9Al/0.6 wt.%CNT	11.8 ± 1.2	230 ± 2	329 ± 1	13 ± 1

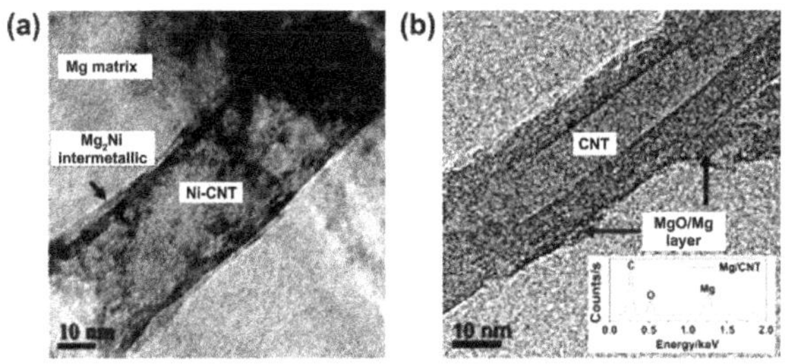

Figure 3.25. TEM micrographs of Mg/Ni-CNT composites showing the presence of (a) intermetallics and (b) an oxide layer (the EDX inset shows the elemental distribution on the surface of the CNT) at the interface of a Ni-CNT embedded in the Mg matrix. Reprinted from [52], Copyright 2014, with permission from Elsevier.

Table 3.20. Properties of Ni-modified Mg/CNT nanocomposites as reported by [52].

Materials	Grain size (μm)	Yield strength (MPa)	Ultimate strength (MPa)	Elongation (%)
Mg	17.9 ± 8.2	126 ± 1	171 ± 2	7.9 ± 0.3
Mg/0.3 wt.%CNT	15.6 ± 6.4	119 ± 4	163 ± 7	5.7 ± 0.2
Mg/0.3 wt.%Ni-CNT	6.1 ± 1.4	206 ± 2	237 ± 1	6.4 ± 0.3

property measurements showed improved microhardness and strength properties. However, the ductility remained largely similar (table 3.20).

A CNT reinforced AZ91 alloy composite was also fabricated by mechanical stirring and high intensity ultrasonic dispersion [53]. From the results of microstructure and mechanical testing (figure 3.26), well dispersed CNTs causing simultaneous improvement in strength and elongation was reported for the AZ91D/1.5CNT nanocomposite. While the grain size was reduced by 72% compared to unreinforced AZ91, the strengths and ductility were improved by 20% and 40%, respectively. In a similar study, AZ61/CNT composites containing (0–1 wt.%) CNTs were fabricated using a stir casting method followed by aging heat treatments [54]. In this study, while the results of grain size measurements showed marginal reduction in the average grain size, the compression strengths were found to increase significantly with the increase in CNT content under the as-cast and aged conditions.

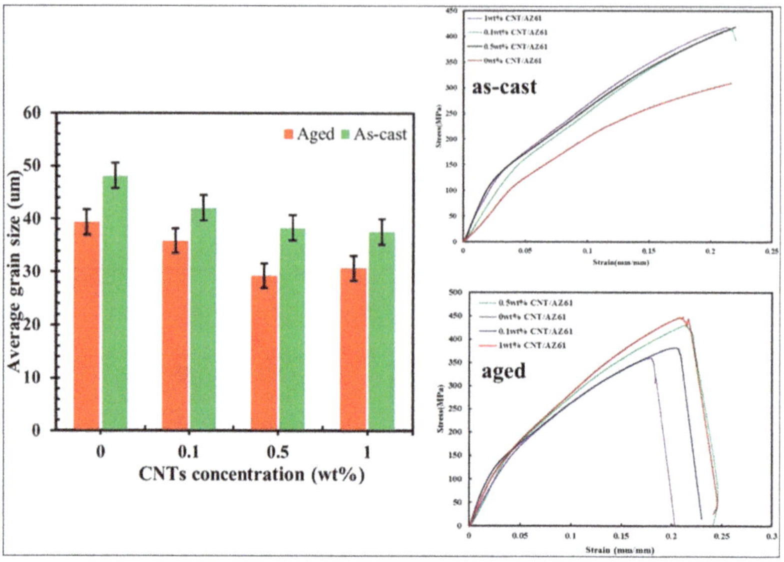

Figure 3.26. Grain size and compressive stress–strain curves of AZ61/CNT nanocomposites in as-cast and aged conditions. Reprinted from [54], Copyright 2019, with permission from Elsevier.

Table 3.21. Tensile properties of Mg/GNP nanocomposites produced by [55–57].

Materials	Yield strength (MPa)	Ultimate strength (MPa)	Fracture strain (%)
Mg	119 ± 5	186 ± 6	9.7 ± 3
Mg–1Al/0.09 wt.%GNP	148 ± 3	206 ± 4	10.5 ± 3.4
Mg–1Al/0.18 wt.%GNP	162 ± 4.1	223 ± 5	15.2 ± 2
Mg–1Al/0.30 wt.%GNP	178 ± 2.9	246 ± 3.5	16.9 ± 3
Mg	162 ± 5	195 ± 4	3.7 ± 2.5
Mg–0.5Al/0.18 wt.%GNP	173 ± 4	230 ± 5.1	10.7 ± 3
Mg–1.0Al/0.18 wt.%GNP	190 ± 5.3	254 ± 3	15.5 ± 3.4
Mg–1.5Al/0.18 wt.%GNP	209 ± 3.9	268 ± 4.5	12.7 ± 2
Mg–1Al-1Sn	161 ± 04	236 ± 5.1	16.7 ± 03
Mg–1Al–1Sn/0.18 wt.%GNPs	208 ± 5.3	269 ± 03	10.9 ± 3.4

The CNT addition did not significantly affect the fracture strain of as-cast composites. However, it was found to improve after thermal aging treatment.

3.3.5 The effect of graphene nanoplatelets

Rashad *et al* [55–57] carried out graphene nanoplatelet (GNP) addition using the PM method to improve the mechanical performance of Mg alloys. In this study, the increment in mechanical properties (table 3.21) was attributed to the uniform dispersion and large specific surface area per volume of GNPs embedded in a magnesium matrix (figure 3.27).

Chen *et al* [58] used a novel nano-processing method (schematic shown in figure 3.28) that combines liquid state ultrasonic processing and solid state stirring to disperse GNPs in a pure magnesium matrix. In this method, GNPs were first fed into molten Mg and treated ultrasonically for 15 min. In the second step, a friction stir process was applied to improve the dispersion of GNPs. The developed nanocomposites showed a uniform dispersion of GNPs and enhanced mechanical properties (figure 3.29). Compared to the unreinforced magnesium, the hardness and strength were increased by 78% and 64%, respectively.

In another study, Xiang *et al* [59] investigated the microstructure and mechanical properties of GNP reinforced Mg matrix composites. In this study, the homogeneous dispersion of GNP using the DMD method significantly improved the overall strength and ductility of Mg composites (table 3.22). The yield strength improvement was mainly attributed to the grain refinement and load transfer capabilities of

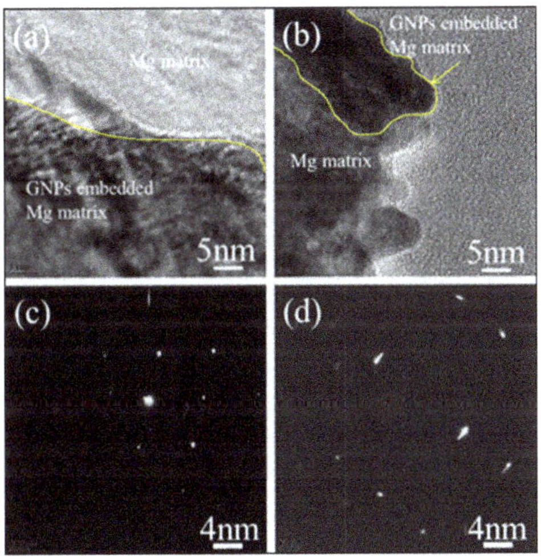

Figure 3.27. TEM micrographs of Mg–1Al–0.3GNPs composite (a)–(b) and SAD patterns of the matrix (c) and GNPs (d). Reprinted from [55], Copyright 2015, with permission from Elsevier.

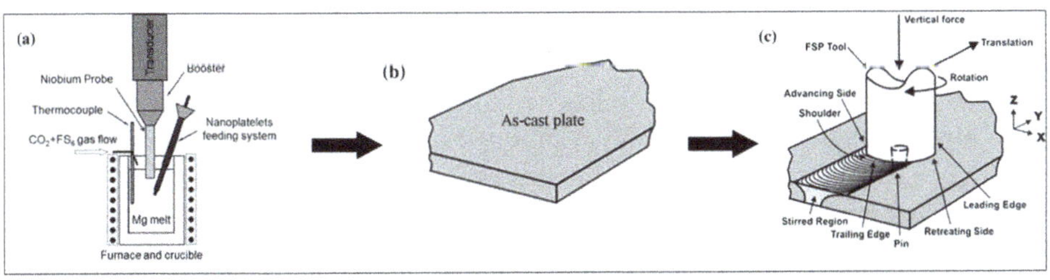

Figure 3.28. Nano-processing method used by Chen *et al* to fabricate Mg/GNP nanocomposites. Reprinted from [58], Copyright 2012, with permission from Elsevier.

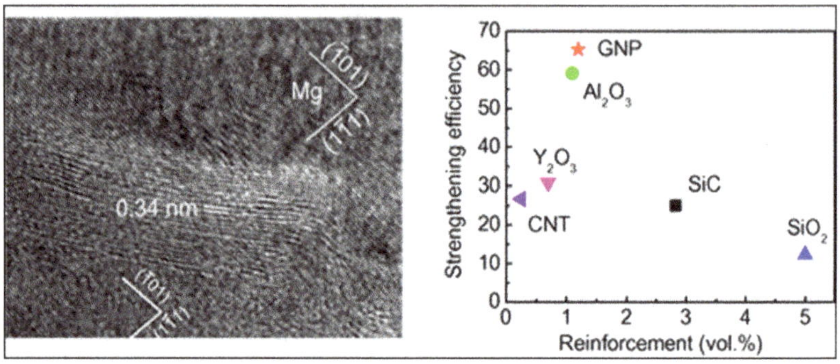

Figure 3.29. (a) TEM micrograph showing GNP distribution and (b) strengthening efficiency of GNP reinforcements as compared to other reinforcements. Reprinted from [58], Copyright 2012, with permission from Elsevier.

Table 3.22. Grain size and tensile properties of Mg/GNP nanocomposites as reported by [59].

Materials	Grain size (μm)	Yield strength (MPa)	Ultimate strength (MPa)	Fracture strain (%)
Mg	7.2	96 ± 4	163 ± 3	7.5 ± 1.5
Mg/0.10 wt.%GNP	5.8	105 ± 1	176 ± 2	10.3 ± 0.6
Mg/0.25 wt.%GNP	4.8	122 ± 2	202 ± 3	14.5 ± 1.2

GNPs. The addition of GNPs also promoted {10–12} extension twins to facilitate favourable orientations for basal slip and prismatic ⟨a⟩ slip initiation for enhanced toughening.

Meng *et al* [60] used a facile strategy to fabricate Mg/GNP nanocomposites with a laminated structure. In this study, the GNPs dispersed in alcohol were sprayed onto Mg foils which are then hot pressed and rolled to form a laminated structure that restricts the dislocation motion and suppresses the strain localization to prolong the crack propagation path. The addition of 0.25 vol.% and 0.75 vol.% GNPs improved the tensile strength of Mg from 136 MPa to 160 MPa and 179 MPa, respectively (figure 3.30). The elongation to failure was slightly compromised. While the composites with 0.25 vol.% GNP exhibited a slight reduction in the elongation to fracture value from 5.5% to 4.9%, it was significantly reduced (2.7%) for 0.75 vol.% GNP addition.

ZK60 alloy was reinforced with low volume fraction GNPs using a cost-effective facile melt stirring and hot extrusion processes [61]. In this study, the GNPs were pre-dispersed with Mg powder and extruded into rods which were then used as a precursor for melting to effectively guarantee the integrity and dispersion of GNPs (schematic shown in figure 3.31). Compared to ZK60 alloy, the developed nanocomposite showed enhanced mechanical properties under tension and compression loads (table 3.23). The theoretical analysis in this study showed load transfer of

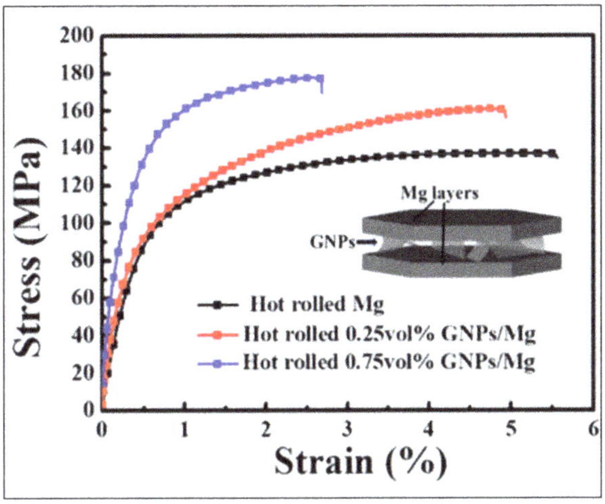

Figure 3.30. Tensile properties of Mg/GNP nanocomposites produced by Meng *et al.* The inset shows the construction of a laminated structure in the Mg/GNP nanocomposite. Reprinted from [60], Copyright 2018, with permission from Elsevier.

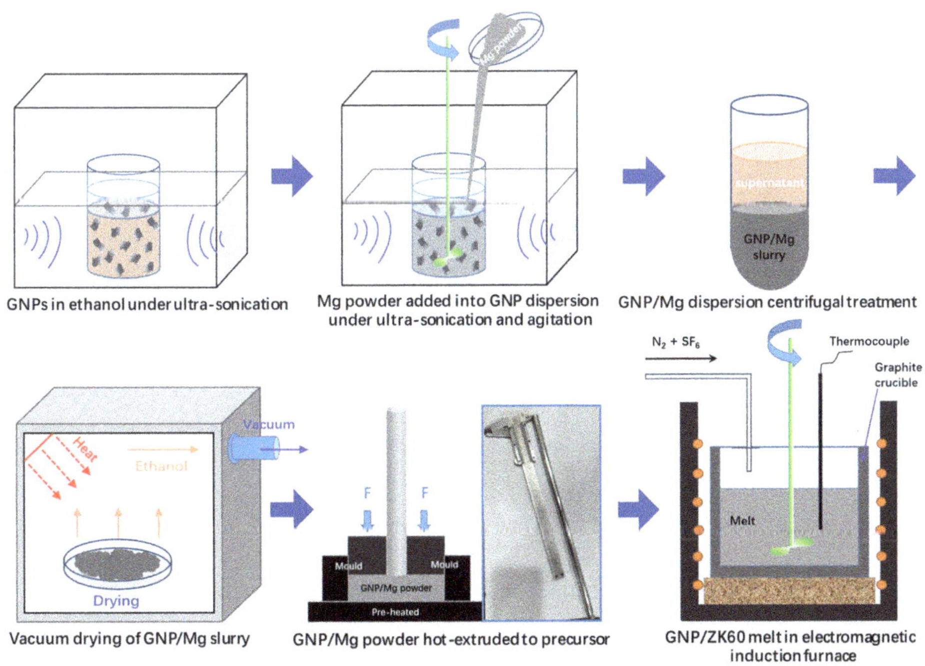

Figure 3.31. Facile melt stirring and hot extrusion processing of Mg/GNP nanocomposites as reported by Du *et al.* Reprinted from [61], Copyright 2018, with permission from Elsevier.

Table 3.23. Mechanical properties of ZK60/GNP nanocomposites [61].

Material	Tensile performances			Compressive performances		
	Yield strength (MPa)	Ultimate strength (MPa)	Fracture strain (%)	Yield strength (MPa)	Ultimate strength (MPa)	Fracture strain (%)
ZK60	282 ± 3.0	158 ± 2.0	11 ± 0.8	364 ± 2.8	126 ± 3.0	9 ± 0.3
ZK60/0.05 wt.%GNP	336 ± 4.0	256 ± 4.0	13 ± 1.2	473 ± 6.2	249 ± 4.0	10 ± 1.0
ZK60/0.10 wt.%GNP	343 ± 3.8	283 ± 3.5	17 ± 2.0	463 ± 5.0	279 ± 3.4	12 ± 1.1

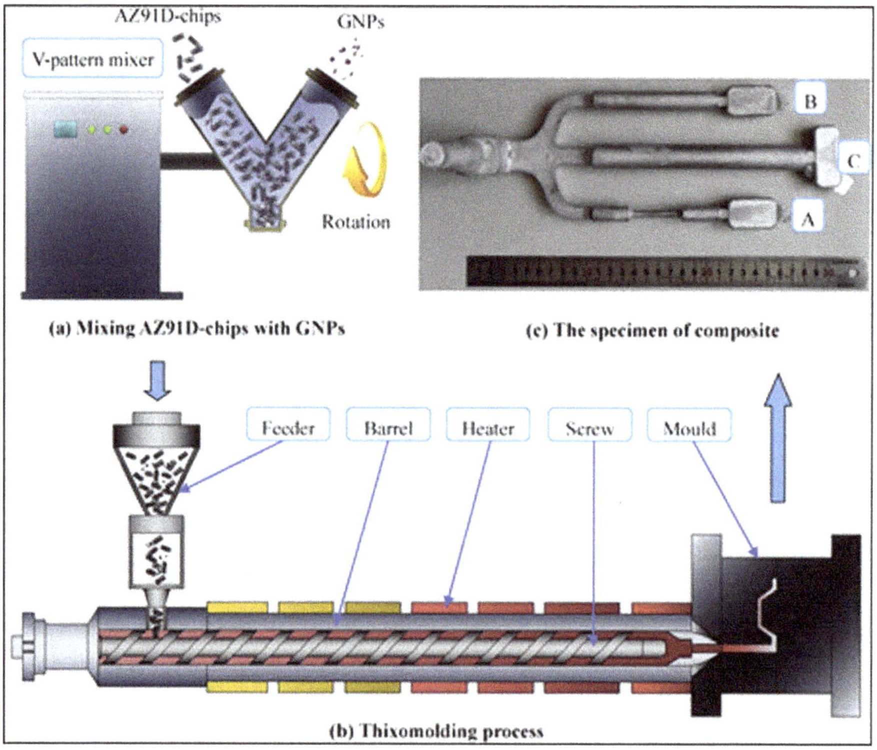

Figure 3.32. Schematic of thixo-moudling used to fabricate AZ91D/GNP nanocomposites. Reprinted from [62], Copyright 2011, with permission from Elsevier.

GNPs due to the 2D interfacial contacting and continuous integration as the prominent strengthening contribution in improving the strength of the GNP/ZK60 composite.

Chen *et al* [62] used the thixo-moudling process (figure 3.32) applicable for mass production to reinforce the AZ91D matrix with GNP reinforcement. This study reported an improvement in the ultimate tensile strength by 38.4% and elongation by 85.7% in AZ9D/GNP composites when compared to AZ91D alloy (figure 3.33). The obtained improvement in mechanical properties was attributed to the uniform dispersion of GNPs and a strong interface between the AZ91 matrix and GNP reinforcements.

Mg–0.3Sr–0.3Ca alloy-based nanocomposites containing different amounts of GNPs were fabricated using a stir casting technique followed by homogenization and extrusion [63]. The results of mechanical property testing under tensile (table 3.24) and compressive loads (table 3.25) showed strength enhancements without a significant effect on the failure strain values and the optimum weight percent of GNP was reported to be 0.2%. Similar improvements in mechanical properties under compressive loads were also reported when Mg was reinforced with

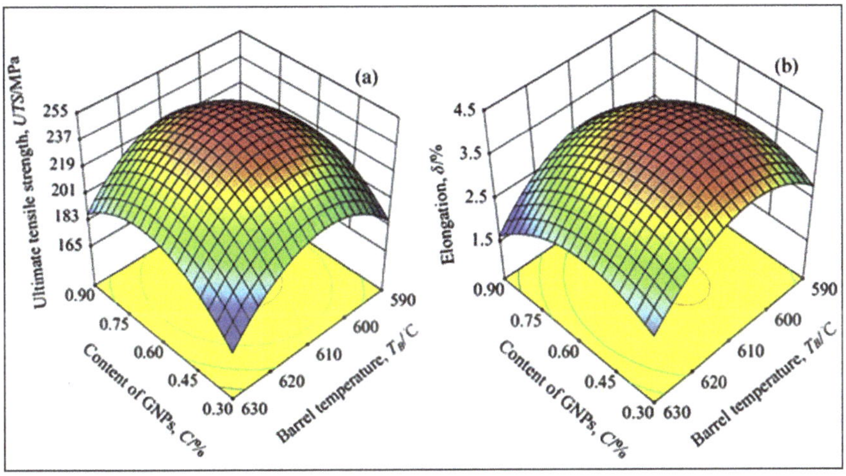

Figure 3.33. Mechanical properties of AZ91D/GNP nanocomposites fabricated using thixo-moudling process. Reprinted from [62], Copyright 2011, with permission from Elsevier.

Table 3.24. Tensile properties of the Mg–0.3Sr–0.3Ca/GNP nanocomposites developed by [63].

Material	Condition	Yield strength (MPa)	Ultimate strength (MPa)	Fracture strain (%)
Mg–0.3Sr–0.3Ca	Homogenized	78 ± 4	102 ± 5	5.4 ± 0.27
	Extruded	171 ± 9	228 ± 11	6 ± 0.3
Mg–0.3Sr–0.3Ca/0.1 wt.%GNP	Homogenized	83 ± 4	108 ± 5	7.2 ± 0.4
	Extruded	184 ± 9	232 ± 11	8.1 ± 0.4
Mg–0.3Sr–0.3Ca/0.2 wt.%GNP	Homogenized	91 ± 5	117 ± 6	8 ± 0.4
	Extruded	210 ± 11	231 ± 12	10 ± 1
Mg–0.3Sr–0.3Ca/0.4 wt.%GNP	Homogenized	96 ± 7	129 ± 9	7.8 ± 0.6
	Extruded	223 ± 13.3	245 ± 15	8.8 ± 0.6

Table 3.25. Compressive properties of the Mg–0.3Sr–0.3Ca/GNP nanocomposites developed by [63].

Material	Condition	Yield strength (MPa)	Ultimate strength (MPa)	Fracture strain (%)
Mg–0.3Sr–0.3Ca	Homogenized	45 ± 2	208 ± 10	31 ± 1.6
	Extruded	65 ± 3	339 ± 17	14.3 ± 0.7
Mg–0.3Sr–0.3Ca/0.1 wt.%GNP	Homogenized	50 ± 3	213 ± 11	29.4 ± 1.5
	Extruded	67 ± 3	335 ± 16	15.5 ± 0.8
Mg–0.3Sr–0.3Ca/0.2 wt.%GNP	Homogenized	62 ± 3	218 ± 10	30.2 ± 1.5
	Extruded	93 ± 5	339 ± 18	18.3 ± 1.0
Mg–0.3Sr–0.3Ca/0.4 wt.%GNP	Homogenized	42 ± 3	227 ± 11	29.1 ± 2.3
	Extruded	82 ± 5	309 ± 17	14.3 ± 0.9

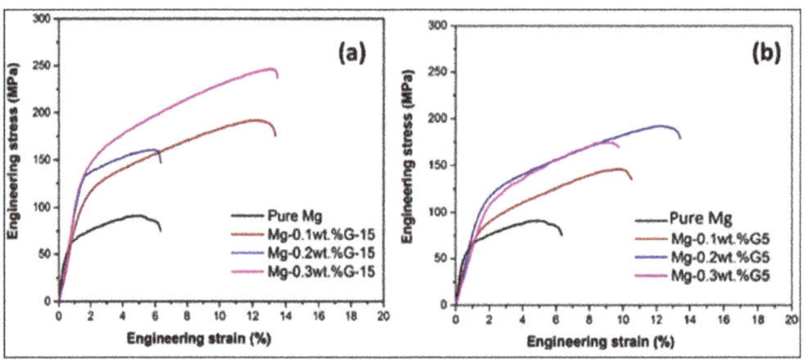

Figure 3.34. Compressive stress–strain curves of (G-15) and(G-5) reinforced Mg nanocomposites. Reprinted from [64], Copyright 2020, with permission from Elsevier.

G-15 and G-5 type GNPs using the ball milling method [64]. From the results, G-15 GNPs were more beneficial in terms of strengthening, although the ductility improvements were comparable for both G-15 and G-5 GNPs addition (figure 3.34).

3.4 Nitride nanoparticle reinforced magnesium composites

3.4.1 The effect of aluminium nitride nanoparticles

Maung *et al* [65] synthesized nanostructured Mg composites of Mg–5 wt.%Al matrix reinforced with different amounts (1–5 wt.%) of nano-sized AlN particles using mechanical milling (MM) and consolidation (sintering + hot extrusion) of the recycled Mg chips and nano-AlN powder. The effects of MM and the amount of nano-AlN reinforcement on the properties of Mg nanocomposites were studied. The property characterization results (figure 3.35) in this study showed better yield strength for MM samples when compared to the un-milled samples. In particular, the ball milled samples with longer milling duration (>10 h) exhibited lower strength and higher ductility alongside prominent strain softening effects.

AlN nanoparticle reinforced Mg composites were also fabricated using a rapid microwave sintering assisted PM method [66]. The extruded samples displayed an

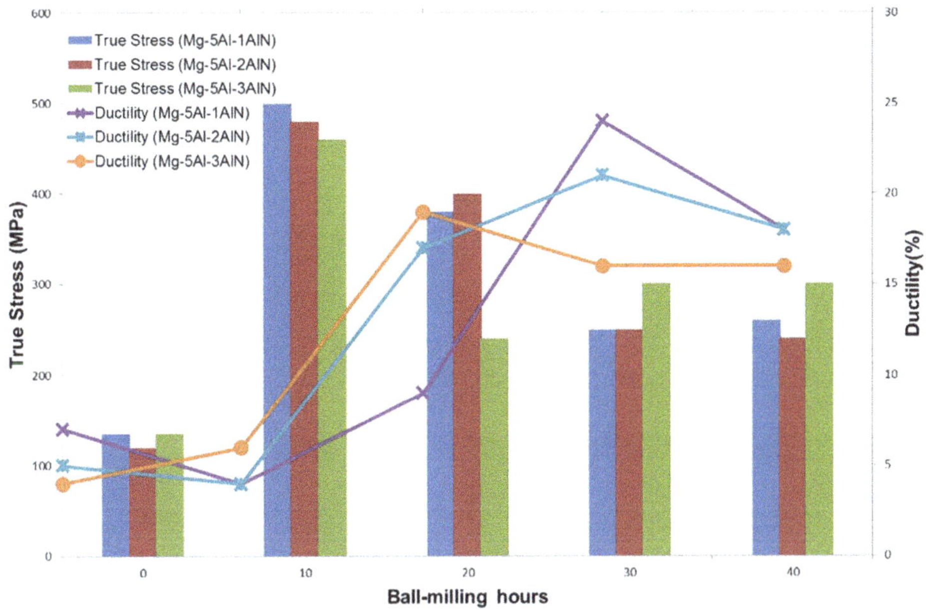

Figure 3.35. Effect of ball milling on the properties of Mg–5Al/AlN nanocomposites. Based on data from [65].

Table 3.26. Properties of Mg/AlN nanocomposites as reported by [66].

		Tensile properties			Compressive properties		
Materials	Grain size (μm)	Yield strength (MPa)	Ultimate strength (MPa)	Fracture strain (%)	Yield strength (MPa)	Ultimate strength (MPa)	Fracture strain (%)
Mg	28.0 ± 7.7	96 ± 6	137 ± 9	6.0 ± 3.0	51 ± 9	268 ± 16	18.9 ± 1.6
Mg/0.1AlN	23.8 ± 7.9	102 ± 6	159 ± 8	11.0 ± 2.2	65 ± 1	284 ± 12	19.3 ± 1.5
Mg/0.4AlN	19.6 ± 6.2	120 ± 1	164 ± 3	8.4 ± 0.9	72 ± 5	314 ± 20	17.5 ± 0.6
Mg/0.8AlN	19.4 ± 5.3	129 ± 5	176 ± 3	6.3 ± 0.4	71 ± 3	307 ± 17	18.3 ± 2.3

overall improvement in tensile and compressive properties (table 3.26). The developed Mg/AlN nanocomposites displayed a reduction in tension–compression yield asymmetry owing to the texture weakening effects.

Cao *et al* [67] used the ultrasonic stir casting method to fabricate nano-AlN reinforced AZ91D alloy composites. The results of mechanical property testing showed significant improvement in tensile strength without any significant changes to the ductility (figure 3.36(a)). While the microstructure of the AZ91D alloy was composed of α-Mg, massive β phase, and some lamellar β phase, the addition of AlN nanoparticles resulted in a finer lamellar β phase which also contributed to the strengthening of the AZ91D alloy matrix (figure 3.37). A similar study also reported

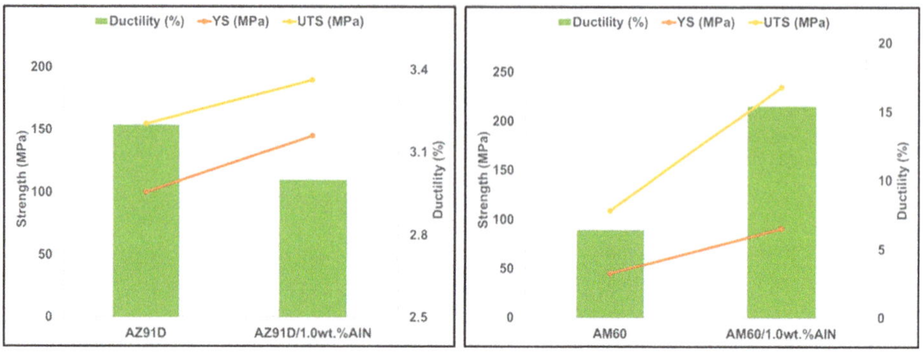

Figure 3.36. Properties of nano-AlN added AZ91D and AM60 alloys. Based on data from [67, 68].

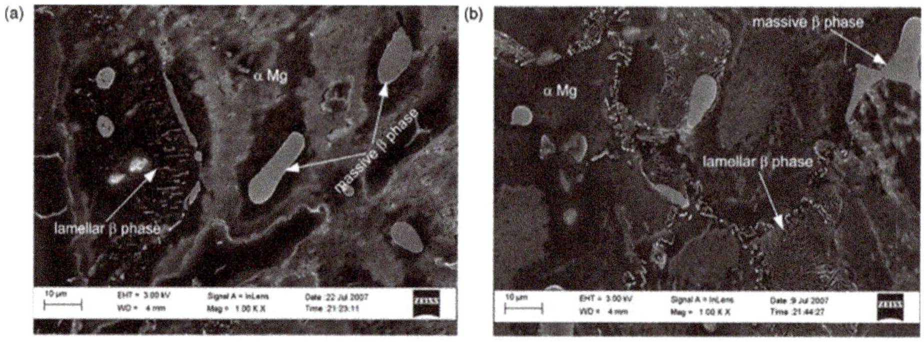

Figure 3.37. Microstructure of AZ91D and its nanocomposite reinforced with AlN nanoparticles as reported by Cao *et al.* Reprinted from [67]. Copyright 2008, with permission from Elsevier.

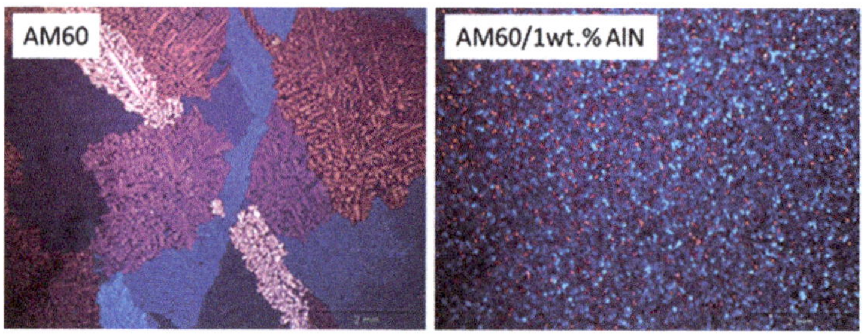

Figure 3.38. Microstructure of AM60 and AM60/1 wt.% AlN nanocomposite. Reprinted from [68], with permission from MDPI.

significant enhancement in the strength and ductility of AM60 alloy due to 1 wt.% nano-AlN addition (figure 3.36(b)) [68]. Here, the improvement in strength and ductility of AM60/AlN nanocomposite was attributed to significant grain refinement from 1227 ± 301 μm to 85 ± 6 μm (figure 3.38) and associated Hall–Petch strengthening. In another study [69], there was no change in the mechanical properties of Elektron 21 alloy when nano-AlN particles were added, although the E21/AlN nanocomposite displayed a marginal refinement in the matrix grain size. Similar results were also reported in the case of DMD processed AZ91-ZK60/AlN nanocomposite [34].

3.4.2 Effect of boron nitride nanoparticles

Mg composites containing BN nanoparticles were developed using the PM and DMD methods followed by hot extrusion [70, 71]. In this study, the results of mechanical testing (table 3.27) conducted on Mg/BN nanocomposites revealed an overall improvement in strength and ductility due to grain refinement and texture weakening (figure 3.39). Among the processing types, the tensile response of DMD processed nanocomposites was superior compared to PM processed composites.

3.4.3 The effect of silicon nitride nanoparticles

Paramsothy *et al* [72] investigated the properties of nano-Si_3N_4 particle reinforced ZK60 alloy processed using the DMD method followed by hot extrusion. In this study, the addition of nano-scale Si_3N_4 particles contributed towards the overall improvement in tensile properties in terms of 0.2YS, UTS, and ductility (table 3.28). The strength and ductility enhancements were attributed to the grain refinement and the activation of the non-basal slip system due to texture randomization. Similar improvement in tensile properties due to nano-Si_3N_4 addition was also reported for AZ31/AZ91 hybrid alloy [73]. Under compression loads, the AZ31/AZ91/1.5 vol.% Si_3N_4 nanocomposite exhibited improved strengths and reduced failure strain. However, the addition of Si_3N_4 nanoparticles improved the ultimate strength and fracture strain of ZK60 alloy at the expense of yield strength.

Table 3.27. Properties of Mg/BN nanocomposites from [70, 71].

		Tensile properties			Compressive properties		
Materials	Grain size (μm)	Yield strength (MPa)	Ultimate strength (MPa)	Fracture strain (%)	Yield strength (MPa)	Ultimate strength (MPa)	Fracture strain (%)
Microwave sintering assisted powder metallurgy method							
Mg	29 ± 6	136 ± 8	170 ± 7	6.1 ± 1.2	70 ± 2	250 ± 7	24.5 ± 2.7
Mg/0.29BN	21 ± 3	127 ± 6	192 ± 8	7.8 ± 0.9	88 ± 6	290 ± 9	20.9 ± 1.8
Mg/0.86BN	22 ± 2	142 ± 4	200 ± 5	8.6 ± 0.5	108 ± 2	312 ± 8	19.9 ± 1.2
Mg/1.44BN	19 ± 3	145 ± 3	217 ± 5	7.2 ± 0.8	115 ± 4	319 ± 4	19.7 ± 1.4
DMD method							
Pure Mg	28 ± 11	120 ± 9	169 ± 11	6.4 ± 0.7	70 ± 8	234 ± 8	20.7 ± 0.9
Mg/0.3BN	18 ± 6	133 ± 4	193 ± 7	8.6 ± 0.8	84 ± 9	275 ± 12	18.9 ± 0.7
Mg/0.6BN	15 ± 4	154 ± 2	223 ± 2	15.3 ± 0.6	97 ± 3	297 ± 8	17.1 ± 1.5
Mg/1.2BN	9 ± 3	178 ± 5	255 ± 3	12.6 ± 1.3	109 ± 4	307 ± 6	17.6 ± 2.0

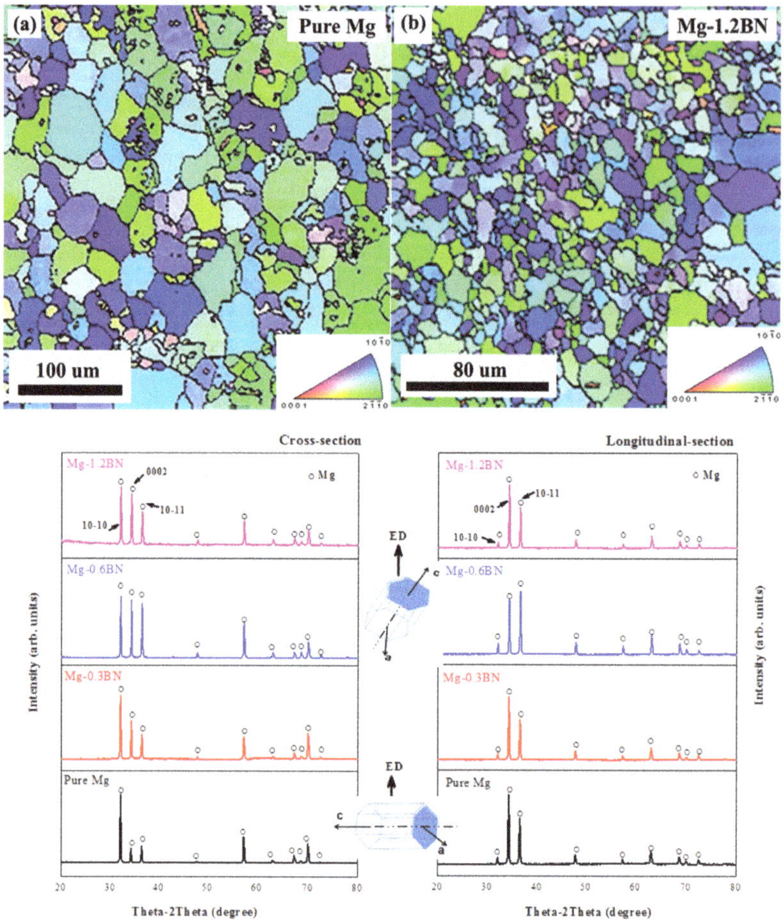

Figure 3.39. EBSD generated IPF maps and XRD results showing grain refinement and texture changes due to nano-BN addition in Mg/BN nanocomposites. Reproduced with permission from [70, 71]. Copyright 2015, 2013 SAGE Publications and MDPI.

3.4.4 The effect of titanium nitride nanoparticles

Mg nanocomposites containing TiN nanoparticles were produced using the DMD method followed by hot extrusion [74]. Grain size analysis conducted on the developed nanocomposites showed significant reduction in grain size from 25.5 ± 2 to 11 ± 3.5 μm with increasing amounts of nano-TiN particles. Mechanical testing results (table 3.29) showed an up to 20% increase in the yield and ultimate strengths under tensile and compressive loads and the strength increment was proportional to the nano-TiN content. The tensile ductility values show an increment under tensile loads up to 0.97 vol.% TiN addition and reduced thereafter. The compressive failure strain values were not significantly affected.

Table 3.28. Properties of nano-Si_3N_4 added ZK60 and AZ31-AZ91 alloys from [72, 73].

		Tensile properties			Compressive properties		
Material	Grain size (μm)	Yield strength (MPa)	Ultimate strength (MPa)	Failure strain (%)	Yield strength (MPa)	Ultimate strength (MPa)	Failure strain (%)
ZK60A	8.9 ± 2.0	163 ± 3	268 ± 3	6.6 ± 0.6	128 ± 11	522 ± 11	19.6 ± 0.9
ZK60A/1.5Si_3N_4	3.4 ± 1.7	198 ± 6	313 ± 4	12.2 ± 0.8	86 ± 11	605 ± 5	29.9 ± 3.1
AZ31-AZ91	5.1 ± 0.7	207 ± 4	316 ± 6	8.0 ± 0.1	117 ± 15	495 ± 13	19.6 ± 1.9
AZ31-AZ91/ 1.5Si_3N_4	5.1 ± 1.0	232 ± 2	331 ± 2	13.1 ± 0.5	158 ± 10	517 ± 2	18.5 ± 1.7

Table 3.29. Properties of Mg/TiN nanocomposites [74].

		Tensile properties			Compressive properties		
Material	Grain size (μm)	Yield strength (MPa)	Ultimate strength (MPa)	Failure strain (%)	Yield strength (MPa)	Ultimate strength (MPa)	Failure strain (%)
Mg	25.5 ± 2	107 ± 5	167 ± 7	10 ± 1	80.4 ± 2.5	347 ± 4	18.5 ± 1.5
Mg/0.58TiN	23 ± 2.5	91 ± 5	151 ± 4	15 ± 1	83.4 ± 2	355 ± 8	21
Mg/0.97TiN	15 ± 2.5	112 ± 2	173 ± 1	15 ± 2	101 ± 3	365.5	20 ± 1
Mg/1.98TiN	13 ± 3.5	130 ± 7	190 ± 11	14.5 ± 1	103 ± 5	385 ± 13	20 ± 1
Mg/2.5TiN	11 ± 3.5	135 ± 8	196 ± 14	10.6 ± 1.2	82 ± 3	345 ± 1	1.5

3.5 Boride nanoparticle reinforced magnesium composites

3.5.1 The effect of titanium diboride nanoparticles

Mg nanocomposites containing nano-TiB_2 particles were developed using the DMD method [75]. The results of mechanical testing (table 3.30) showed significant improvements in strength and ductility due to nano-TiB_2 addition under tension and compression loads. While the best properties were recorded for Mg 1.98 vol.% TiB_2 nanocomposite, with 54% and 79% increments in yield strength and ductility, the ductility was maximum for 0.97 vol.% TiB_2 addition. The increase in strengths and tensile ductility was attributed to the grain refinement and the change in basal plane orientation of Mg achieved upon the addition of nano-TiB_2 particulates. In another study, Choi *et al* [76] used the solidification nano-processing method to disperse TiB_2 nanoparticles in AZ91D melt. In this study, the results of tensile property testing showed marginal improvement in strength alongside an ~50% increase in ductility due to nano-TiB_2 addition.

Table 3.30. The effect of nano-TiB_2 addition on the properties of pure Mg and AZ91D alloy from [75, 76].

Material	Grain size (μm)	Tensile properties: Yield strength (MPa)	Tensile properties: Ultimate strength (MPa)	Tensile properties: Failure strain (%)	Compressive properties: Yield strength (MPa)	Compressive properties: Ultimate strength (MPa)	Compressive properties: Failure strain (%)
Pure Mg	45 ± 2.4	92 ± 5	157 ± 5	8.2 ± 0.2	57 ± 3	332 ± 10	18
$Mg/0.58TiB_2$	37 ± 2.1	93 ± 7	149 ± 9	13 ± 0.5	78 ± 3	352 ± 4	17 ± 1
$Mg/0.97TiB_2$	28 ± 3.6	110 ± 3	173 ± 8	16 ± 0.5	75 ± 7	333 ± 9	21 ± 1
$Mg/1.98TiB_2$	16.5 ± 2	140 ± 9	186 ± 1	14 ± 1	81 ± 7	365 ± 10	19 ± 1
AZ91D	100	88 ± 3	162 ± 5	2.88 ± 0.12	—	—	—
$AZ91D/1TiB_2$	50	104 ± 1	180 ± 0	3.33 ± 0.05	—	—	—
$AZ91D/2.7TiB_2$	50	107 ± 1	188 ± 7	4.27 ± 0.61	—	—	—

Figure 3.40. Microstructure and properties of $Mg_{94}Al_2Gd_2Y_1Zn_1/ZrB_2$ (AZR6M) nanocomposites as reported by Paramsothy and Gupta. Reproduced from [77], with permission © Springer Nature.

3.5.2 The effect of zirconium diboride nanoparticles

Paramsothy and Gupta [77] used the DMD method and thermal aging treatment (200 °C for 24 h) to develop a nanocomposite (referred to as the AZR6M nanocomposite in this study) by adding 1.5 vol.% ZrB_2 nanoparticles to $Mg_{94}Al_2Gd_2Y_1Zn_1$ alloy (referred to as the AZR6 alloy in this study, subscripts represent the composition in at.%). The composition was adjusted by melting the appropriate quantities of AZ80 and ZK60 commercial alloys together with Mg–30Gd and Mg–30Y master alloys and ZrB_2 nanoparticles. In this study, the addition of ZrB_2 nanoparticles induced the formation of nano-LPSO layers which strengthened the matrix alloy in the as-extruded and thermal aged (denoted as TA) conditions (figure 3.40).

3.6 Calcium phosphate nanoparticle reinforced magnesium composites

Calcium phosphates (CaP) are the major constituents of bone and tooth enamel. The high chemical stability and low solubility of CaP make them ideal for use in regenerative medicine and orthopaedics [78]. For example, apatites, which are the CaP compounds with a Ca/P atomic ratio between 1.5 and 1.67, are widely used as bone substitutes for the restoration of cavities in maxillofacial surgery, toothpaste components, and implant coatings in dental applications [78]. For these reasons, calcium phosphate (CAP) compounds are increasingly being used as reinforcements in biodegradable magnesium composites [79].

For the first time, Khalil [80] investigated the effects of nano-hydroxyapatite (HA) particles on the microstructural and mechanical properties of Mg nanocomposites. Mg/HA nanocomposites containing (0–10 wt.%) HA nanoparticles were prepared using a high frequency induction heat sintering method. From the results of microstructural analysis, the authors confirmed the presence of nano-HA particles along the particle boundaries. While the uniform distribution of less than 2 wt.% nano-HA particles improved the relative density and compressive strength (from 173 MPa to 204 MPa), larger amounts of nano-HA addition resulted in decreased strength (127 MPa for 10 wt.% nano-HA addition) due to the agglomeration of nano-HA particles.

Khodaei *et al* [81] also investigated the properties of Mg nanocomposites containing up to 10 wt.% nano-HA particles. In this study, the Mg/HA nanocomposites were densified using both conventional sintering (CS) and spark plasma sintering methods with an aim of understanding the effects of the sintering methods. The results (table 3.31) revealed improved mechanical properties for spark plasma sintered (SPS) materials when compared to CS. While the compression strength of SPS processed Mg/HA nanocomposite was improved from 160 MPa to 221 MPa, no significant strengthening effects were reported for the conventionally sintered Mg/HA nanocomposite.

Using a two-pass friction stir processing method, WE43/HA nanocomposites were fabricated by Cao *et al* [82]. In this study, while the nano-HA particles along with FSP processing reduced the average grain size of WE43 matrix alloy, the yield and tensile strength were slightly affected due to the localized agglomeration of nano-HA particles and their poor interfacial bonding characteristics (figure 3.41).

Table 3.31. Compressive strength of Mg and Mg/HA nanocomposites from [81].

Material	Processing method	Compressive strength (MPa)
Mg	Conventional sintering	130 ± 4
	Spark plasma sintering	160 ± 3
Mg/10HA	Conventional sintering	136 ± 4
	Spark plasma sintering	221 ± 3

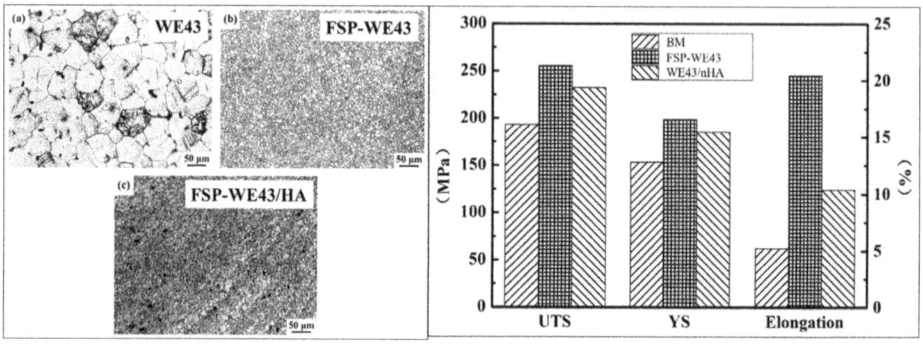

Figure 3.41. Microstructure and mechanical properties of WE43 alloy before FSP (a), after FSP (b), and WE43/HA nanocomposite (c). Reprinted from [82], with permission from MDPI.

Table 3.32. Properties of AZ91 magnesium alloy and AZ91/FA nanocomposites [83–85].

Materials	Relative density (%)	Elastic modulus (GPa)	Compressive yield strength (MPa)	Failure strain (%)
AZ91	92.7	30.2	105.3	6.15
AZ91/10 wt.% FA	94.1	34.1	116.2	5.78
AZ91/20 wt.% FA	95.8	37.5	123.2.	5.32
AZ91/30 wt.% FA	93.2	42.3	112.4	4.51

Razavi *et al* [83] used a planetary ball milling process to synthesize fluorapatite (FA) nanoparticles which were used to develop AZ91/FA nanocomposites by the PM method. The results of mechanical testing (table 3.32) showed marginal improvement in strength properties under compression loads. While the maximum strength was reported for 20 wt.% nano-FA addition due to the improved uniform distribution of FA nanoparticles, the severe agglomeration of nano-FA particles within the grain and grain boundaries of AZ91/30FA nanocomposite resulted in strength reduction in the case of 30 wt.% nano-FA composites. The failure strain was adversely affected for all the composite formulations. The same authors also carried out (20 wt.%) nano-FA addition to improve the compression strength of the AZ91 alloy [84, 85].

Mg-based nanocomposites containing nano-β-tricalcium phosphate (TCP) particles were synthesized using the hybrid PM technique, followed by hot extrusion [86]. The addition of (0.5–1.5 vol.%) TCP nanoparticles refined the average grain size and enhanced the strength and fracture strain of pure magnesium under compression loads (figure 3.42). Similar improvements in grain refinement and mechanical properties were also reported for induction melted Mg–3Zn–0.8Zr/β-TCP nanocomposites by He *et al* [87].

In a similar study, Liu *et al* [88] investigated the effects of conventional mechanical stirring and nano-sized β-TCP melt shearing on the properties of high

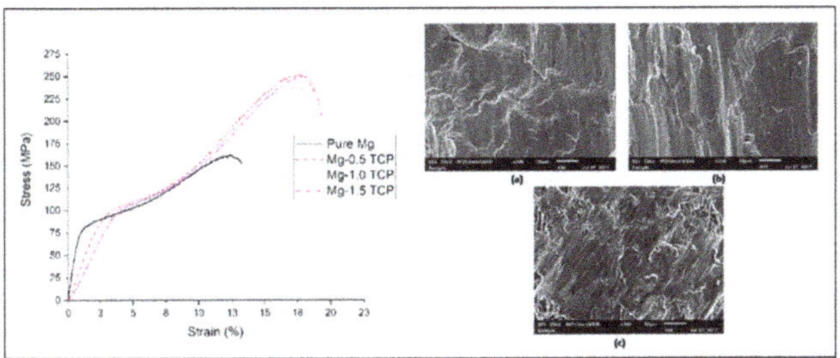

Material	Grain Size (μm)	Compressive Yield Strength (MPa)	Ultimate Compressive Strength (MPa)	Compressive Fracture Strain (%)
Pure Mg	34 ± 2	77 ± 5	156 ± 7	15.8 ± 0.3
Mg/0.5TCP	18 ± 2	92 ± 1	258 ± 4	18.5 ± 0.6
Mg/1.0TCP	13 ± 1	96 ± 2	223 ± 7	17.2 ± 0.7
Mg/1.5TCP	10 ± 1	103 ± 7	240 ± 7	19.3 ± 0.5

Figure 3.42. Compressive properties of Mg/TCP nanocomposites. Reprinted from [86], with permission from MDPI.

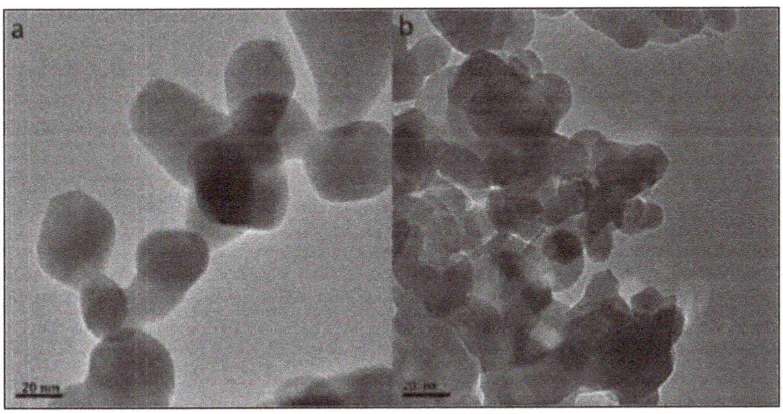

Figure 3.43. Microstructure of (a) MgO modified and (b) unmodified TCP particles. Adapted from [89] © KeAi.

pressure die cast Mg–3Zn–1Ca alloy reinforced with particles. From the results, it was reported that the agglomeration tendency of β-TCP particles was reduced by melt shearing. While the yield strength and ultimate tensile strength of the nanocomposite were found to be 125.4 MPa and 150.0 MPa, respectively, the elongation was reported to be 2.85%. Zheng *et al* [89] added MgO coated β-TCP nanoparticles (figure 3.43) by melt shearing to improve the properties of Mg–3Zn–0.8Zr alloy. In this study, the surface modification of β-TCP nanoparticles reduced their clustering tendency (figure 3.44) and improved their uniform distribution in the Mg matrix of as-cast and extruded nanocomposites. The results of mechanical testing revealed better strength properties for the nanocomposite reinforced with MgO coated β-TCP nanoparticles.

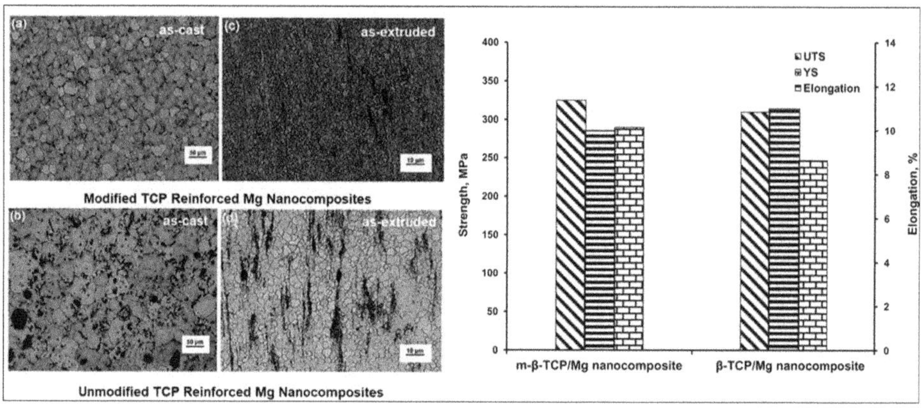

Figure 3.44. Microstructure and mechanical properties of MgO modified and unmodified TCP reinforced Mg nanocomposites. Adapted from [89] © KeAi.

Table 3.33. Grain size and tensile properties of Mg nanocomposites reinforced with elemental Cu and Al nanoparticles [91].

Materials	Grain size (μm)	Yield strength (MPa)	Ultimate strength (MPa)	Fracture strain (%)
Mg	27 ± 7	116 ± 11	168 ± 10	6.1 ± 2.0
Mg/0.3Cu	17 ± 5	188 ± 13	218 ± 11	5.9 ± 1.1
Mg/0.6Cu	15 ± 4	237 ± 24	286 ± 8	5.4 ± 1.2
Mg/1.0Cu	15 ± 4	194 ± 17	221 ± 17	2.9 ± 0.4
Mg/0.00Al	23 ± 6	134 ± 11	190 ± 10	4.6 ± 0.6
Mg/0.25Al	17 ± 5	181 ± 14	221 ± 15	4.8 ± 0.4
Mg/0.50Al	17 ± 8	218 ± 16	271 ± 11	6.2 ± 0.9
Mg/0.75Al	16 ± 7	202 ± 7	261 ± 10	5.0 ± 1.6
Mg/1.00Al	16 ± 6	185 ± 9	226 ± 12	3.3 ± 1.0

3.7 Metallic nanoparticle reinforced magnesium composites

The incorporation of metallic particles in the nano-lengthscale also improves the mechanical performance of Mg. Using a microwave sintering based PM method, Wong and Gupta [90] developed Mg/Cu nanocomposites and investigated the properties under tensile roads. The results (table 3.33) of mechanical characterization showed that the addition of nano-sized Cu particulates led to an increase in hardness, 0.2% YS, UTS, and work of fracture (except for Mg/1.0Cu) of the Mg matrix. The largest improvement in strength (an ~104% increase in 0.2% YS and an ~70% increase in UTS) was reported in the case of 0.6 vol.% nano-Cu addition. However, the ductility of Mg progressively reduced with the increasing addition of Cu due to the formation of the Mg_2Cu intermetallic phase.

Table 3.34. Properties of DMD processed Mg/Ti nanocomposites [92].

		Tensile properties			Compressive properties		
Material	Grain size (μm)	Yield strength (MPa)	Ultimate strength (MPa)	Fracture strain (%)	Yield strength (MPa)	Ultimate strength (MPa)	Fracture strain (%)
Pure Mg	16 ± 4	73.5 ± 5.4	130.3 ± 4.4	13.82 ± 1.42	86 ± 1	326 ± 1	20.8 ± 1.7
Mg/0.58Ti	4.65 ± 1.5	134 ± 7	190 ± 7	6.3 ± 0.6	129 ± 2	431 ± 8	17.4 ± 0.3
Mg/0.97Ti	2.5 ± 1.5	135 ± 3	197 ± 8	8.3 ± 0.6	130 ± 8	413 ± 15	18.5 ± 0.6
Mg/1.98Ti	1.25 ± 0.25	162 ± 5	231 ± 12	7.7 ± 0.1	120 ± 5	415 ± 4	17.7 ± 0.8

In a similar study, the effects of nano-sized Al particle integration on the properties of magnesium were investigated by Zhong *et al* [91]. For this purpose, the authors produced and tested Mg nanocomposites containing 0.25, 0.5, 0.75, and 1 vol.% nano-Al particles. The reported results (table 3.33) indicated significant enhancement in strength. Similarly, the average ductility of the composites (Mg⩽0.75 vol.% Al) was higher when compared to pure Mg and decreased thereafter.

The DMD method was used to synthesize nano-Ti reinforced Mg nanocomposites [92]. The results of microstructural measurements showed significant grain refinement from 16 μm to 1 μm with increasing nano-Ti addition (table 3.34). Similarly, the results of tensile testing also showed significant improvement in strength properties (0.2% YS up to <112% and UTS up to <81%) with increasing amounts of nano-Ti particles, however, the ductility was adversely affected (table 3.34). The DMD method was also used to fabricate Mg nanocomposites containing up to 3 wt.% nano-sized NiTi particles [93]. In this study, a significant refinement in the average grain size (from 40 μm to 9 μm) was reported upon nano-NiTi addition which in turn improved the strength properties of Mg. However, the ductility was marginally affected (figure 3.45).

3.8 Summary

A critical investigation of open literature shows that a number of ceramic particles including Al_2O_3, B_4C, Y_2O_3, AlN, TiC, TiN, SiC, Si_3N_4, ZnO, TiO_2, and CaO along with CNTs, graphene platelets (GNP), and CAP are used as nano-reinforcement for Mg and Mg alloys. Some of the recent publications also highlight the effectiveness of nano-sized metallic particles.

While most nano-scale reinforcements offered strength enhancements under tensile loads, the oxide nanoparticles including Al_2O_3 and Y_2O_3 were extensively used owing to their prominent strengthening and ductilization benefits (figure 3.46(a)). One of the latest works also reported the exceptional performance

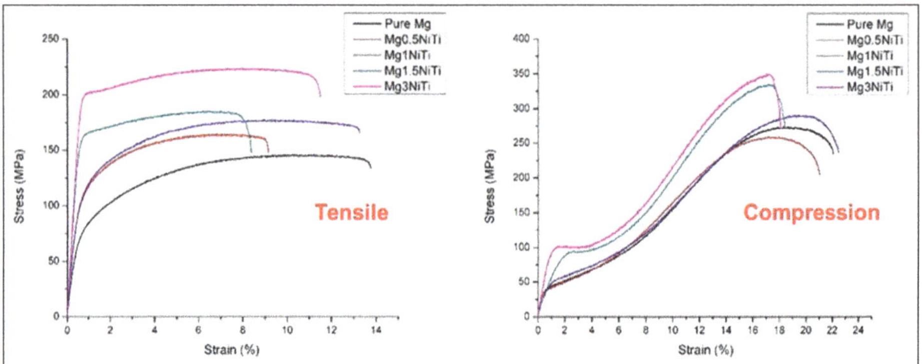

Material	Grain Size (μm)	Tensile Yield Strength (MPa)	Ultimate Tensile Strength (MPa)	Tensile Fracture Strain (%)	Compressive Yield Strength (MPa)	Ultimate Compressive Strength (MPa)	Compressive Fracture Strain (%)
Pure Mg	40 ± 9.4	84 ± 6	148 ± 6	13 ± 0.9	46 ± 2	274 ± 4	23 ± 0.9
Mg/0.5wt.%NiTi	17 ± 2.2	106 ± 12	167 ± 14	9 ± 0.8	42 ± 4	265 ± 15	23 ± 0.3
Mg/1.0wt.%NiTi	13 ± 1	123 ± 2	176 ± 9	9 ± 4.2	52 ± 6	290 ± 4	23 ± 0.7
Mg/1.5wt.%NiTi	10 ± 3.3	163 ± 9	187 ± 8	9 ± 0.4	78 ± 5	337 ± 4	18 ± 0.4
Mg/3.0wt.%NiTi	9 ± 1.8	193 ± 17	217 ± 16	11 ± 0.2	94 ± 3	345 ± 5	18 ± 0.9

Figure 3.45. Mechanical properties of Mg/NiTi nanocomposites. Reprinted from [93], with permission from MDPI.

of Mg–1.8Y/CaO nanocomposite which had superior strength and ductility than most commercial magnesium alloys. Under compressive loads, the addition of nano-scale reinforcements improves the compressive strengths of Mg and Mg alloys (figure 3.46(b)). However, their failure strain remains either unaffected or showed a slight decrement except for nanocomposites reinforced with CNTs, GNPs, Al_2O_3, Si_3N_4, and TiO_2 nanoparticles.

In the case of metallic reinforcements, the addition of nitinol (NiTi) nanoparticles improved the tensile and compressive yield strengths of Mg by ~130% and ~100%, respectively, without significant adverse effects on the failure strain. Among the processing techniques, melt processed Mg nanocomposites displayed a better combination of strength and ductility while the PM methods produced modest improvement in strength as in the case of Mg/Y_2O_3 and Mg/Al_2O_3 nanocomposites.

Regardless of the processing methods, the inherent high strength and high modulus of these finely dispersed reinforcing phases offered distinct advantages in terms of effective load sharing, enhanced dislocation density, particle (Orowan) strengthening, and grain refinement (figure 3.46). The presence of hard and strong reinforcements with good wettability and interfacial characteristics actively contributes towards effective load sharing. Further, the mismatch in elastic modulus and thermal expansion coefficients between the matrix and the nano-reinforcements contribute towards an increase in dislocations around the interfacial region for

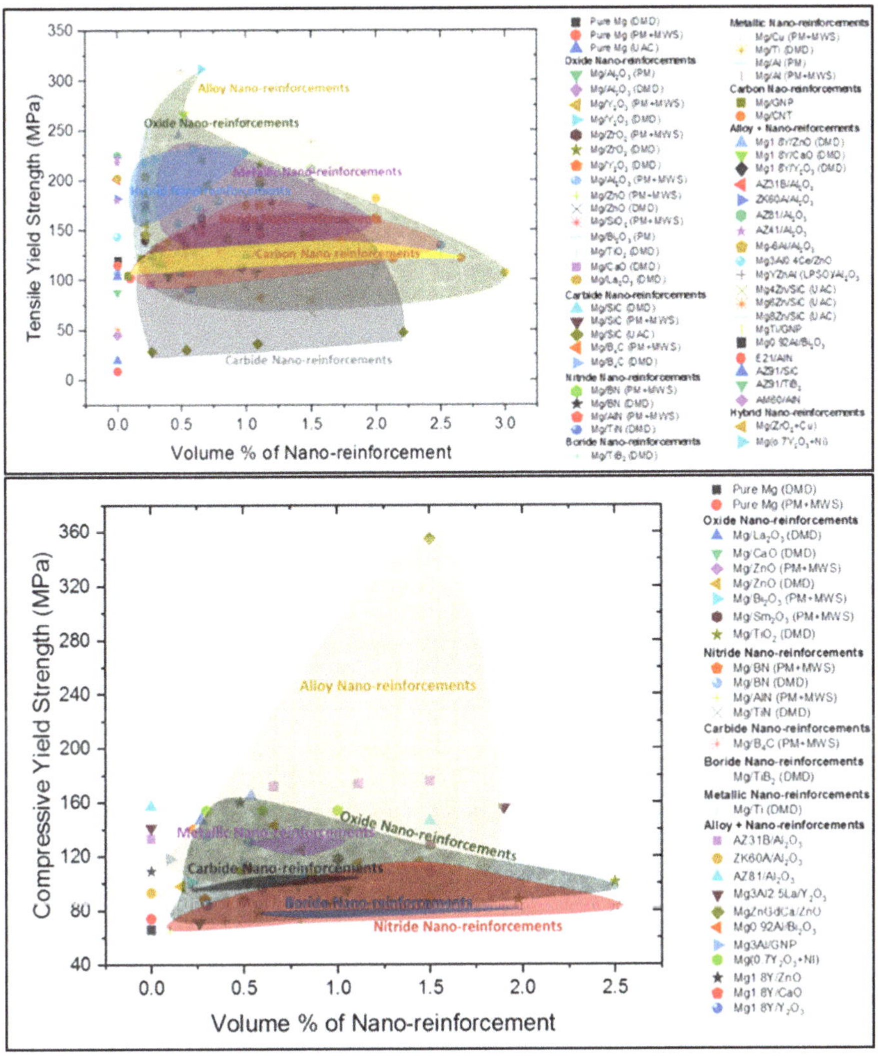

Figure 3.46. (a) Tensile and (b) compressive properties of Mg nanocomposites.

matrix strengthening. During solidification and recrystallization, the effective dispersion of nano-sized reinforcing phases of reasonable size (a few nanometres to a hundred nanometres) and volume fraction facilitates multiple sites for grain nucleation and Zener pinning of matrix grain boundaries. With regards to ductility enhancement, the grain refinement and the activation of non-basal slip systems by texture modification are identified as the responsible mechanisms.

Because of the preferred crystallographic orientation, the extruded Mg nanocomposites displayed different yield strengths under tension and compression loads. In wrought Mg alloys and composites, the basal planes tend to align parallel to the

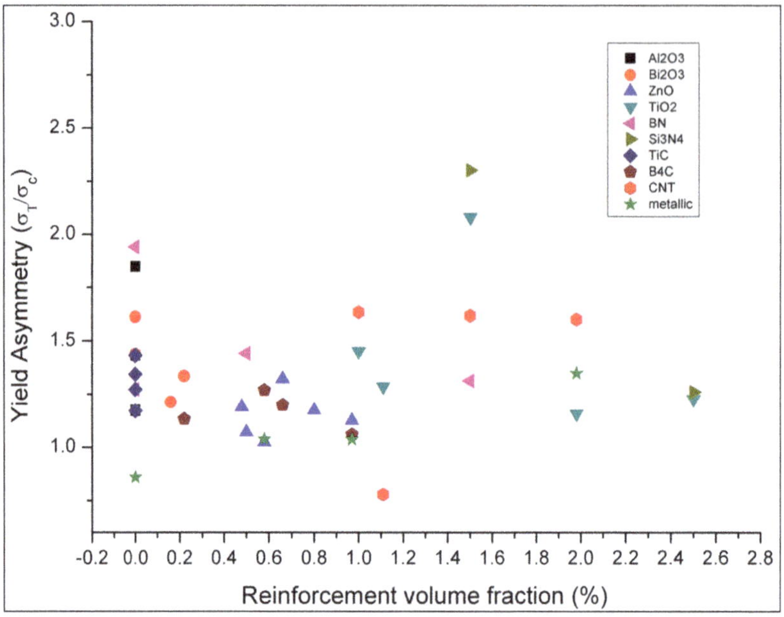

Figure 3.47. Tension–compression yield asymmetry of Mg nanocomposites.

extrusion direction which affects the deformation behaviour. While the twinning process with relatively low critical resolved shear stress (CRSS) governs the compressive deformation and results in lower yield strength, the inability of twinning and the relatively larger CRSS value of basal/non-basal slips results in larger (~1.2 to 2 times larger than CYS) tensile yield strength (figure 3.47).

References

[1] Kainer K U 2006 *Basics of Metal Matrix Composites* (Weinheim: Wiley)

[2] Clyne T W and Withers P J 1995 *An Introduction to Metal Matrix Composites* (Cambridge: Cambridge University Press)

[3] Kumar S, Dieringa H and Kainer K U 2007 Thermal cycling behaviour of the magnesium alloy based hybrid composites in the transverse direction *Mater. Sci. Eng.* A **36** 321–5

[4] Gupta M and Wong W L E 2015 Magnesium-based nanocomposites: lightweight materials of the future *Mater. Charact.* **105** 30–46

[5] Hassan S F 2006 Creation of new magnesium-based material using different types of reinforcements *PhD Thesis* National University of Singapore

[6] Hassan S F and Gupta M 2005 Enhancing physical and mechanical properties of Mg using nanosized Al_2O_3 particulates as reinforcement *Metall. Mater. Trans.* A **36** 2253–8

[7] Wong W L E and Gupta M 2007 Improving overall mechanical performance of magnesium using nano-alumina reinforcement and energy efficient microwave assisted processing route *Adv. Eng. Mater.* **9** 902–9

[8] Nguyen Q B and Gupta M 2008 Increasing significantly the failure strain and work of fracture of solidification processed AZ31B using nano-Al_2O_3 particulates *J. Alloys Compd.* **459** 244–50

[9] Paramsothy M, Tan X H, Chan J, Kwok R and Gupta M 2012 Al_2O_3 nanoparticle addition to concentrated magnesium alloy AZ81: enhanced ductility *J. Alloys Compd.* **545** 12–8
[10] Paramsothy M, Chan J, Kwok R and Gupta M 2012 Nanoparticle interactions with the magnesium alloy matrix during physical deformation: tougher nanocomposites *Mater. Chem. Phys.* **137** 472–82
[11] Tan X H, Chee W K H, Chan J K W, Kwok R W O and Gupta M 2016 Stretching the engineering strain of high strength LPSO quaternary Mg-Y-Zn-Al alloy via integration of nano-Al_2O_3 *J. Mater. Sci.* **51** 4160–8
[12] Habibi M K, Hamouda A S and Gupta M 2013 Using hierarchical composite approach to improve mechanical response of Mg and Mg-Bi_2O_3 nano-composites *Mater. Des.* **49** 627–7
[13] Tekumalla S, Gupta M and Min K H 2018 Using CaO nanoparticles to improve mechanical and ignition response of magnesium *Curr. Nanomater.* **3** 44–51
[14] Tekumalla S, Nandigam Y, Bibhanshu N, Rajashekara S, Yang C, Suwas S and Gupta M 2018 A strong and deformable *in-situ* magnesium nanocomposite igniting above 1000 °C *Sci. Rep.* **8** 7038
[15] Johanes M, Tekumalla S and Gupta M 2019 Fe_3O_4 nanoparticle-reinforced magnesium nanocomposites processed via disintegrated melt deposition and turning-induced deformation techniques *Metals* **9** 1225
[16] Kujur M S, Mallick A, Manakari V, Parande G, Tun K S and Gupta M 2017 Significantly enhancing the ignition/compression/damping response of monolithic magnesium by addition of Sm_2O_3 nanoparticles *Metals* **7** 357
[17] Lee C J, Huang J C and Hsieh P J 2006 Mg based nano-composites fabricated by friction stir processing *Scr. Mater.* **54** 1415–20
[18] Parande G, Manakari V, Meenashisundaram G K and Gupta M 2017 Enhancing the tensile and ignition response of monolithic magnesium by reinforcing with silica nanoparticulates *J. Mater. Res.* **32** 2169–78
[19] Meenashisundaram G K, Nai M H, Almajid A and Gupta M 2015 Development of high performance Mg–TiO_2 nanocomposites targeting for biomedical/structural applications *Mater. Des.* **65** 104–14
[20] Meenashisundaram G K, Nai M H and Gupta M 2015 Effects of primary processing techniques and significance of Hall–Petch strengthening on the mechanical response of magnesium matrix composites containing TiO_2 nanoparticulates *Nanomaterials* **5** 1256–83
[21] Meenashisundaram G K, Nai M H, Almajid A, Khalil K A, Abdo H S and Gupta M 2016 Effects of TiO_2 powder morphology on the mechanical response of pure magnesium: 1D nanofibers versus 0D nanoparticulates *J. Alloys Compd* **664** 45–58
[22] Tun K S and Gupta M 2007 Improving mechanical properties of magnesium using nano-yttria reinforcement and microwave assisted powder metallurgy method *Compos. Sci. Technol.* **67** 2657–64
[23] Tun K S and Gupta M 2008 Effect of heating rate during hybrid microwave sintering on the tensile properties of magnesium and Mg/Y_2O_3 nanocomposite *J. Alloys Compd.* **466** 140–5
[24] Tun K S and Gupta M 2008 Effect of extrusion ratio on microstructure and mechanical properties of microwave-sintered magnesium and Mg/Y_2O_3 nanocomposite *J. Mater. Sci.* **43** 4503–11
[25] Kumar A, Tun K, Kohadkar A and Gupta M 2017 Improved compressive, damping and coefficient of thermal expansion response of Mg–3Al–2.5La alloy using Y_2O_3 nano reinforcement *Metals* **7** 104

[26] Chang C I, Wang Y N, Pei H R, Lee C J, Du X H and Huang J C 2007 Microstructure and mechanical properties of nano-ZrO_2 and nano-SiO_2 particulate reinforced AZ31-Mg based composites fabricated by friction stir processing *Key Eng. Mater.* **351** 114–9

[27] Tun K S, Jayaramanavar P, Nguyen Q B, Chan J, Kwok R and Gupta M 2012 Investigation into tensile and compressive responses of Mg–ZnO composites *Mater. Sci. Technol.* **28** 582–8

[28] Sankaranarayanan S, Pranav Nayak U, Sabat R K, Suwas S, Almajid A and Gupta M 2014 Nano-ZnO particle addition to monolithic magnesium for enhanced tensile and compressive response *J. Alloys Compd.* **615** 211–9

[29] Tekumalla S, Farhan N, Srivatsan T and Gupta M 2016 Nano-ZnO particles' effect in improving the mechanical response of Mg-3Al-0.4Ce alloy *Metals* **6** 276

[30] Tekumalla S, Shabadi R, Yang C, Seetharaman S and Gupta M 2017 Strengthening due to the *in-situ* evolution of $\beta 1'$ Mg-Zn rich phase in a ZnO nanoparticles introduced Mg-Y alloy *Scr. Mater.* **133** 29–32

[31] Chen Y, Tekumalla S, Guo Y B and Gupta M 2016 Introducing Mg-4Zn-3Gd-1Ca/ZnO nanocomposite with compressive strengths matching/exceeding that of mild steel *Sci. Rep.* **6** 32395

[32] Sankaranarayanan S, Sabat R K, Jayalakshmi S, Suwas S and Gupta M 2014 Effect of nanoscale boron carbide particle addition on the microstructural evolution and mechanical response of pure magnesium *Mater. Des.* **56** 428–36

[33] Habibi M K, Hamouda A S and Gupta M 2013 Hybridizing boron carbide (B4C) particles with aluminum (Al) to enhance the mechanical response of magnesium based nano-composites *J. Alloys Compd.* **550** 83–93

[34] Paramsothy M and Gupta M 2013 The opposing nanoscale and macroscale effects of selected nanoparticle addition to AZ91/ZK60A hybrid magnesium alloy *J. Nanoparticle Res.* **15** 1938

[35] Tekumalla S, Yuan N J, Haghshenas M and Gupta M 2019 Enhancing properties of aerospace alloy Elektron 21 using boron carbide nanoparticles as reinforcement *Appl. Sci.* **9** 5470

[36] Ferkel H and Mordike B L 2001 Magnesium strengthened by SiC nanoparticles *Mater. Sci. Eng.* A **298** 193–9

[37] Wong W L E and Gupta M 2006 Simultaneously improving strength and ductility of magnesium using nano-size SiC particulates and microwaves *Adv. Eng. Mater.* **8** 735–40

[38] Nie K B, Wang X J, Xu L, Wu K, Hu X S and Zheng M Y 2012 Influence of extrusion temperature and process parameter on microstructures and tensile properties of a particulate reinforced magnesium matrix nanocomposite *Mater. Des.* **36** 199–205

[39] Cao G, Konishi H and Li X 2008 Mechanical properties and microstructure of Mg/SiC nanocomposites fabricated by ultrasonic cavitation based nanomanufacturing *J. Manuf. Sci. Eng. Trans. ASME* **130** 031105

[40] Cao G, Choi H, Konishi H, Kou S, Lakes R and Li X 2008 Mg-6Zn/1.5%SiC nano-composites fabricated by ultrasonic cavitation-based solidification processing *J. Mater. Sci.* **43** 5521–6

[41] De Cicco M, Konishi H, Cao G, Choi H S, Turng L S, Perepezko J H, Kou S, Lakes R and Li X 2009 Strong, ductile magnesium-zinc nanocomposites *Metall. Mater. Trans.* A **40** 3038

[42] Wang Z H, Wang X D, Zhao Y X and Du W B 2010 SiC nanoparticles reinforced magnesium matrix composites fabricated by ultrasonic method *Trans. Nonferrous Met. Soc. China (English Ed.)* **20** s1029–32

[43] Shiying L, Feipeng G, Qiongyuan Z and Wenzhen L 2009 Mechanical properties and microstrutures of nano-sized SiC particles reinforced AZ91D nanocomposites fabricated by high intensity ultrasonic assisted casting *Mater. Sci. Forum* **618-619** 449–52
[44] Paramsothy M, Chan J, Kwok R and Gupta M 2011 Adding TiC nanoparticles to magnesium alloy ZK60A for strength/ductility enhancement *J. Nanomater.* **2011** 642980
[45] Vidu R, Rahman M, Mahmoudi M, Enachescu M, Poteca T D and Opris I 2014 Nanostructures: a platform for brain repair and augmentation *Front. Syst. Neurosci.* **8** 91
[46] Goh C S, Wei J, Lee L C and Gupta M 2006 Development of novel carbon nanotube reinforced magnesium nanocomposites using the powder metallurgy technique *Nanotechnology* **17** 7
[47] Li Q and Tian B 2013 Compression behavior of magnesium/carbon nanotube composites *J. Mater. Res.* **28** 1877–84
[48] Zhou M, Qu X, Ren L, Fan L, Zhang Y, Guo Y, Quan G, Tang Q, Liu B and Sun H 2017 The effects of carbon nanotubes on the mechanical and wear properties of AZ31 alloy *Materials* **10** 1385
[49] Paramsothy M, Tan X H, Chan J, Kwok R and Gupta M 2013 Carbon nanotube addition to concentrated magnesium alloy AZ81: enhanced ductility with occasional significant increase in strength *Mater. Des.* **45** 15–23
[50] Paramsothy M, Chan J, Kwok R and Gupta M 2011 Addition of CNTs to enhance tensile/compressive response of magnesium alloy ZK60A *Composites* A **42** 180–8
[51] Hou J, Du W, Parande G, Gupta M and Li S 2019 Significantly enhancing the strength + ductility combination of Mg-9Al alloy using multi-walled carbon nanotubes *J. Alloys Compd* **790** 974–82
[52] Nai M H, Wei J and Gupta M 2014 Interface tailoring to enhance mechanical properties of carbon nanotube reinforced magnesium composites *Mater. Des.* **60** 490–5
[53] Liu S Y, Gao F P, Zhang Q Y, Zhu X and Li W Z 2010 Fabrication of carbon nanotubes reinforced AZ91D composites by ultrasonic processing *Trans. Nonferrous Met. Soc. China (English Ed.)* **20** 1222
[54] Huang S J, Abbas A and Ballóková B 2019 Effect of CNT on microstructure, dry sliding wear and compressive mechanical properties of AZ61 magnesium alloy *J. Mater. Res. Technol.* **8** 4273
[55] Rashad M, Pan F, Hu H, Asif M, Hussain S and She J 2015 Enhanced tensile properties of magnesium composites reinforced with graphene nanoplatelets *Mater. Sci. Eng.* A **630** 36–44
[56] Rashad M, Pan F, Tang A, Asif M, Hussain S, Gou J and Mao J 2015 Improved strength and ductility of magnesium with addition of aluminum and graphene nanoplatelets (Al + GNPs) using semi powder metallurgy method *J. Ind. Eng. Chem.* **23** 243–50
[57] Rashad M, Pan F, Asif M and Tang A 2014 Powder metallurgy of Mg-1%Al-1%Sn alloy reinforced with low content of graphene nanoplatelets (GNPs) *J. Ind. Eng. Chem.* **20** 4250–5
[58] Chen L Y, Konishi H, Fehrenbacher A, Ma C, Xu J Q, Choi H, Xu H F, Pfefferkorn F E and Li X C 2012 Novel nanoprocessing route for bulk graphene nanoplatelets reinforced metal matrix nanocomposites *Scr. Mater.* **67** 29–32
[59] Xiang S L, Gupta M, Wang X J, Wang L D, Hu X S and Wu K 2017 Enhanced overall strength and ductility of magnesium matrix composites by low content of graphene nanoplatelets *Composites* A **100** 183–93

[60] Meng L, Hu X, Wang X, Zhang C, Shi H, Xiang Y, Liu N and Wu K 2018 Graphene nanoplatelets reinforced Mg matrix composite with enhanced mechanical properties by structure construction *Mater. Sci. Eng.* A **733** 414–8
[61] Du X, Du W, Wang Z, Liu K and Li S 2018 Ultra-high strengthening efficiency of graphene nanoplatelets reinforced magnesium matrix composites *Mater. Sci. Eng.* A **711** 633–42
[62] Chen L, Zhao Y, Hou H, Zhang T, Liang J, Li M and Li J 2019 Development of AZ91D magnesium alloy-graphene nanoplatelets composites using thixomolding process *J. Alloys Compd.* **788** 359–74
[63] Ramezanzade S, Ebrahimi G R, Parizi M T and Ezatpour H R 2020 Microstructure and mechanical characterizations of graphene nanoplatelets-reinforced Mg–Sr–Ca alloy as a novel composite in structural and biomedical applications *J. Compos. Mater.* **54** 711–28
[64] Munir K, Wen C and Li Y 2020 Graphene nanoplatelets-reinforced magnesium metal matrix nanocomposites with superior mechanical and corrosion performance for biomedical applications *J. Magnes. Alloy.* **8** 269–90
[65] Thein M A, Lu L and Lai M O 2006 Mechanical properties of nanostructured Mg-5 wt%Al-x wt%AlN composite synthesized from Mg chips *Compos. Struct.* **75** 206–12
[66] Sankaranarayanan S, Habibi M K, Jayalakshmi S, Jia Ai K, Almajid A and Gupta M 2015 Nano-AlN particle reinforced Mg composites: microstructural and mechanical properties *Mater. Sci. Technol.* **31** 1122–31
[67] Cao G, Choi H, Oportus J, Konishi H and Li X 2008 Study on tensile properties and microstructure of cast AZ91D/AlN nanocomposites *Mater. Sci. Eng.* A **494** 127–31
[68] Dieringa H, Katsarou L, Buzolin R, Szakács G, Horstmann M, Wolff M, Mendis C, Vorozhtsov S and StJohn D 2017 Ultrasound assisted casting of an AM60 based metal matrix nanocomposite, its properties, and recyclability *Metals* **7** 388
[69] Katsarou L, Mounib M, Lefebvre W, Vorozhtsov S, Pavese M, Badini C, Molina-Aldareguia J M, Jimenez C C, Pérez Prado M T and Dieringa H 2016 Microstructure, mechanical properties and creep of magnesium alloy Elektron21 reinforced with AlN nanoparticles by ultrasound-assisted stirring *Mater. Sci. Eng.* A **659** 84–92
[70] Sankaranarayanan S, Sabat R K, Jayalakshmi S, Suwas S, Almajid A and Gupta M 2015 Mg/BN nanocomposites: nano-BN addition for enhanced room temperature tensile and compressive response *J. Compos. Mater.* **49** 3045–55
[71] Seetharaman S, Subramanian J, Tun K S, Hamouda A S and Gupta M 2013 Synthesis and characterization of nano boron nitride reinforced magnesium composites produced by the microwave sintering method *Materials* **6** 1940–55
[72] Paramsothy M, Chan J, Kwok R and Gupta M 2011 Enhanced mechanical response of magnesium alloy ZK60A containing Si_3N_4 nanoparticles *Composites* A **42** 2093–100
[73] Paramsothy M, Chan J, Kwok R and Gupta M 2011 Enhanced mechanical response of hybrid alloy AZ31/AZ91 based on the addition of Si_3N_4 nanoparticles *Mater. Sci. Eng.* A **528** 6545–51
[74] Meenashisundaram G K, Nai M H, Almajid A and Gupta M 2016 Reinforcing low-volume fraction nano-tin particulates to monolithical, pure MG for enhanced tensile and compressive response *Materials* **9** 134
[75] Meenashisundaram G K, Seetharaman S and Gupta M 2014 Enhancing overall tensile and compressive response of pure Mg using nano-TiB_2 particulates *Mater. Charact.* **94** 178–88
[76] Choi H, Sun Y, Slater B P, Konishi H and Li X 2012 AZ91D/TiB_2 nanocomposites fabricated by solidification nanoprocessing *Adv. Eng. Mater.* **14** 291–5

[77] Paramsothy M and Gupta M 2013 ZrB_2 nanoparticle induced nano-LPSO-grain and nano-LPSO-layer reinforced ultra-high strength Mg-RE alloy *J. Mater. Sci.* **48** 8368–76

[78] Eliaz N and Metoki N 2017 Calcium phosphate bioceramics: a review of their history, structure, properties, coating technologies and biomedical applications *Materials* **10** 334

[79] Haghshenas M 2017 Mechanical characteristics of biodegradable magnesium matrix composites: a review *J. Magnes. Alloy.* **5** 189–201

[80] Khalil K A 2012 A new-developed nanostructured Mg/HAp nanocomposite by high frequency induction heat sintering process *IOP Conf. Ser.: Mater. Sci. Eng.* **40** 012031

[81] Khodaei M, Nejatidanesh F, Shirani M J, Iyengar S, Sina H and Savabi O 2020 Magnesium/nano-hydroxyapatite composite for bone reconstruction: the effect of processing method *J. Bionic Eng.* **17** 92–9

[82] Cao G, Zhang L, Zhang D, Liu Y, Gao J, Li W and Zheng Z 2019 Microstructure and properties of nano-hydroxyapatite reinforced WE43 alloy fabricated by friction stir processing *Materials* **12** 2994

[83] Razavi M, Fathi M H and Meratian M 2010 Microstructure, mechanical properties and bio-corrosion evaluation of biodegradable AZ91-FA nanocomposites for biomedical applications *Mater. Sci. Eng.* A **527** 6938–44

[84] Razavi M, Fathi M H and Meratian M 2010 Fabrication and characterization of magnesium–fluorapatite nanocomposite for biomedical applications *Mater. Charact.* **61** 1363–70

[85] Razavi M, Fathi M H and Meratian M 2010 Bio-corrosion behavior of magnesium–fluorapatite nanocomposite for biomedical applications *Mater. Lett.* **64** 2487–90

[86] Parande G, Manakari V, Gupta H and Gupta M 2018 Magnesium-β-tricalcium phosphate composites as a potential orthopedic implant: a mechanical/damping/immersion perspective *Metals* **8** 343

[87] He S Y, Sun Y, Chen M F, Liu D B and Ye X Y 2011 Microstructure and properties of biodegradable β-TCP reinforced Mg-Zn-Zr composites *Trans. Nonferrous Met. Soc. China (English Ed.)* **21** 814–9

[88] Liu D, Zuo Y, Meng W, Chen M and Fan Z 2012 Fabrication of biodegradable nano-sized β-TCP/Mg composite by a novel melt shearing technology *Mater. Sci. Eng.* C **32** 1253–8

[89] Zheng H R, Li Z, You C, Liu D B and Chen M F 2017 Effects of MgO modified β-TCP nanoparticles on the microstructure and properties of β-TCP/Mg-Zn-Zr composites *Bioact. Mater.* **2** 1–9

[90] Wong W L E and Gupta M 2007 Development of Mg/Cu nanocomposites using microwave assisted rapid sintering *Compos. Sci. Technol.* **67** 1541–52

[91] Zhong X L, Wong W L E and Gupta M 2007 Enhancing strength and ductility of magnesium by integrating it with aluminum nanoparticles *Acta Mater.* **55** 6338–44

[92] Meenashisundaram G K and Gupta M 2014 Low volume fraction nano-titanium particulates for improving the mechanical response of pure magnesium *J. Alloys Compd* **593** 176–83

[93] Parande G, Manakari V, Wakeel S, Kujur M S and Gupta M 2018 Enhancing mechanical response of monolithic magnesium using nano-NiTi (Nitinol) particles *Metals* **8** 1014

IOP Publishing

Magnesium-Based Nanocomposites
Advances and applications
Sankaranarayanan Seetharaman, Sravya Tekumalla and Manoj Gupta

Chapter 4

Temperature and strain rate effects on the tensile and compressive properties of magnesium nanocomposites

In this chapter, the effects of temperature and strain rates on the tensile and compressive properties of magnesium-based nanocomposites are summarized. While the room temperature mechanical properties of Mg nanocomposites have been extensively studied under quasi-static tension and compression loads, only very few research works report their high temperature response and these are presented in the first section. Similarly, the strain rate effects and the dynamic response of Mg nanocomposites at room temperature are discussed in the second section. Finally, the processing maps that are constructed to describe the combined effects of temperature and strain rate variations on the mechanical properties of Mg nanocomposites are also reviewed and summarized.

4.1 Introduction

Magnesium, as an ideal light-weight structural metal with high specific strength, damping capacity, and electromagnetic shielding capability, is an excellent prospect for applications in automobiles, aircrafts, consumer electronics, medical implants, and other fields [1, 2]. While the limited availability of slip systems at room temperature limits the cold formability of HCP structured Mg, thermal softening at elevated temperatures (as shown in the representative stress–strain curve of AZ31 alloy, figure 4.1) along with poor creep and oxidation resistance compared to aluminium alloys restrict its wider application [3–6].

Similarly, the variation in strain rate also affects the mode of deformation in Mg [8]. While most Mg materials show negligible sensitivity to strain rate at lower strains, the strain hardening rates are found to be high for larger strain rates. In this regard, heat processing maps that are constructed using the Arrhenius model and constitutive equations attract enormous interest in understanding the flow

doi:10.1088/978-0-7503-3535-5ch4

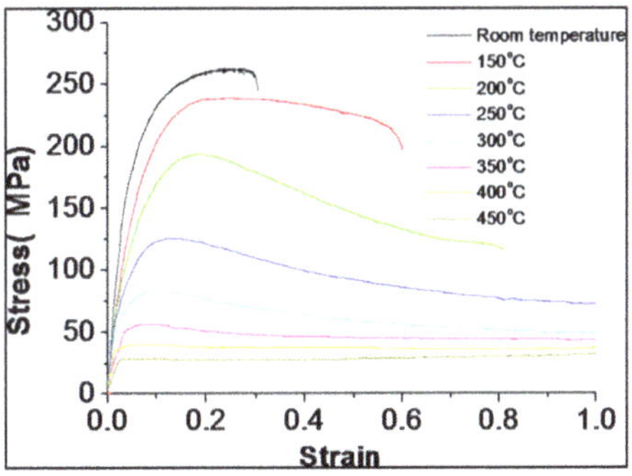

Figure 4.1. The effect of temperature on the mechanical properties of AZ31 magnesium alloy. Adapted from [7].

characteristics [9–11]. In view of this, some recent research works explored the effects of nanoscale reinforcement addition on the deformation behaviour of Mg materials and an attempt is made in this chapter to summarize the high temperature and high strain rate deformation properties of Mg nanocomposites. This chapter is divided into three sections. In the first section the temperature effects on the quasi-static mechanical properties of Mg nanocomposites are presented. The strain rate effects and the dynamic response at room temperature are discussed in the second section. The third section presents the details of processing maps that are constructed for Mg nanocomposites.

4.2 Temperature effects on the quasi-static mechanical properties of magnesium nanocomposites

Hassan *et al* [12] used the DMD method followed by hot extrusion to develop a Mg/Al_2O_3 nanocomposite reinforced with 1.11 vol.% Al_2O_3 nanoparticles. When tested for tensile properties at different temperatures from 25 °C to 200 °C, it was found that the Mg/1.11Al_2O_3 nanocomposite exhibited strength retention and matrix strain hardening up to 150 °C (see table 4.1). Above 150 °C the strength was reduced drastically. Similar results were also reported for AZ31 alloy nanocomposite reinforced with 1.5 vol.% nano-Al_2O_3 [13].

In a similar study, Habibi *et al* [14] investigated the high temperature mechanical properties of Mg nanocomposites containing a hierarchical configuration of Al_2O_3 nanoparticles embedded in micron-sized Al. In this study, the materials were synthesized using a microwave sintering assisted powder metallurgy method followed by hot extrusion. The extruded materials were tested for tension and compression properties at different temperatures between 25 °C and 200 °C and the results (table 4.2) showed a deterioration in the strength properties with increasing temperature. The results also highlighted the dismal strengthening effects from hierarchical (Al–Al_2O_3) reinforcement addition at temperatures above 100 °C. The increased ductility and improved tension–compression yield asymmetry of the tested Mg materials confirmed the activation of non-basal slips and dynamic recovery at

Table 4.1. The effect of temperature on the tensile properties of Mg/Al_2O_3 nanocomposites from [12, 13].

Materials	Testing temp.	Yield strength (MPa)	Ultimate strength (MPa)	Fracture strain (%)
$Mg/0.00Al_2O_3$	25 °C	97	173	7.4
$Mg/1.11Al_2O_3$	25 °C	175	246	14.0
	150 °C	162	224	45.0
	200 °C	95	110	44.7
AZ31	25 °C	172	263	10.4
$AZ31/1.5Al_2O_3$	25 °C	204	317	22.2
	150 °C	145	—	82.9
	200 °C	76	—	140.8
	250 °C	38	—	184.1

Table 4.2. The effect of temperature on the mechanical properties of $Mg–Al/Al_2O_3$ nanocomposites, from [14].

		Tensile properties			Compressive properties		
Material	Test temp.	Yield strength (MPa)	Ultimate strength (MPa)	Fracture strain (%)	Yield strength (MPa)	Ultimate strength (MPa)	Fracture strain (%)
Mg	25 °C	93 ± 01	153 ± 07	7.9 ± 3.4	91 ± 03	239 ± 11	19.8 ± 1.7
	100 °C	46 ± 02	103 ± 06	18.7 ± 2.5	80 ± 06	145 ± 10	23.4 ± 4.0
	150 °C	42 ± 10	77 ± 12	21.8 ± 3.2	55 ± 09	82 ± 12	22.5 ± 4.2
	200 °C	41 ± 08	57 ± 08	23.9 ± 2.9	45 ± 02	53 ± 02	21.1 ± 3.6
$Mg/0.972Al–0.22Al_2O_3$	25 °C	160 ± 07	225 ± 05	9.5 ± 0.4	151 ± 10	430 ± 11	13.1 ± 1.5
	100 °C	42 ± 07	136 ± 02	30.3 ± 5.2	126 ± 19	261 ± 29	18.3 ± 1.5
	150 °C	39 ± 05	88 ± 02	31.9 ± 2.6	60 ± 06	127 ± 08	20.2 ± 2.1
	200 °C	45 ± 02	67 ± 03	27.7 ± 2.7	42 ± 07	66 ± 07	20.0 ± 1.7
$Mg/0.972Al–0.66Al_2O_3$	25 °C	182 ± 10	276 ± 08	11.2 ± 1.5	164 ± 10	429 ± 05	10.2 ± 1.4
	100 °C	66 ± 09	153 ± 17	29.5 ± 3.4	119 ± 14	258 ± 15	27.0 ± 5.2
	150 °C	58 ± 02	109 ± 10	28.7 ± 4.6	89 ± 09	146 ± 18	27.8 ± 3.4
	200 °C	40 ± 05	70 ± 06	30.6 ± 2.6	54 ± 08	69 ± 09	24.1 ± 1.6
$Mg/0.972Al–1.11Al_2O_3$	25 °C	172 ± 12	236 ± 19	4.0 ± 0.3	120 ± 11	421 ± 04	13.2 ± 0.9
	100 °C	70 ± 08	143 ± 18	22.2 ± 1.7	104 ± 07	251 ± 25	16.5 ± 2.7
	150 °C	45 ± 06	103 ± 06	27.2 ± 5.1	66 ± 03	91 ± 05	25.0 ± 4.3
	200 °C	45 ± 05	64 ± 03	25.7 ± 3.3	49 ± 05	62 ± 08	22.1 ± 2.4

elevated temperature while the room temperature deformation under tensile and compressive loads were dominated by basal slip and tensile twinning, respectively.

The influence of extrusion temperature ratio on the high temperature tensile behaviour of AZ91/SiC nanocomposites was investigated by Nie *et al* [15]. In this study, the nanocomposites were produced using an ultrasonic assisted semisolid stirring method followed by hot extrusion. From the tensile test results (figure 4.2), the authors confirmed the effectiveness of nano-SiC reinforcements up to 175 °C. Similar observations were also reported (see table 4.3) for a CNT reinforced AZ31 and Elektron21/AlN nanoparticles reinforced alloy [13, 16, 17].

Mallick *et al* [18] investigated the mechanical behaviour of pure Mg and Mg/0.7Y_2O_3 nanocomposite in the temperature range 25 °C–250 °C and the results (figure 4.3) revealed strength loss and ductility enhancement in both pure Mg and Mg/0.7Y_2O_3 nanocomposites with increasing test temperature up to 200 °C. The ductility of the Mg/Y_2O_3 nanocomposite was reduced thereafter and the results were attributed to the change in crystallographic texture under the influence of

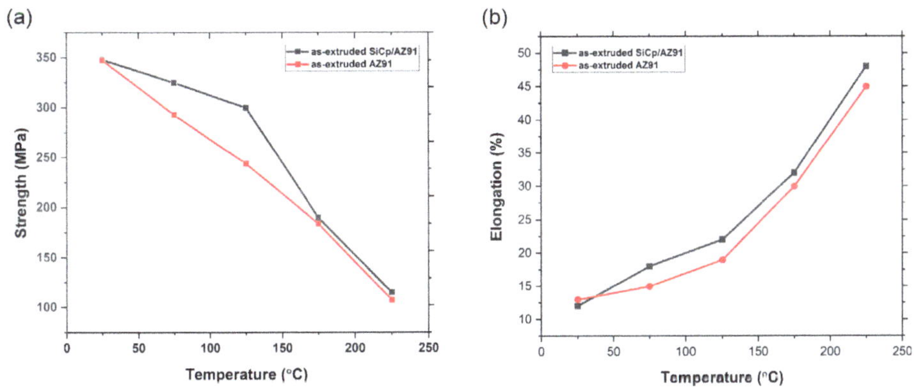

Figure 4.2. Mechanical properties of AZ91/SiC nanocomposites. Reprinted from [15], Copyright 2012, with permission from Elsevier.

Table 4.3. The effect of temperature on the mechanical properties of E21/AlN and AZ31/CNT nanocomposites from [16, 17].

Material	Temperature	Yield strength (MPa)	Ultimate strength (MPa)	Elongation (%)
Elektron21	25 °C	115 ± 5	229 ± 10	12 ± 3
	240 °C	86 ± 7	214 ± 15	28 ± 1
Elektron21/1 wt.%	25 °C	107 ± 3	226 ± 3	14 ± 1
AlN	240 °C	80 ± 2	217 ± 8	27 ± 1
AZ31/1.0CNT	25 °C	190	—	17.5
	150 °C	134	—	60.3
	200 °C	88	—	70.6
	250 °C	43	—	131.5

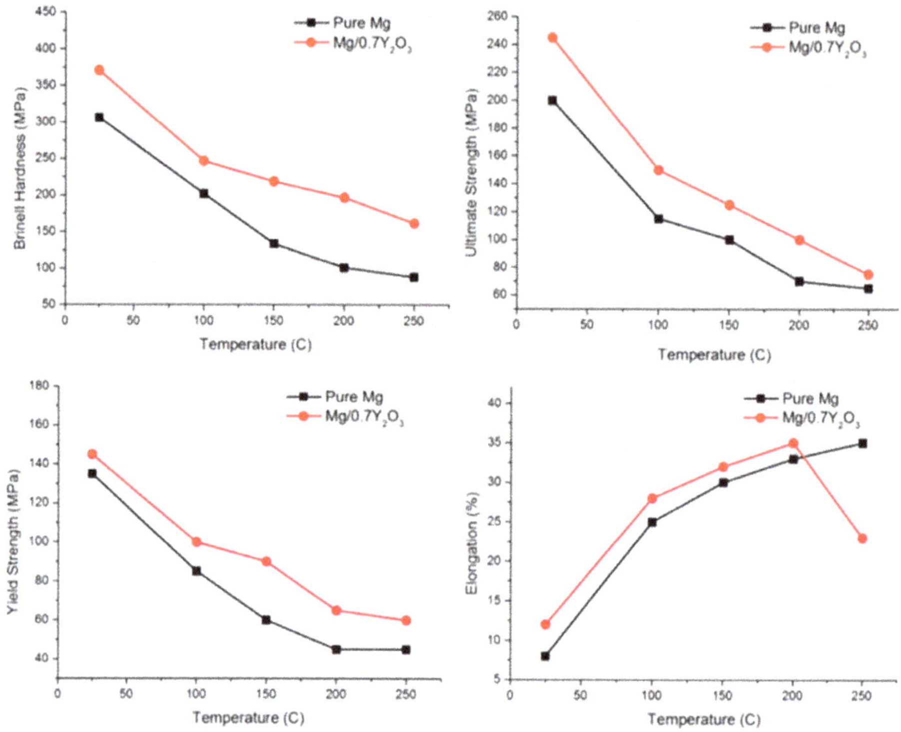

Figure 4.3. Properties of Mg/Y_2O_3 nanocomposites at different temperatures. Based on data from [18].

temperature and test loads. At all temperatures, the overall mechanical response of the nanocomposite was found to be superior compared to pure Mg.

The effects of copper addition (0.3–1.0 vol.%) on the high temperature mechanical properties of microwave sintered Mg/Y_2O_3 nanocomposites were also investigated [19]. In this study, although the mechanical property results showed strength loss at elevated temperatures, the presence of hybrid (Y_2O_3–Cu) reinforcement resulted in better yield strength compared to pure Mg (see figure 4.4).

4.3 Strain rate effects on room temperature mechanical properties

In order to evaluate the feasibility of magnesium nanocomposite usage in vehicle, aircraft, and armour applications, it is extremely important that the effects of dynamic loading conditions on the properties of these nanocomposites be investigated and understood thoroughly. However, the available literature suggests that only limited studies have been carried out so far.

Guo *et al* [20, 21] investigated the dynamic mechanical properties of DMD processed Mg nanocomposites reinforced with 0.22 vol.% Al_2O_3 nanoparticles. In this study, the results of mechanical testing under dynamic tensile loads (table 4.4) revealed significantly improved strengths and ductility compared to the quasi-static condition. Under compression loads, the developed materials displayed a

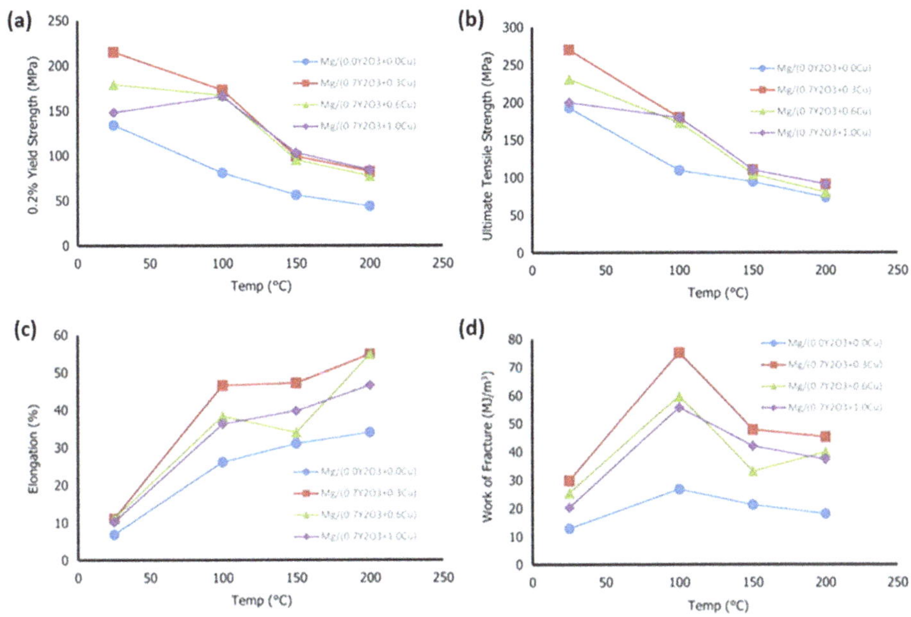

Figure 4.4. Mechanical properties of Mg/Y_2O_3–Cu nanocomposites. Reproduced with permission from [19] © Taylor & Francis.

Table 4.4. Dynamic tensile and compressive properties of Mg–6Al/Al_2O_3 nanocomposites [20, 21].

Material	Tensile properties				Compressive properties			
	Strain rate (s^{-1})	Yield strength (MPa)	Ultimate strength (MPa)	Failure strain (%)	Strain rate (s^{-1})	Yield strength (MPa)	Ultimate strength (MPa)	Failure train (%)
Mg–6Al	0.0012	221	289	7.7	0.001	183	453	12.9
	1000	286	314	13.8	1000–1250	—	484	18.5
	1600	301	313	13.4				
Mg–6Al/ 0.66Al_2O_3	0.0012	227	291	7.2	0.001	184	470	13.7
	1000	277	311	13.3	1000–1250	—	491	18.1
	1600	289	328	15.2				

notable increase in ductility and compressive strength due to strain rate variation. On the other hand, the influence of nanoparticle addition was not significantly pronounced. The tension–compression yield asymmetry was also maintained at different strain rates.

In a similar study [22], AZ31 alloy and its composites, reinforced by different volume fractions (1 vol.%, 1.4 vol.%, and 3 vol.%) of 50 nm Al_2O_3 nanoparticles, were fabricated using the DMD method and tested for tensile properties at different strain rates between 10^{-4} and 10^3 s^{-1}. Results from this study (see figure 4.5) showed

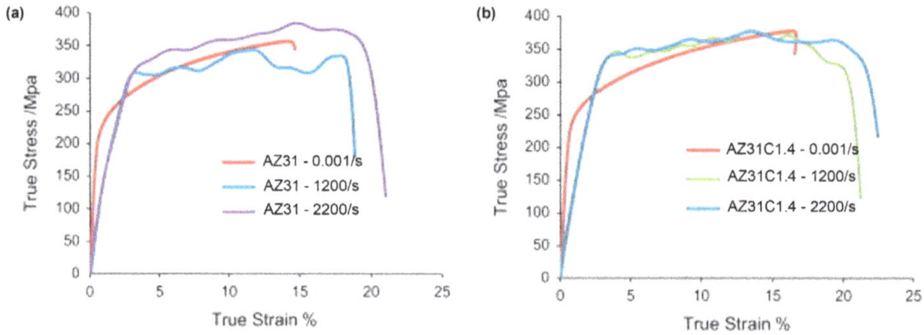

Figure 4.5. Tensile stress–strain curves of (a) AZ31 and (b) AZ31/Al_2O_3 nanocomposites. Reprinted with permission from [22]. Copyright 2013 Elsevier.

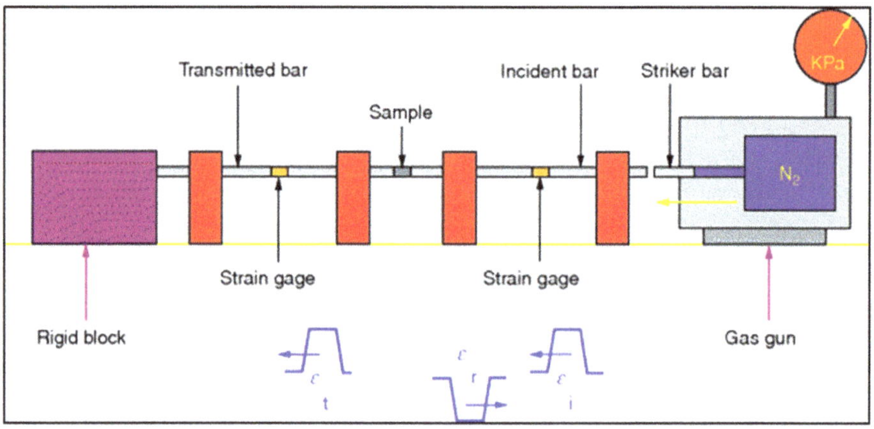

Figure 4.6. Schematic of a split-Hopkinson bar. Reproduced with permission from [24] © John Wiley & Sons.

significant improvements to yield strength and ductility due to nano-reinforcement addition at all strain rates. While the strain rate sensitivity (SRS) of nanocomposites was decreased, the strain hardening behaviour of AZ31 matrix alloy remained unchanged at higher strain rates. In this study as well, the enhanced dislocation movement and improved ductility of the nanocomposite were attributed to the confirmed texture changes due to nanoparticle addition.

Chen *et al* [23] studied the dynamic compressive response of DMD processed AZ31/Al_2O_3 for strain rates up to $2.5 \times 10^3\ s^{-1}$. In this study, the dynamic properties were investigated using a split-Hopkinson pressure bar (schematic in figure 4.6) in which a test sample was sandwiched between an input and the output bar and the values of displacement, velocity, and stress conditions were measured to evaluate the dynamic properties [24].

From the test results (as shown in figure 4.7) it was reported that the nanocomposites displayed a negative SRS due to the localized stresses caused by the nano-Al_2O_3 reinforcements which in turn reduced the required stress to activate

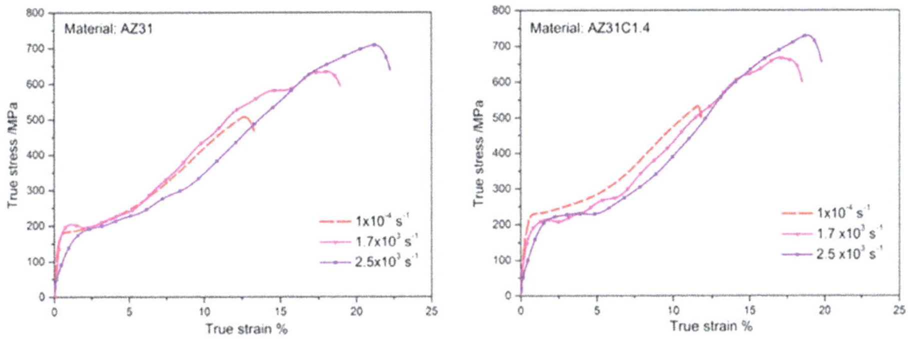

Figure 4.7. Compressive stress–strain curves of (a) AZ31 and (b) AZ31/Al_2O_3 nanocomposite. Reprinted from [23], Copyright 2016, with permission from Elsevier.

tension twinning at high strain rates. The results also confirmed that the nano-Al_2O_3 particle addition did not have any influence on the compressive failure strain of AZ31, unlike under tensile loads wherein a significant enhancement was reported which again could be attributed to the difference in deformation modes.

Chen *et al* [25] also studied the dynamic compressive properties of DMD processed Mg–4Zn–3Gd–1Ca alloy and its nano-ZnO particle reinforced composite at strain rates up to 1×10^3 s^{-1}. While both materials displayed a negative SRS (figure 4.8), the Mg–4Zn–3Gd–1Ca/2ZnO nanocomposite displayed enhanced plastic flow in the early stage of deformation (by ~100 MPa) due to significant grain refinement and the ability of nanoparticles to retard twin nucleation and growth at grain boundaries. However, the effectiveness of nanoparticles was found to be weakened at a higher strain rate. It was also reported that the strong solute strengthening from Gd and Ca enhanced the critical resolved shear stress for slip and maintained tension twinning as the dominant deformation mechanism for yielding. In a similar study, Habibi *et al* [26] investigated the effects of strain rate change on the properties of Mg/Al-CNT nanocomposites under tension and compression loads. In this study, the nanocomposites were fabricated using a microwave assisted powder metallurgy method and were subjected to quasi-static and dynamic loads. From the results (table 4.5) it was found that the presence of hybrid Al-CNT particles significantly improved the mechanical response of Mg in both the quasi-static and dynamic regimes under both tension and compression loads.

4.4 Combined effects of temperature and strain rate on the mechanical properties of magnesium nanocomposites

Xiao *et al* [27] investigated the effects of strain rate and temperature on the compressive behaviour of AZ31B alloy nanocomposites containing different amounts (0.5 and 1 vol.%) of SiC nanoparticles. In this study, the AZ31B matrix alloy and its nanocomposites were fabricated using an ultrasonic vibration assisted casting process. Hopkinson pressure bar experiments were used to quantify the strain rate/temperature sensitivity and strain hardening behaviour and the results

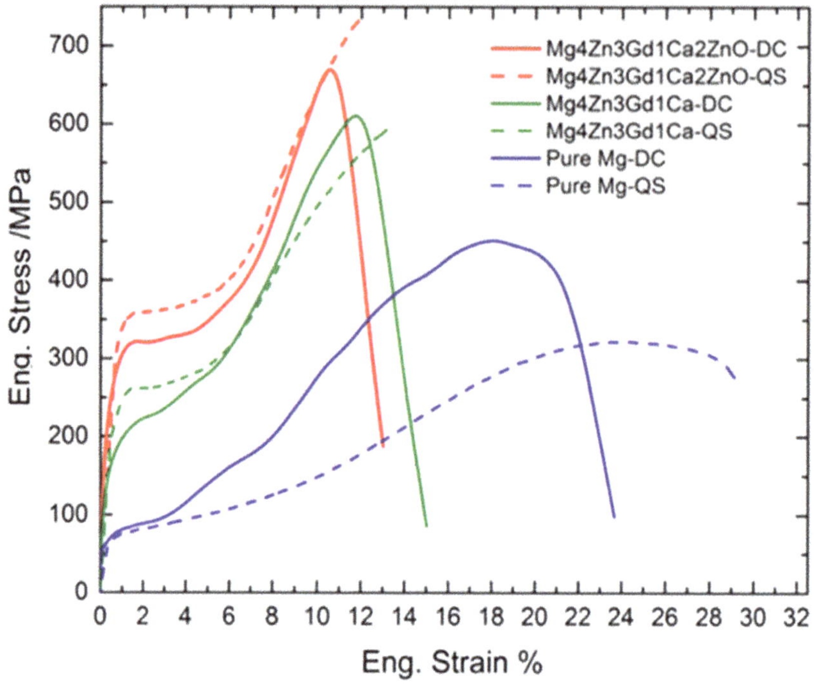

Material	Load*	Yield Strength (MPa)	Compressive Strength (MPa)	Failure Strain (%)	Initial Grain Size (μm)
Pure Mg	QS	88 ± 11	335 ± 14	25.6 ± 2.2	24.6 ± 4.0
	D	91 ± 11	443 ± 6.1	20 ± 0.8	
Mg-4Zn-3Gd-1Ca	QS	259 ± 3	585 ± 19	12.6 ± 0.3	1.42 ± 0.20
	D	214 ± 25	602 ± 26	12 ± 0.4	
Mg-4Zn-3Gd-1Ca/2ZnO	QS	359 ± 2	703 ± 40	10.6 ± 0.4	0.91 ± 0.47
	D	312 ± 16	654 ± 31	11.4 ± 0.2	

*QS denotes loading at a low strain rate (5×10^{-3} s^{-1}); D denotes high strain rate compression (10^3 s^{-1}).

Figure 4.8. Compressive properties of Mg–4Zn–3Gd–1Ca alloy and its nano-ZnO reinforced composite under quasi-static and dynamic loading. Reprinted from [25], Copyright 2017, with permission from Elsevier.

(figures 4.9 and 4.10) revealed superior strengthening and a strong rate dependence. While the performance of nanocomposites was better at any given strain rate and temperature, the flow stress was found to increase with an increase in the strain rate (figure 4.9) and its significance was more pronounced at room temperature. Similarly, at a constant strain rate, the flow stress was found to decrease with increasing temperatures and the temperature had a significant effect at quasi-static loading (figure 4.10).

The DMD processed AZ31 alloy and AZ31/Al_2O_3 nanocomposites were tested for compression properties at different temperatures (250 °C–400 °C) and strain rates (0.01–1.0 s^{-1}) and the results showed comparable properties, as shown in

Table 4.5. Quasi-static and dynamic mechanical properties of Mg/Al-CNT nanocomposites at room temperature as reported by [26].

Material	Load[a]	Ultimate tensile strength (MPa)	Ultimate compressive strength (MPa)	Tensile failure strain (%)	Compressive failure strain (%)
Mg	QS	153 ± 07	239 ± 11	7.9 ± 3.4	19.8 ± 1.7
	D	215 ± 08	341 ± 10	10.0 ± 0.6	18.4 ± 1.3
Mg/1.00Al-0.09CNT	QS	206 ± 04	337 ± 14	9.1 ± 0.6	10.1 ± 1.5
	D	230 ± 17	390 ± 11	8.1 ± 0.8	9.5 ± 1.5
Mg/1.00Al-0.18CNT	QS	208 ± 08	421 ± 13	11.2 ± 2.9	12.5 ± 0.4
	D	241 ± 12	462 ± 10	9.0 ± 1.3	11.1 ± 1.7
Mg/1.00Al-0.30CNT	QS	227 ± 14	424 ± 12	8.6 ± 0.4	12.4 ± 1.7
	D	270 ± 10	471 ± 25	7.5 ± 1.0	11.0 ± 1.0
Mg/1.00Al-0.50CNT	QS	220 ± 06	401 ± 15	4.6 ± 0.5	10.6 ± 1.5
	D	278 ± 11	479 ± 10	4.3 ± 1.0	11.2 ± 0.6

[a] QS denotes loading at a low strain rate ($\times 10^{-4}$ s^{-1}); D denotes high strain rate compression (2×10^3 s^{-1}).

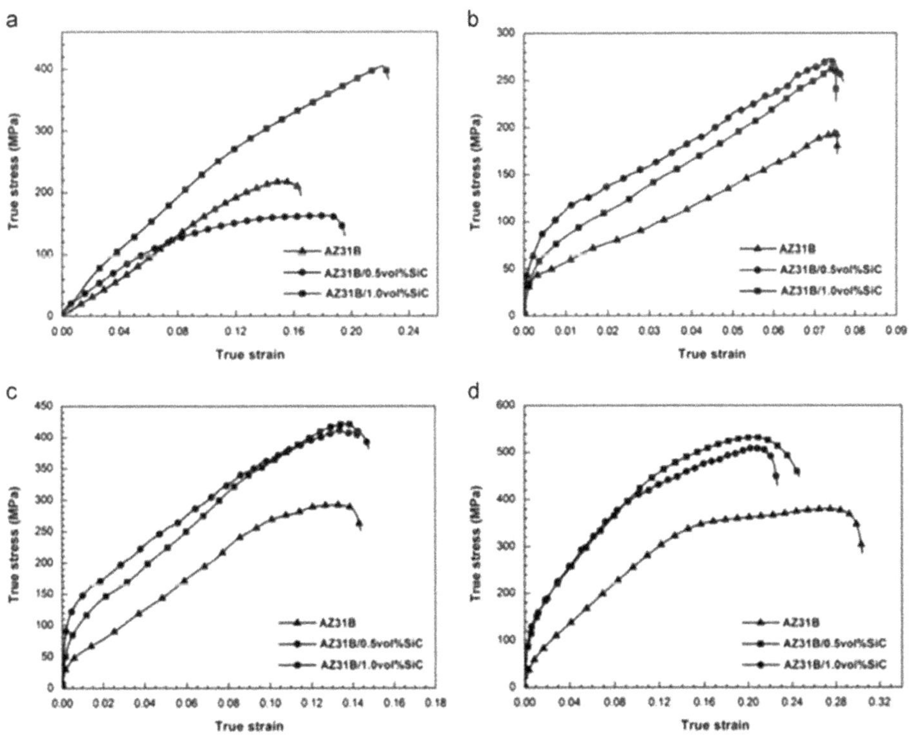

Figure 4.9. Room temperature mechanical behaviour of AZ31B and AZ31B/SiC nanocomposites at strain rates of (a) 0.001 s^{-1}, (b) 700 s^{-1}, (c) 1400 s^{-1}, and (d) 3300 s^{-1}. Reprinted from [27], Copyright 2014, with permission from Elsevier.

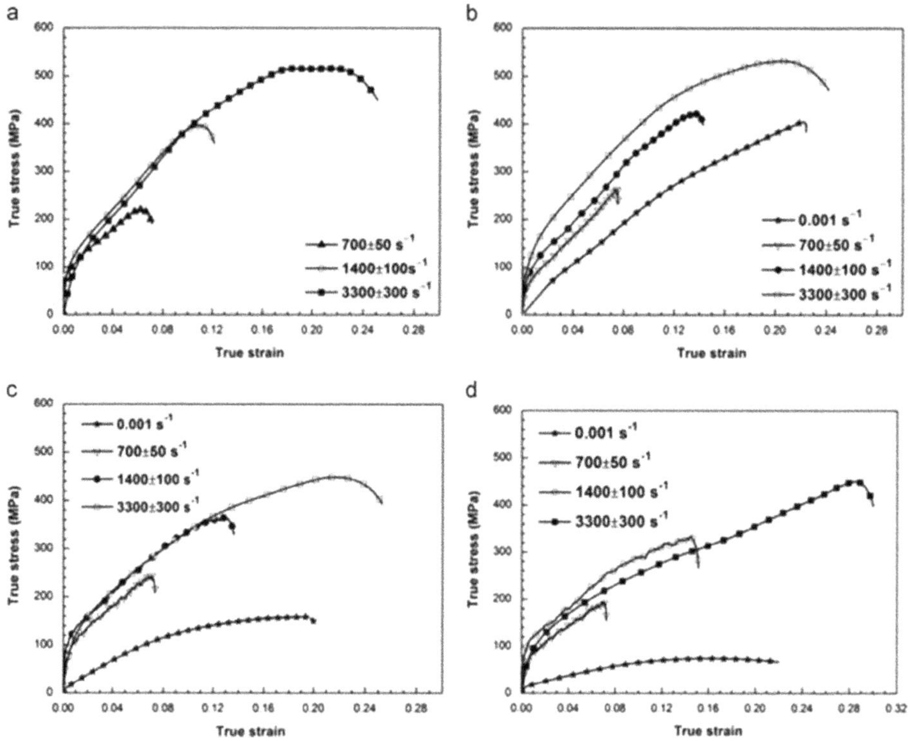

Figure 4.10. Stress–strain curves of AZ31B/1SiC nanocomposite at different strain rates and temperatures of (a) −30 °C, (b) 25 °C, (c) 100 °C, and (d) 200 °C. Reprinted from [27], Copyright 2014, with permission from Elsevier.

figure 4.11 [28]. The mechanical testing results at different temperatures and strain rates, as shown in figure 4.11, were used to construct a three-dimensional processing map (see figure 4.12) to describe the material's response to hot working process parameters, i.e. temperature (T) and strain rate ($\varepsilon°$). In general, processing maps are constructed on the basis of a dynamic materials model by superimposing the power dissipation and instability maps. While the power dissipation map presents various domains of microstructural mechanisms such as dynamic recrystallization (DRX) and recovery (DRY), the instability map separates out the regions of unstable material flow. In the processing map, while the high strain rate–high temperature zone was identified as the instability regime as characterized by shear bands, flow localization, and twinning, the low strain rate regime with dominant DRX was identified as the stable domain. The processing map also highlighted 0.01 s^{-1} and 400 °C as the optimum strain rate and temperature to process the Mg alloy nanocomposite.

In a similar study, Zhong *et al* [29] also developed a processing map (see figure 4.12) for the DMD processed $AZ31/1.5Al_2O_3$ nanocomposite using the results of mechanical testing performed at different temperatures (250 °C–500 °C) and strain rates (0.0003–10 s^{-1}). The processing map (figure 4.12) of both the AZ31 alloy

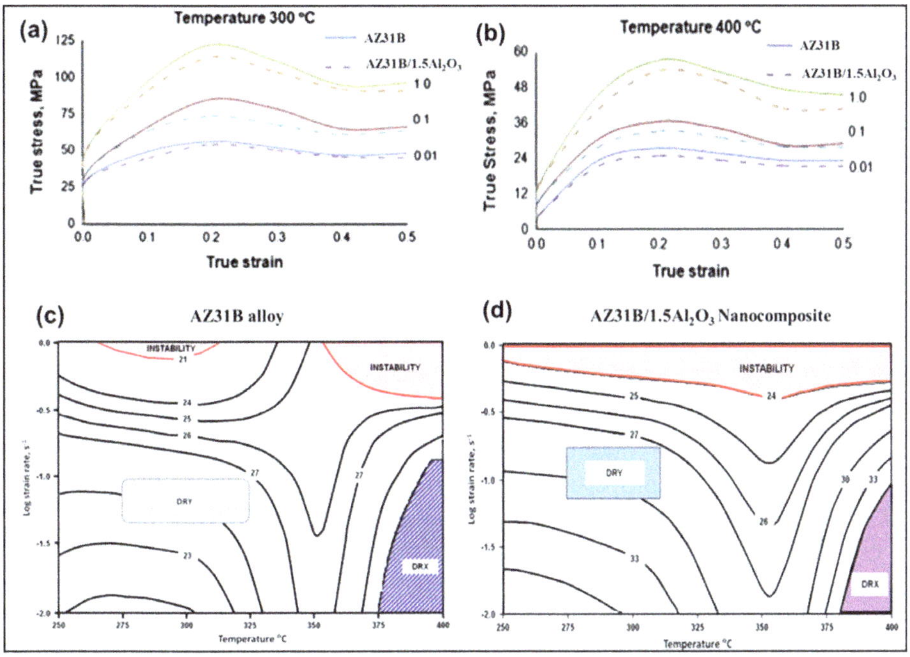

Figure 4.11. True stress–strain curves for AZ31B alloy and AZ31B/Al_2O_3 nanocomposite at temperatures (a) 300 °C and (b) 400 °C at different strain rates, and processing maps for the (c) AZ31B alloy and (d) AZ31B/Al_2O_3 nanocomposite. Reprinted from [28], Copyright 2013, with permission from Elsevier.

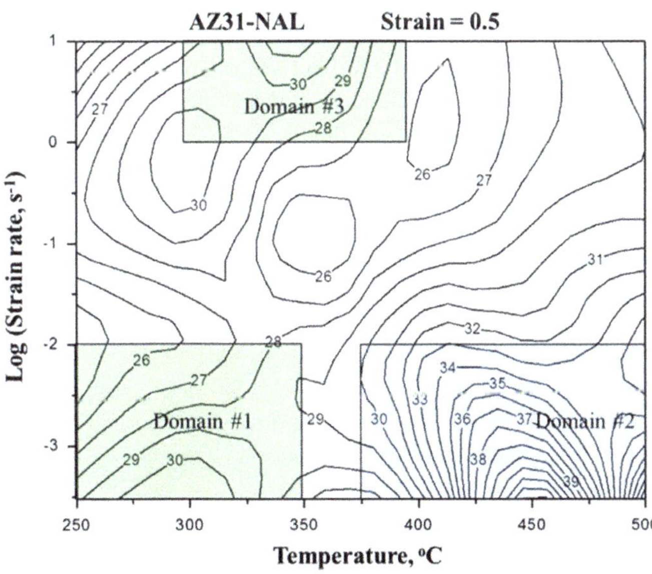

Figure 4.12. Processing map for AZ31/1.5Al_2O_3 nanocomposite as proposed by Zhong *et al.* Reprinted from [29], Copyright 2014, with permission from Elsevier.

and its nanocomposite consisted of three domains: (i) 250 °C–350 °C/0.0003–0.01 s^{-1}, (ii) 300 °C–400 °C/1–10 s^{-1}, and (iii) 375 °C–500 °C/0.0003–0.01 s^{-1}. While domains 1 and 2 were identified with fine-grained microstructure due to DRX, domain 3 was dominated by grain boundary sliding which resulted in wedge and inter-crystalline cracking under compression and tension loads, respectively. Although domain 1 in the AZ31/Al_2O_3 nanocomposite occurred under similar conditions as the AZ31 alloy, the peak efficiency was found to decrease from 43% to 30% due to the weaker initial texture of the nanocomposite. Similarly, domain 2 moved to a lower temperature (by 50 °C) as the fine grains were found to promote the grain boundary diffusion. In the case of domain 3 it moved slightly to a higher temperature.

Wang *et al* [3] and Nie *et al* [30] investigated the hot deformation behaviour of as-cast AZ91 magnesium alloy reinforced with SiC nanoparticles. In these studies, the nanocomposites were produced using the ultrasonic assisted stir casting method and tested for compression properties in the temperature and strain rate ranges between 250 °C–400 °C and 0.001 s^{-1} to 1 s^{-1}, respectively. While the low temperature compression occurred by incoordinate deformation compression due to profuse twinning and basal slip, the <*c*+*a*> glide was reported as dominant at high temperatures (>350 °C) as the SiC nanoparticles reduced the activation energy for hot deformation. From the processing map (figure 4.13), while the ranges of temperature (250 °C–400 °C) and strain rate (0.018 s^{-1}–0.001 s^{-1}) were identified as the optimized workability regions for both the AZ91 alloy and AZ91/SiC nanocomposite, the instability region for AZ91/SiC was reduced at high temperatures.

Nie *et al* [31] also investigated the hot deformation processing of AZ91/SiC nanocomposite prepared by 12 passes of multi-pass forging. In this study, the as-cast coarse-grained nanocomposites were subjected to multi-pass forging. The results of compression testing at different temperatures (250 °C–400 °C) and strain rates (0.001–1 s^{-1}) were used to calculate the activation energy and the stress factor to understand the effects of grain refinement and second phase precipitates on the hot deformation behaviour. Microstructural analysis revealed an increase in the average

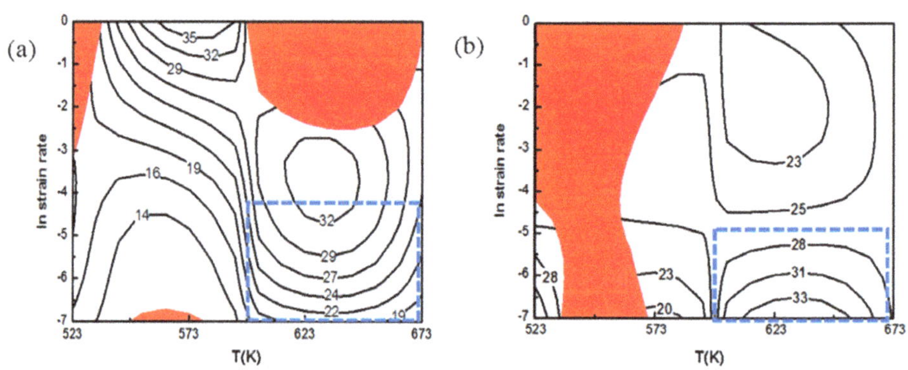

Figure 4.13. Processing maps for (a) AZ91 and (b) AZ91/SiC nanocomposite at a strain of 0.5. Reprinted from [30], with permission from MDPI.

grain size with increasing temperatures alongside a reduction in the volume of second phases which in turn affected the deformation zones. While the fine-grained nanocomposites with an average grain size of 1.5 μm recorded no dynamic softening post-peak stress, its critical stress was reported as low compared to the coarse-grained nanocomposite (average grain size –96 μm). Hence, the authors concluded that grain boundary slip assisted dislocation climb was the dominant deformation mechanism for the fine-grained nanocomposite while the deformation of the coarse-grained nanocomposite was reported as being controlled by dislocation climb resulting from lattice diffusion. Based on the processing map (figure 4.14), the optimal temperature for the hot compression of fine-grained nanocomposite was reported as 300 °C, which is significantly lower compared to the coarse-grained nanocomposite (350 °C–400 °C).

In a recent study, hot deformation tests were conducted on Mg/ZnO nanocomposites [32] in the temperature and strain rate range (250 °C–400 °C) and (0.01–1.5 s^{-1}), respectively. While the processing map (figure 4.15) showed DRX formation and better workability at temperatures between 375 °C and 400 °C, at a strain rate of 0.01 s^{-1} DRY formation occurred between 300 °C and 350 °C at a strain rate of 0.01 s^{-1}. Defects such as cracks and pores which are indicative of instability were observed at high strain rates (>0.1 s^{-1}) at all hot working temperatures.

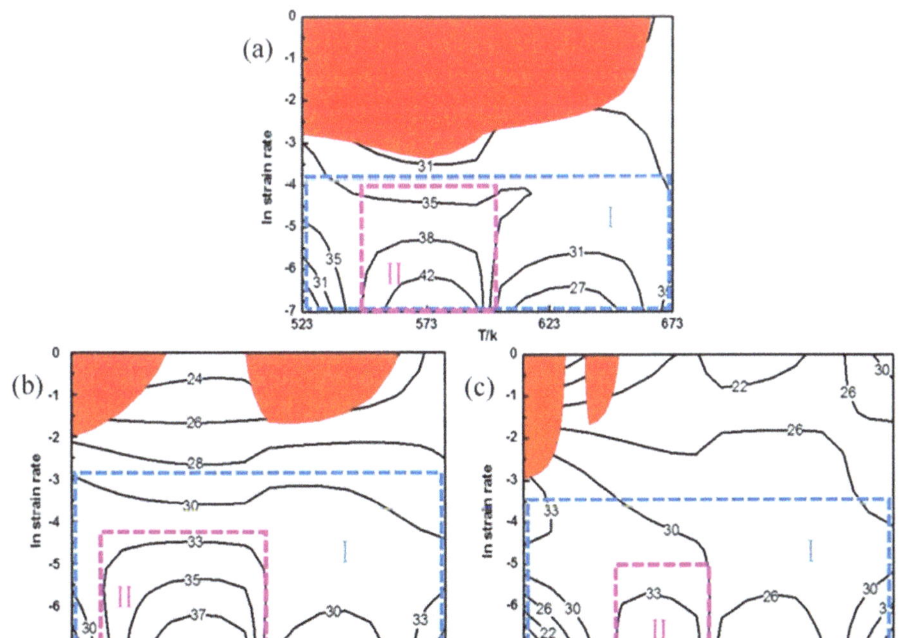

Figure 4.14. Processing maps of the fine-grained nanocomposite compressed at different strains: (a) 0.3, (b) 0.4, and (c) 0.5. Reprinted from [31], with permission from MDPI.

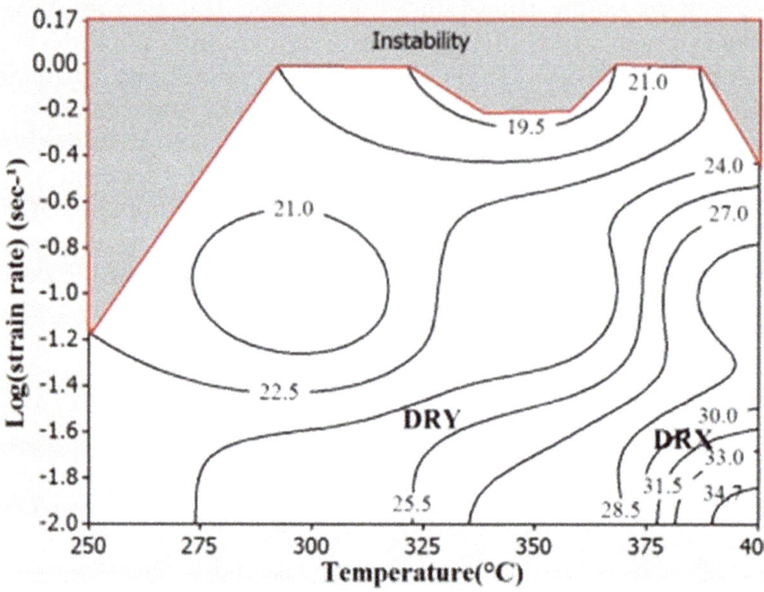

Figure 4.15. Processing map of the Mg/ZnO nanocomposite. Reprinted from [32], Copyright 2015, with permission from Elsevier.

4.5 Summary

As the lightest structural metal, magnesium has excellent application potential in automobile, aerospace, and electronic sectors. However, the limitations such as low modulus, loss of strength at high temperature, poor creep resistance, and rampant corrosion restrict the applications of Mg. Most of these limitations can be circumvented through alloying and reinforcement additions. In this regard, Mg-based nanocomposites developed through nanoscale reinforcement addition have received increased attention as the addition of only a very small amount of nanoscale reinforcements results in significant enhancements to mechanical properties. While most of the published literature on Mg nanocomposites reports the room temperature mechanical properties, only very few research works have focussed on the high temperature mechanical properties of Mg nanocomposites. From these studies, it is evident that the addition of nanoscale reinforcement is beneficial at low temperatures as they help in grain refinement and texture modification to bring in additional non-basal slip systems. However, at elevated temperatures, the non-basal slip systems are naturally active, which result in comparable deformation behaviour. These studies also highlighted the strength loss with increasing temperatures due to thermal softening and grain growth of the Mg matrix. With respect to strain rate, no noticeable effects were reported, i.e. the Mg nanocomposites exhibited more or less similar mechanical properties under different strain rates. From the processing maps that are constructed from the results of mechanical testing at different temperatures and strain rates, most Mg nanocomposites were found to exhibit an optimum hot working temperature range between 300 °C to 400 °C at a strain rate of 0.01 s^{-1}.

While any increase in strain rate above 0.01 s^{-1} requires a working temperature of more than 400 °C, the high temperature–high strain combination would more often result in enhanced cracking and porosity.

References

[1] Gupta M and Sharon N M L 2010 *Magnesium, Magnesium Alloys, and Magnesium Composites* (New York: Wiley)

[2] Gupta M and Wong W L E 2015 Magnesium-based nanocomposites: lightweight materials of the future *Mater. Charact.* **105** 30–46

[3] Wang T, Nie K B, Deng K K and Liang W 2016 Analysis of hot deformation behavior and microstructure evolution of as-cast SiC nanoparticles reinforced magnesium matrix composite *J. Mater. Res.* **31** 3437–47

[4] Pekguleryuz M, Kainer K and Kaya A A 2013 *Fundamentals of Magnesium Alloy Metallurgy* (Cambridge: Woodhead)

[5] Miranda R 2013 Advances in wrought magnesium alloys *Int. J. Environ. Stud.* **70** 336

[6] Yamashita A, Horita Z and Langdon T G 2001 Improving the mechanical properties of magnesium and a magnesium alloy through severe plastic deformation *Mater. Sci. Eng.* A **300** 142–7

[7] National Engineering Research Center for Magnesium Alloys n.d. Stress–strain curves of AZ31 Mg alloy under different temperatures *National Engineering Research Center for Magnesium Alloys* (Accessed: 12 June 2020) http://chinaccma.net/ccmg/common/images/image06.gif

[8] Singh A and Saal J E 2014 Dynamic properties of magnesium alloys *JOM* **66** 275–6

[9] Li J, Liu J and Cui Z 2014 Characterization of hot deformation behavior of extruded ZK60 magnesium alloy using 3D processing maps *Mater. Des.* **56** 889–97

[10] Guo Y, Xuanyuan Y, Lia C and Yang S 2019 Characterization of hot deformation behavior and processing maps of Mg-3Sn-2Al-1Zn-5Li magnesium alloy *Metals* **9** 1262

[11] Liao H M, Zeng M, Long S Y and Cao H X 2011 High temperature deformation behaviour and processing maps of AZ81 magnesium alloy *Mater. Sci. Forum* **686** 168–75

[12] Hassan S F, Tan M J and Gupta M 2008 High-temperature tensile properties of Mg/Al_2O_3 nanocomposite *Mater. Sci. Eng.* A **486** 56–62

[13] Hassan S F, Paramsothy M, Patel F and Gupta M 2012 High temperature tensile response of nano-Al_2O_3 reinforced AZ31 nanocomposites *Mater. Sci. Eng.* A **558** 278–84

[14] Habibi M K, Qian M, Hamouda A S and Gupta M 2012 Differentiating the mechanical response of hierarchical magnesium nano-composites as a function of temperature *Mater. Des.* **42** 102–10

[15] Nie K B, Wang X J, Xu L, Wu K, Hu X S and Zheng M Y 2012 Effect of hot extrusion on microstructures and mechanical properties of SiC nanoparticles reinforced magnesium matrix composite *J. Alloys Compd.* **512** 355–60

[16] Hassan S F, Paramsothy M, Yilbas B S and Gupta M 2014 Study of comparative effectiveness of thermally stable nanoparticles on high temperature deformability of wrought AZ31 alloy *J. Mater. Res.* **29** 1264–9

[17] Katsarou L, Mounib M, Lefebvre W, Vorozhtsov S, Pavese M, Badini C, Molina-Aldareguia J M, Jimenez C C, Pérez Prado M T and Dieringa H 2016 Microstructure, mechanical properties and creep of magnesium alloy Elektron21 reinforced with AlN nanoparticles by ultrasound-assisted stirring *Mater. Sci. Eng.* A **659** 84–92

[18] Mallick A, Tun K S, Vedantam S and Gupta M 2010 Mechanical characteristics of pure Mg and a Mg/Y_2O_3 nanocomposite in the 25–250 °C temperature range *J. Mater. Sci.* **45** 3058–66

[19] Fida Hassan S, Al-Aqeeli N, Gasem Z M, Tun K S and Gupta M 2016 Magnesium nanocomposite: increasing copperisation effect on high temperature tensile properties *Powder Metall.* **59** 66–72

[20] Guo Y B, Shim V P W and Tan B W F 2012 Dynamic tensile properties of magnesium nanocomposite *Mater. Sci. Forum* **706–709** 780–5

[21] Guo Y B, Shim V P W and Tan B W F 2010 Response of magnesium nanocomposites to quasi static and dynamic loading *Proc. IMPLAST 2010 Conf. (Providence, RI, USA, 12-14 October 2010)*

[22] Chen Y, Guo Y B, Gupta M and Shim V P W 2013 Dynamic tensile response of magnesium nanocomposites and the effect of nanoparticles *Mater. Sci. Eng.* A **582** 359–67

[23] Chen Y, Guo Y B, Gupta M and Shim V P W 2016 A study of the dynamic compressive response of AZ31/Al_2O_3 nanocomposites and the influence of nanoparticles *Int. J. Impact Eng.* **89** 114–23

[24] Gray G T R III 2012 High-strain-rate testing of materials: the split-Hopkinson pressure bar *Characterization of Materials* (New York: Wiley) pp 1–15

[25] Chen Y, Tekumalla S, Guo Y B, Shabadi R, Shim V P W and Gupta M 2017 The dynamic compressive response of a high-strength magnesium alloy and its nanocomposite *Mater. Sci. Eng.* A **702** 65–72

[26] Habibi M K, Pouriayevali H, Hamouda A M S and Gupta M 2012 Differentiating the mechanical response of hybridized Mg nano-composites as a function of strain rate *Mater. Sci. Eng.* A **545** 51–60

[27] Xiao J, Shu D W and Wang X J 2014 Effect of strain rate and temperature on the mechanical behavior of magnesium nanocomposites *Int. J. Mech. Sci.* **89** 381–90

[28] Srinivasan M, Loganathan C, Narayanasamy R, Senthilkumar V, Nguyen Q B and Gupta M 2013 Study on hot deformation behavior and microstructure evolution of cast-extruded AZ31B magnesium alloy and nanocomposite using processing map *Mater. Des.* **47** 449–55

[29] Zhong T, Rao K P, Prasad Y V R K, Zhao F and Gupta M 2014 Hot deformation mechanisms, microstructure and texture evolution in extruded AZ31–nano-alumina composite *Mater. Sci. Eng.* A **589** 41–9

[30] Nie K, Kang X, Deng K, Wang T, Guo Y and Wang H 2018 Effect of SiC nanoparticles on hot deformation behavior and processing maps of magnesium alloy AZ91 *Nanomaterials* **8** 82

[31] Nie K, Zhu Z, Deng K, Wang T and Han J 2019 Hot deformation behavior and processing maps of SiC nanoparticles and second phase synergistically reinforced magnesium matrix composites *Nanomaterials* **9** 57

[32] Selvam B, Marimuthu P, Narayanasamy R, Senthilkumar V, Tun K S and Gupta M 2015 Effect of temperature and strain rate on compressive response of extruded magnesium nanocomposite *J. Magnes. Alloy.* **3** 224–30

IOP Publishing

Magnesium-Based Nanocomposites

Advances and applications

Sankaranarayanan Seetharaman, Sravya Tekumalla and Manoj Gupta

Chapter 5

The fatigue response of magnesium-based nanocomposites

As magnesium is approximately 33% lighter than aluminium, magnesium-based materials with comparable mechanical properties are highly suitable for weight critical applications in transportation industries. The biodegradable nature of Mg-based materials also makes them ideal candidates for biomedical applications. Some of the limitations of Mg include its poor ductility, low elastic modulus, and poor creep and corrosion resistance, which can be circumvented through the addition of nano-sized reinforcements. As most engineering and biomedical applications involve cyclic loads, it is vital to understand the fatigue performance of Mg-based nanocomposites. However, only very few detailed studies have been conducted in the last two decades and an attempt is made in this chapter to summarize the key findings in the published literature.

5.1 Introduction

Lightweight magnesium alloys are excellent candidates for weight-critical applications. However, they suffer from serious limitations such as poor room temperature formability, loss of strength at high temperature, and low resistance to wet corrosion. With the latest developments in Mg-based nanocomposites, most of these limitations can be circumvented and the extended use of these materials is foreseen in automobiles, aircraft structures, consumer electronics, sporting equipment, and biomedical implants, etc [1]. Since many of these applications also involve cyclic loads, it is essential to understand the fatigue behaviour of Mg-based nanocomposites. From the available literature, it can be seen that a number of recent research works have focussed on the fatigue crack growth behaviour of magnesium alloys [2–4], however, only a few research works have focussed on the effects of nanoscale reinforcement addition and their key findings are summarized in this chapter.

doi:10.1088/978-0-7503-3535-5ch5

5.2 The mechanism of fatigue deformation

Fatigue deformation consists of four stages (see figure 5.1): (i) crack-free damage or cyclic hardening, (ii) crack initiation, (iii) crack propagation, and (iv) rupture [5]. Crack-free damage or cyclic hardening is the initial stage of fatigue deformation. It involves microplastic deformation to generate complex dislocation structures which upon interaction with microstructural constituents such as grain boundaries and precipitates result in increased dislocation densities. Since the formation of complex dislocation structures is unlikely in HCP Mg materials, the crack-free fatigue damage of Mg depends on the interaction between dislocations and microstructural constituents [2–5]. While fatigue cracks develop from the slip bands in defect-free materials, they also initiate from the sites of defects, inclusions, and surface/edge layers. Therefore, the crack initiation stage in defect-free materials is significantly greater than the crack propagation stage and constitutes ~90% of total fatigue life [2].

Post-initiation, the short cracks advance in an intergranular or trans-granular manner as per the microstructure. Although the crack propagation rate is generally faster in the case of Mg alloys, it is strongly influenced by the microstructural constituents. For example, the precipitates in age hardened Mg–Li, Mg–Nd, and Mg–Zn alloys exhibit a deceleration effect. On the other hand, the $Mg_{17}Al_{12}$ phases accelerate crack propagation in age hardened AZ91 alloy which can be improved by Y and Nd addition [2]. Similarly, surface treatments also have significant influence on fatigue crack propagation.

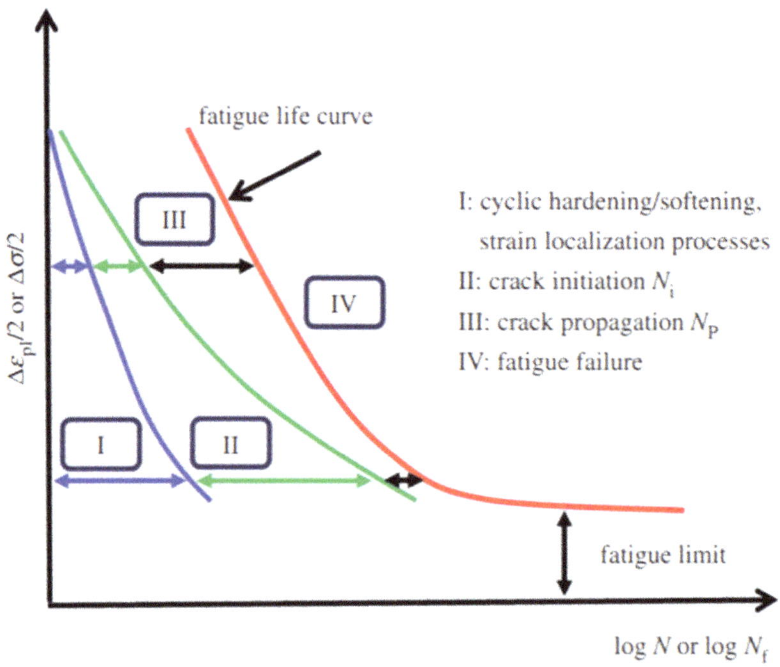

Figure 5.1. Schematic illustration of the four stages of fatigue. Reproduced from [5], Royal Society Publishing.

As the propagation continues, the cross-section of the sample is reduced which weakens the part and results in complete failure, which can be either ductile (with a dimpled fracture surface) or brittle (with a cleavage, or perhaps even intergranular, fracture surface) or in combination depending on the composition and microstructure [5].

5.3 Evaluation of fatigue properties

On the basis of fatigue life, the fatigue analysis can be broadly classified as either low cycle or high cycle fatigue (HCF). While fatigue failure between 10^3 and 10^5 cycles is called low cycle fatigue, HCF refers to the fatigue failure between 10^5 and 10^7 cycles [6]. In general, the fatigue behaviour can be evaluated by either stress-controlled (*S–N*) or strain-controlled (ε-*N*) methods. Since the stress at low load levels can be easily controlled, HCF tests are usually conducted under stress-controlled conditions [6]. On the other hand, strain-controlled tests involve a wide range of strain levels covering up to 10^7 cycles. However, the heating of the sample due to internal friction is a concern in strain-controlled tests at high strain levels [6]. In the stress-controlled method, the cyclic loads are applied using different methods such as rotating bend, cantilever bend, axial push–pull, and torsion, and the number of cycles to failure is measured after subjecting the individual test samples to different stresses. The test results are usually represented as a logarithmic *S–N* (stress versus number of cycles) plot (similar to figure 5.2(a)) from which the fatigue strength and endurance limit are evaluated. Fatigue strength is the stress below which the failure is not expected in less than a given number of cycles, called the endurance limit. In the strain life approach, strain based calculations as opposed to stresses are used to evaluate the fatigue response of a material in the low cycle. Experimentation involves the tension loading of a test sample to a given strain level and then fully reversing the strain to an equal and opposite compressive strain. This procedure is repeated until the specimen fractures and the results are recorded as a logarithmic plot (similar to figure 5.2(b)) indicating the variation of average strain amplitude ($\Delta\varepsilon/2$) with respect to the number of strain reversals to failure; there are two reversals in each cycle.

5.4 The fatigue response of magnesium nanocomposites

Goh *et al* [7] investigated the fatigue behaviour of a DMD processed Mg nano-composite containing 1.3 vol.% CNTs. In this study, cylindrical samples of pure Mg

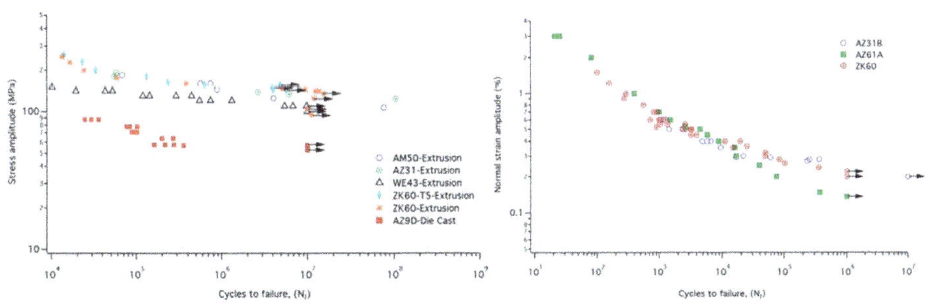

Figure 5.2. Representative *S–N* and ε–*N* test results for Mg alloys. Reproduced from [6], IntechOpen.

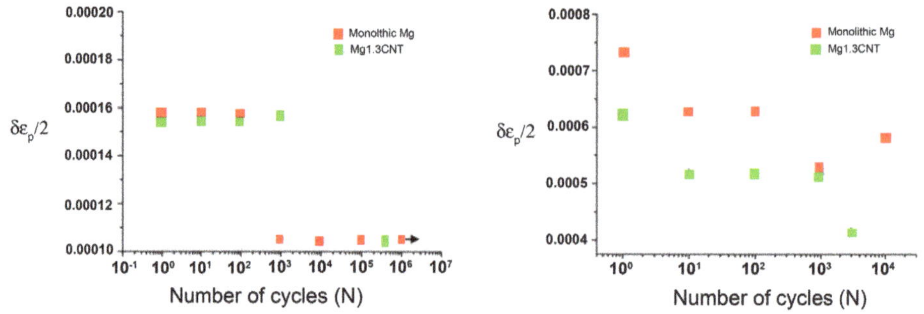

Figure 5.3. Strain-controlled fatigue properties of Mg and Mg/1.3CNT nanocomposites at constant stress amplitudes (a) ±35 MPa and (b) ±70 MPa. Reprinted from [7], Copyright 2008, with permission from Elsevier.

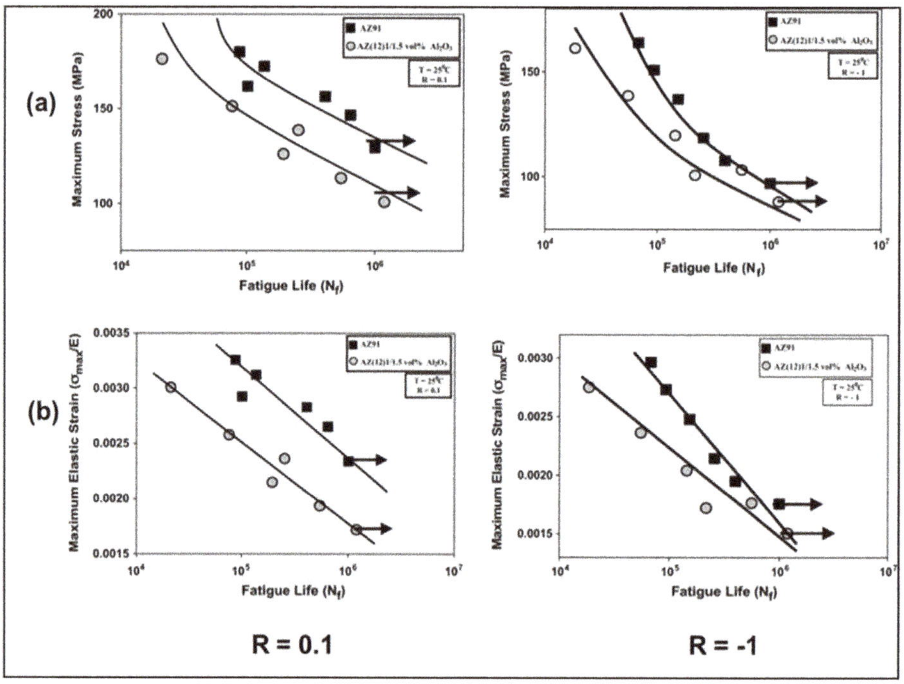

Figure 5.4. Variation of (a) maximum stress and (b) maximum elastic strain versus fatigue life for AZ(12)l and AZ(12)1/1.5Al_2O_3 at load ratios of (i) 0.1 and (ii) −1. Reprinted from [9], Copyright 2012, with permission from Elsevier.

and Mg/1.3CNT nanocomposite were tested for strain-controlled fatigue response at stress amplitudes of ±35 and ±70 MPa. While the results (figure 5.3) showed significant hardening due to the formation of sessile forest dislocations in both samples, the fatigue life of Mg/1.3CNT was compromised compared to pure Mg. The same authors also reported comparable fatigue life and significant hardening due to forest dislocations in Mg/Y_2O_3 nanocomposites [8].

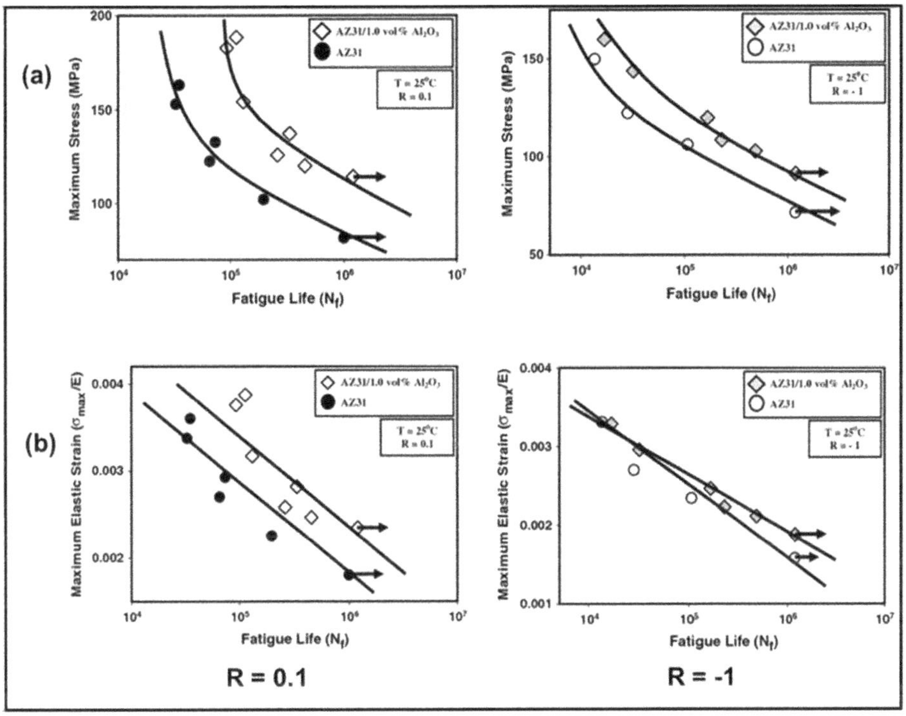

Figure 5.5. Variation of (a) maximum stress and (b) maximum elastic strain versus fatigue life for AZ31 and AZ31/1.5Al_2O_3 at load ratios of (i) 0.1 and (ii) −1. Reproduced from [10], with permission © Springer Nature.

The DMD method was used to develop an AZ(12)1/1.5Al_2O_3 alloy nanocomposite by adding Al 1050 alloy lumps and nano-Al_2O_3 particles to AZ91 alloy. The developed nanocomposite was investigated for HCF response at two different stress ratios (0.1 and −1.0) and a constant frequency (5 Hz) in comparison to AZ91 alloy [9]. In this study, a range of stress values were employed to establish the maximum stress (σ_{max}) and the cyclic fatigue life (N_f). From the results (figure 5.4) it was evident that the nanocomposite displayed a relatively poor fatigue response due to the early initiation of very fine microcracks and their concurrent rapid propagation through the composite microstructure.

Srivatsan *et al* [10, 11] also investigated influence of nano-Al_2O_3 and CNT reinforcements on the cyclic stress-controlled fatigue behaviour of AZ31 alloy. Unlike the AZ(12)l/Al_2O_3 nanocomposite, the addition of nano-Al_2O_3 and CNT reinforcements significantly enhanced the cyclic fatigue life and the endurance limit of the AZ31 matrix alloy (figures 5.5 and 5.6). The improved fatigue behaviour of AZ31/Al_2O_3 and AZ31/CNT nanocomposites was attributed to the influence of the nanoparticles in delaying the crack initiation and its propagation/growth. Similar improvement was also reported for the AZ31 alloy upon the combined addition of nano-Al_2O_3 and micro-Ni particles (as shown in figure 5.7) [12].

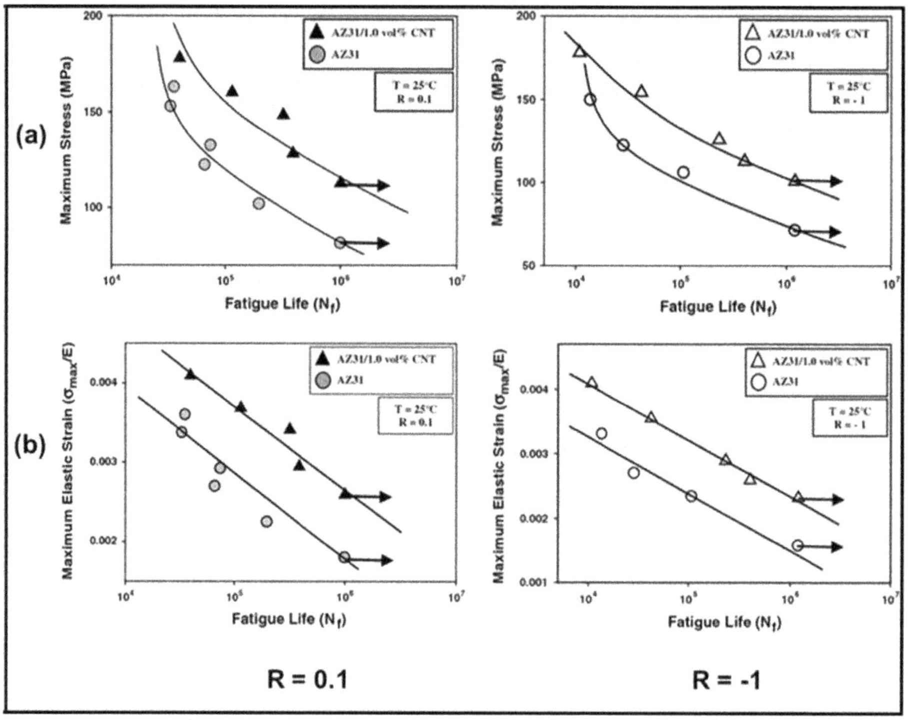

Figure 5.6. Variation of (a) maximum stress and (b) maximum elastic strain versus fatigue life for AZ31 and AZ31/1CNT at load ratios of (i) 0.1 and (ii) −1. Reproduced from [11], with permission © Springer Nature.

Recently, the effects of texture and tension–compression yield asymmetry on the fatigue behaviour of an *in situ* Mg–1.8Y/1.53ZnO nanocomposite were investigated [13]. In this study, the authors attributed the improved fatigue properties (figure 5.8) of an *in situ* Mg–1.8Y/1.53ZnO nanocomposite to a very strong two component texture (basal and prismatic) and a high tension–compression asymmetry (TCA) of 1.63 when compared to the near symmetric TCA (~1.07) of the Mg–1.8Y alloy.

5.5 Summary

Magnesium is a promising material for lightweight applications. Because most of the engineering and biomedical applications involve cyclic loading, fatigue life estimation of Mg materials is critical for their efficient design and implementation. Under cyclic loads, the crack initiation in Mg materials is related to basal slip and the crack growth follows an intergranular or trans-granular micromechanism. In general, the crack propagation is faster in the case of Mg alloys which can be strongly affected by the microstructural constituents. A significant amount of stress-life (*S–N*) highlights the role of alloying elements and heat treatment on the fatigue behaviour of Mg. For example, the precipitates in age hardened Mg–Li, Mg–Nd, and Mg–Zn alloys retard the fatigue crack growth while the $Mg_{17}Al_{12}$ phases are known to accelerate the crack propagation in age hardened AZ91 alloy. Similar improvements in fatigue

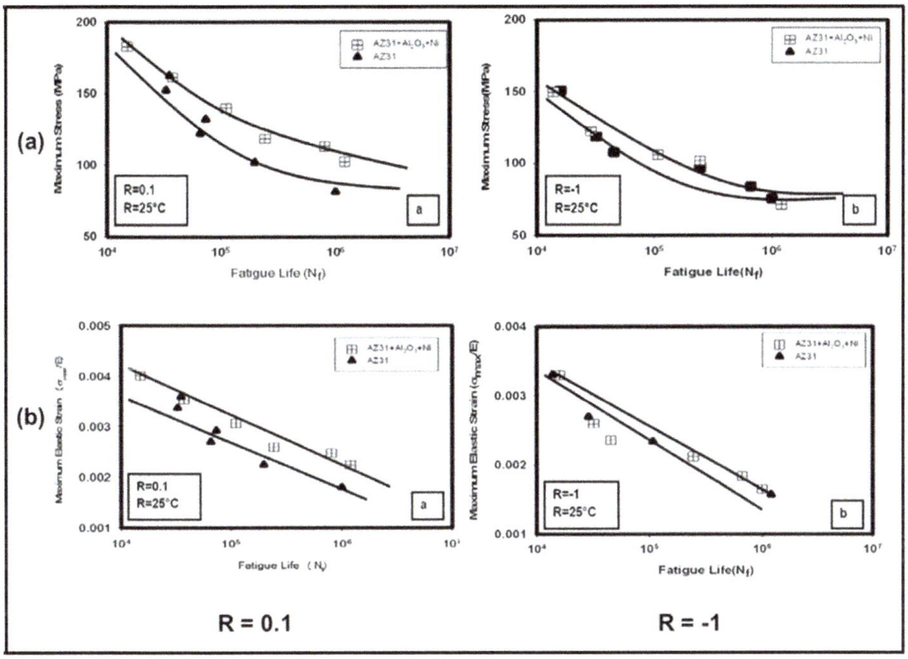

Figure 5.7. Variation of (a) maximum stress and (b) maximum elastic strain versus fatigue life for AZ31 and AZ31/Al_2O_3–Ni at load ratios of (i) 0.1 and (ii) −1. Reprinted from [12], with permission from MDPI.

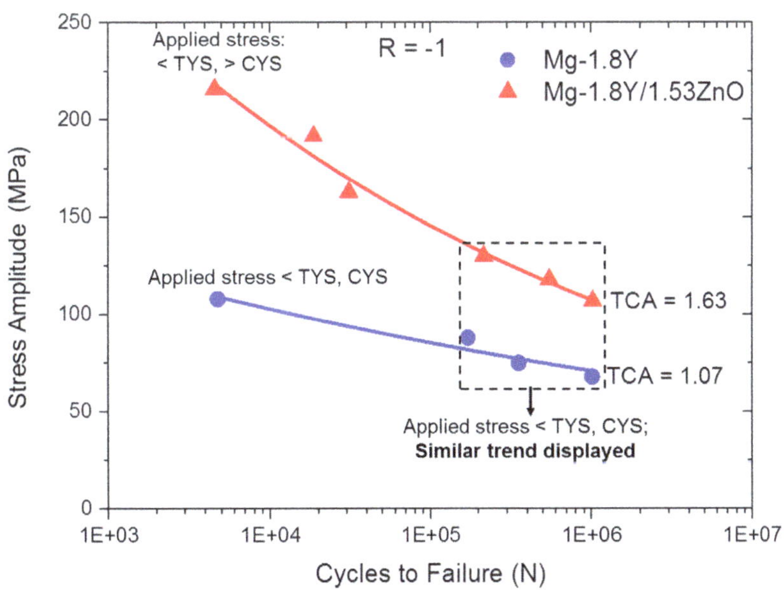

Figure 5.8. The *S–N* curve of the Mg–1.8Y and Mg–1.8Y/1.53ZnO nanocomposite from fully reversed fatigue tests. Reprinted from [13], Copyright 2013, with permission from Elsevier.

Table 5.1. Cycle fatigue characteristics of Mg nanocomposites.

Material	Test frequency (Hz)	Load ratio	No. of cycles	Endurance strength (MPa)[a]	Reference
AZ91	5	0.1	1×10^6	128	Srivatsan *et al* [9]
AZ91	5	−1	1.2×10^6	97	
AZ(12)1/Al_2O_3	5	0.1	1×10^6	100	
AZ(12)1/Al_2O_3	5	−1	1.2 × 106	88	
AZ31	5	0.1	1×10^6	81	Srivatsan *et al* [10]
AZ31	5	−1	1.2×10^6	71	
AZ31/CNT	5	0.1	1×10^6	**113**	
AZ31/CNT	5	−1	1.2×10^6	**100**	
AZ31/1.5Al_2O_3	5	0.1	1×10^6	**190**	Srivatsan *et al* [11]
AZ31/1.5Al_2O_3	5	−1	1.2×10^6	**91**	
AZ31/1.5Al_2O_3–Ni	5	0.1	1×10^6	100	Srivatsan *et al* [12]
AZ31/1.5Al_2O_3–Ni	5	−1	1×10^6	95	
Mg–1.8Y	20	−1	1×10^6	68	Tekumalla *et al* [13]

[a]The number in bold indicates the positive effect of nano-reinforcement addition in increasing the endurance strength of monolithic materials.

strength were also common upon the addition of hard and strong ceramic reinforcements that exhibit better interfacial integrity with a Mg matrix. In this regard, a few interesting research works were undertaken to understand the effectiveness of nanoscale reinforcements on the fatigue response of Mg alloys. Although only limited research works have been published so far, the findings clearly establish the influence of nano-reinforcements (table 5.1). However, it is essential to note that the nano-reinforcement addition must be incorporated using efficient processing methods that are capable of producing a sound, defect-free microstructure.

References

[1] Gupta M and Wong W L E 2015 Magnesium-based nanocomposites: lightweight materials of the future *Mater. Charact.* **105** 30–46

[2] Potzies C and Kainer K U 2004 Fatigue of magnesium alloys *Adv. Eng. Mater.* **6** 281–9

[3] Kuffova M 2011 Fatigue endurance of magnesium alloys *Magnesium Alloys—Design, Processing and Properties* (London: IntechOpen)

[4] Li Z, Wang Q, Luo A A, Peng L and Zhang P 2015 Fatigue behavior and life prediction of cast magnesium alloys *Mater. Sci. Eng.* A **647** 113–26

[5] Mughrabi H 2015 Microstructural mechanisms of cyclic deformation, fatigue crack initiation and early crack growth *Philos. Trans. R. Soc.* A **373** 20140132

[6] Jafar A 2020 *Magnesium—The Wonder Element for Engineering/Biomedical Applications* ed M Gupta (Rijeka: IntechOpen) ch 4

[7] Goh C S, Wei J, Lee L C and Gupta M 2008 Ductility improvement and fatigue studies in Mg-CNT nanocomposites *Compos. Sci. Technol.* **68** 1432–9

[8] Goh C S, Wei J, Lee L C and Gupta M 2007 Properties and deformation behaviour of Mg–Y_2O_3 nanocomposites *Acta Mater.* **55** 5115–21

[9] Srivatsan T S, Godbole C, Paramsothy M and Gupta M 2012 The role of aluminum oxide particulate reinforcements on cyclic fatigue and final fracture behavior of a novel magnesium alloy *Mater. Sci. Eng.* A **532** 196–211

[10] Srivatsan T S, Godbole C, Paramsothy M and Gupta M 2012 Influence of nano-sized carbon nanotube reinforcements on tensile deformation, cyclic fatigue, and final fracture behavior of a magnesium alloy *J. Mater. Sci.* **47** 3621–38

[11] Srivatsan T S, Godbole C, Quick T, Paramsothy M and Gupta M 2013 Mechanical behavior of a magnesium alloy nanocomposite under conditions of static tension and dynamic fatigue *J. Mater. Eng. Perform.* **22** 439–53

[12] Srivatsan T S, Manigandan K, Godbole C, Paramsothy M and Gupta M 2012 Influence of nickel particle reinforcement on cyclic fatigue and final fracture behavior of a magnesium alloy composite *Metals* **2** 143–69

[13] Tekumalla S, Bibhanshu N, Shabadi R, Suwas S, Ha T M H and Gupta M 2018 Evolution of texture and asymmetry and its impact on the fatigue behaviour of an *in-situ* magnesium nanocomposite *Mater. Sci. Eng.* A **727** 61–9

IOP Publishing

Magnesium-Based Nanocomposites

Advances and applications

Sankaranarayanan Seetharaman, Sravya Tekumalla and Manoj Gupta

Chapter 6

The tribological behaviour of magnesium nanocomposites

One of the primary objectives of composite making is to improve the wear resistance of soft metallic materials. While the reinforcements, in general, enhance the mechanical properties of magnesium, the delamination of micron-sized particles promotes wear loss when compared to unreinforced materials. In this regard, the superior particle strengthening and grain refinement benefits of nano-sized reinforcements are particularly interesting. As tribological investigations of magnesium-based nanocomposites have been gaining attention in recent years, this chapter summarizes the outcomes of the various research works carried out so far on the wear characteristics of magnesium-based metal matrix composites containing nano-lengthscale reinforcements.

6.1 Introduction

Wear can be defined as the progressive loss of material from a surface due to its interaction with a mating surface in relative motion [1]. Together with fatigue and corrosion, wear constitutes one of the three major factors affecting the performance of any engineering system [1]. Hence, improving the wear resistance of engineering materials has become one of the primary objectives of materials development. In this regard, the integration of hard and strong reinforcing phases is expected to promote the wear resistance of soft metallic matrices [2]. Amongst the metallic materials, magnesium, with its extremely low density, qualifies for structural weight savings. While the composites made of magnesium alloys are known to exhibit enhanced strength and better thermal stability, the delamination of large-sized particles results in severe wear when compared to unreinforced materials [3]. Recently, nano-sized reinforcements with better strengthening capabilities have become an attractive option [4]. Therefore, the tribological investigations of magnesium-based nanocomposites are gaining attention to promote the deployment of Mg nanocomposites

doi:10.1088/978-0-7503-3535-5ch6

in tribological applications such as automobile brake pads, biomedical implants, etc [4, 5]. This chapter summarizes the outcomes of the various research works carried out so far on the friction and wear characteristics of magnesium-based metal matrix composites containing nano-lengthscale reinforcements.

6.2 Wear mechanisms in magnesium nanocomposites

The available literature suggests that the wear loss in Mg nanocomposites can be due to (i) abrasive wear, (ii) adhesive wear, or (iii) oxidation wear. While abrasive wear occurs due to the presence of hard particles and other surface irregularities, adhesive wear results from the cold-welding and subsequent rupture of contact surfaces. Oxidation wear, on the other hand, refers to the progressive oxidation and removal of weak surface oxide films due to the relative motion between mating surfaces. Figure 6.1 illustrates the schematic of these wear mechanisms in magnesium composites.

6.3 Evaluation of sliding wear properties

While tribometers of different kinds can be used to evaluate the friction and wear characterization of materials, simple pin-on-disk experiments are the most common. Figure 6.2 shows the schematic of a simple pin-on-disk wear tester. It consists of a stationary pin holder and a rotating disk and the tests involve loading the samples of preferred shape and dimension into the stationary pin and pressing it against the rotating disk at a recommended load. The results are recorded in terms of friction coefficient and wear rate for different loads and sliding distances, as per the ASTM G99 standard [6]. Microscopic and surface topography analyses are conducted to understand the wear mechanisms. For this purpose, modern tribometers are equipped with inline imaging systems such as optical profilometers and atomic force microscopes.

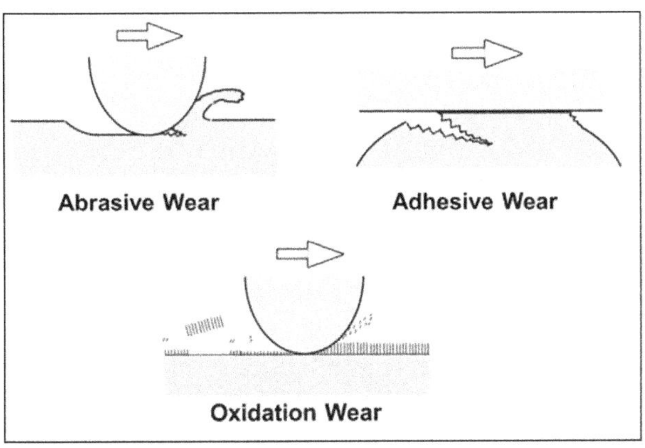

Figure 6.1. Schematic of the wear mechanisms in magnesium nanocomposites [1].

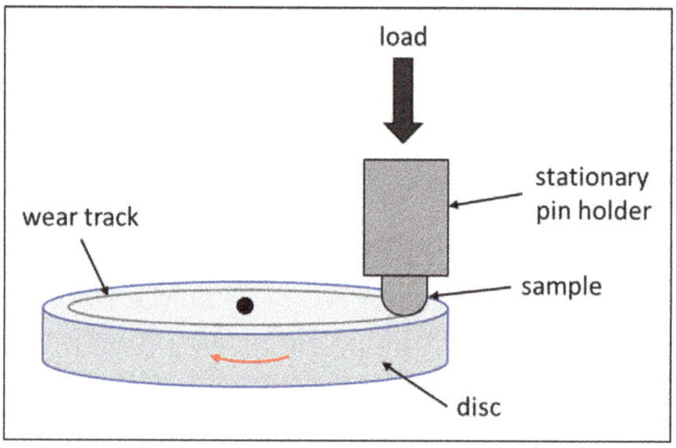

Figure 6.2. Schematic showing a pin-on-disc wear tester set-up [7].

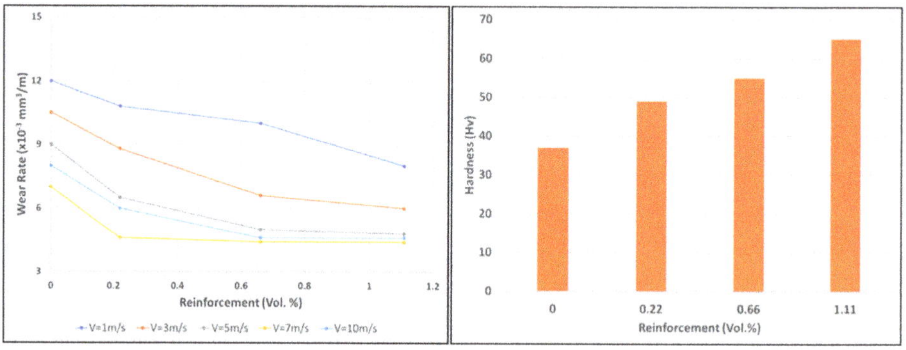

Figure 6.3. Wear rate and hardness of Mg/Al_2O_3 nanocomposites. Based on data from [8].

6.4 Wear response of magnesium nanocomposites

Lim *et al* [8] investigated the wear characteristics of DMD processed Mg composites containing various amounts (up to 1.11 vol.%) of nano-sized alumina particulates using pin-on-disc dry sliding tests. In this study, pin specimens (d = 5 mm, l = 15 mm) were machined from the extruded rods and tests were conducted against the rotating AISI-O1 tool-steel disc at a constant normal load of 10 N (applied using dead weights) for a range of sliding speeds from 1 to 10 m s^{-1} (1, 3, 5, 7, and 10 m s^{-1}). The weight loss data were used to calculate the volumetric wear rates (as shown in figure 6.3) which indicated consistent improvement with increasing amounts of nano-alumina reinforcement and the improvements were particularly effective at higher sliding speeds. The authors also carried out investigation on the worn out surface to identify the dominant wear mechanism. Micrographs showed evidence of abrasive wear at lower sliding speeds. However, a change from abrasive to adhesive wear was reported at higher sliding speeds and the same was attributed to the enhanced deformation due to frictional heating and localized thermal softening. The wear rates trend reported in this study was also consistent with Archard's wear relationship ($Q = kW/H$; where Q is the volume worn per unit sliding distance, W is the applied load, H is the hardness, and k is the wear coefficient), i.e. material wear is inversely related to its hardness (see figure 6.3). It is also worth noting that the Mg nanocomposites investigated in this study did not exhibit delamination wear, as in the case of most metal matrix composites.

Improvement in the wear resistance due to nano-Al_2O_3 addition was also reported for the DMD processed Ca modified AZ31B alloy nanocomposites [9]. While the tested samples revealed abrasive wear at low sliding speeds (less than 5 m s^{-1}), a transition to adhesion type wear occurred when the sliding speeds were between 5 and 7 m s^{-1} (see figure 6.4(a)). In this study, the addition of Ca and nano-alumina delayed the onset of thermal softening and improved the wear resistance of the nanocomposite by ~30%. Similar enhancement in wear resistance due to Ca and nano-Al_2O_3 addition was also reported for different sliding speed ranges between 0.60 and 1.2 m s^{-1} for distances up to 2000 m (see figure 6.4(a)) [10].

Kaviti *et al* [11] also studied the effects of nano-Al_2O_3 particles on the dry sliding wear behaviour of AZ31/Al_2O_3 nanocomposites. In this study, the AZ31 alloy was reinforced with different amounts (0.66, 1.11, and 1.5 wt.%) of Al_2O_3 nanoparticles using the powder metallurgy method. The developed nanocomposites were tested for their sliding wear properties at different loads (5, 7, and 10 N), sliding velocities (0.6, 0.9, and 1.2 m s^{-1}), and sliding distances (500, 1000, and 1500 m). From the results (see figure 6.4(b)) it was reported that the AZ31/1.5Al_2O_3 nanocomposite had the best wear resistance, although the wear rate of all the nanocomposites was found to increase with increasing loads and sliding velocities. The analysis of worn out surfaces showed evidence of abrasive wear and insignificant delamination and the severity increased with increasing load.

In another study [12], nano-Al_2O_3 addition was performed to improve the mechanical properties of cast pure Mg and AZ31 alloys. The developed nanocomposites were investigated for friction and wear properties in the as-cast condition using a pin-on-disc set-up for a sliding distance of up to 2000 m under different loads (12, 24, and 36 N) and sliding speeds (0.5–1.5 m s^{-1}). While the results (figure 6.5) showed material softening and poor wear characteristics for pure Mg, the AZ31–Al_2O_3 nanocomposite displayed better wear resistance (low wear rate) due to the increased work-hardening capabilities. From the reported results, it is also clear that

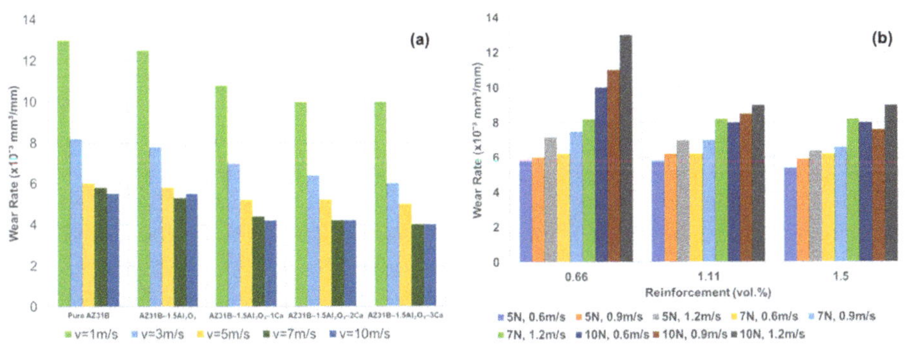

Figure 6.4. Wear rates of (a) nano-Al_2O_3 reinforced Ca modified AZ31B and (b) AZ31 alloy nanocomposites. Based on data from [9, 11].

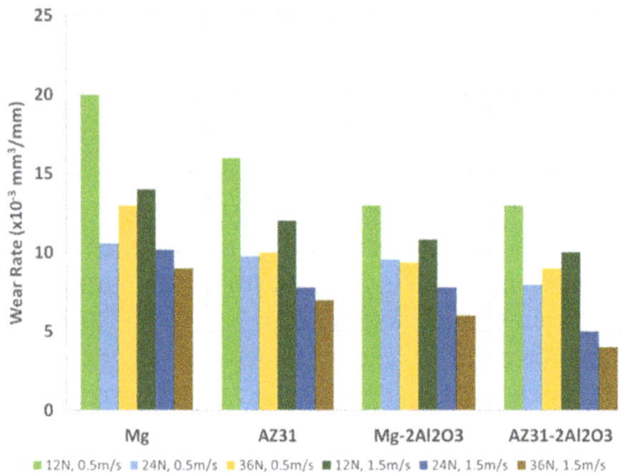

Figure 6.5. Wear properties of nano-Al_2O_3 reinforced Mg and AZ31 alloy nanocomposites. Based on data from [12].

the wear characteristics of pure Mg, AZ31 alloy, and the nanocomposites were profoundly influenced by the test conditions and the wear mechanism was modified from abrasion to oxidation in the case of nanocomposites.

AZ91/Al_2O_3 nanocomposites were also fabricated using the metal casting approach involving mechanical stirring and ultrasonic processing [13]. In this study, the results of dry sliding wear tests conducted at loads between 4.9 N and 14.7 N showed reduced wear rates due to an increasing amount of reinforcements and the analysis of worn out surfaces revealed abrasion and oxidation as the primary wear mechanisms. Similar improvements in sliding wear rates were also reported for friction stir processed AZ31 alloy nanocomposites reinforced with nano-Al_2O_3 particles [14]. This study showed a direct relation between the FSP process parameters and the nanoparticle distribution, which in turn affected the wear characteristics of Mg nanocomposites. While nano-Al_2O_3 addition modified the wear mechanism from adhesive- to abrasive-dominant in the nanocomposite, the optimum tool rotational speed of 1400 rpm was found to be effective for better wear resistance.

Dry sliding wear tests were performed on the powder metallurgy processed Mg nanocomposites reinforced with nanoscale Y_2O_3 and Cu reinforcements at a constant sliding speed of 1 m s^{-1} for loads between 5 and 30 N, and the results showed comparable wear resistance with abrasion wear as the primary wear mechanism [15, 16]. In this study, the nanocomposites with higher copper content showed significant oxidation which positively affected the wear resistance. At higher loads, limited thermal softening was reported along with delamination. In this study, the authors also identified the dominant mechanism for different loads and copper content, as shown in figure 6.6. In a similar study, Mondal and Kumar [17, 18] also reported lower wear rates for AE42 alloy nanocomposites due to hybrid (saffil fibre + nano-Al_2O_3) reinforcements.

Mg nanocomposites containing up to 15 vol.% nano-SiC particles were produced using the powder ball-milling and consolidation method [19]. The friction

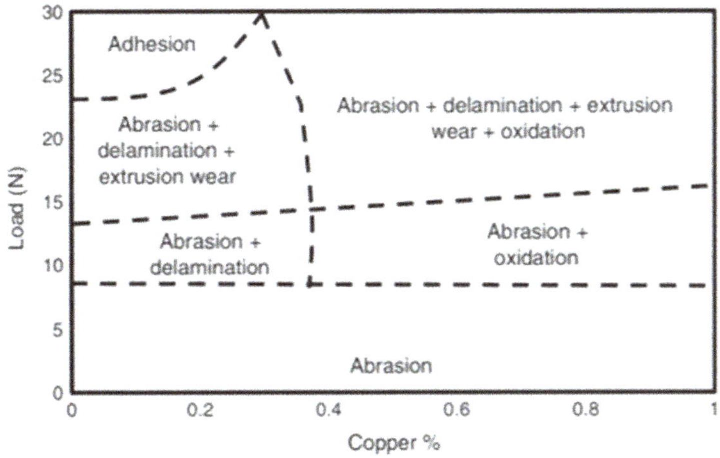

Figure 6.6. Wear mechanisms for Mg/Y_2O_3–Cu nanocomposites. Reproduced from [16], with permission © Springer Nature.

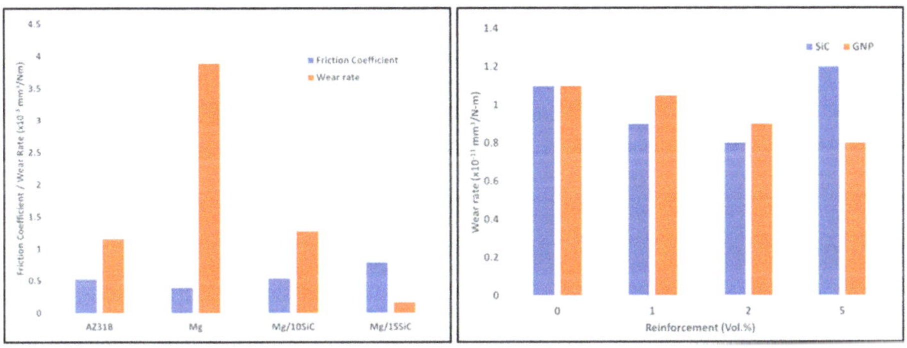

(a) (b)

Figure 6.7. The wear behaviour of Mg/SiC and Mg/GNP nanocomposites. Based on data from [19, 20].

coefficients and wear rates of the developed Mg/SiC nanocomposites were measured and compared with those of the AZ31 alloy (see figure 6.7(a)). While Mg/10SiC exhibited a similar friction coefficient (30% higher than pure Mg) as the AZ31B Mg alloy, a significantly higher friction coefficient was reported for the Mg/15SiC nanocomposite. The wear rate of the Mg/15SiC nanocomposite (0.17×10^{-3} mm^3 Nm^{-1}) was about 23 times lower than that for pure Mg. With respect to wear mechanisms, oxidation was reported as the dominant mechanism in these nanocomposites.

In another study, Mg nanocomposites reinforced with nanoscale SiC particles and graphene nanoplates (GNPs) were fabricated using the ball-milling and spark plasma sintering process [20]. When dry sliding tests were carried out with an applied load of 1 N and a sliding velocity of 150 r min^{-1}, the results (see figure 6.7(b)) showed marked improvements in the wear resistance of Mg nanocomposites due to the presence of the GNPs and SiC reinforcements. The wear rate was reduced by

23% and 28%, respectively, for 2 vol.% GNP and nano-SiC addition. Similarly, an improvement in wear rate of 30% was also observed in AZ91–2.0Ca–0.3Sb reinforced with 2% SiC nanoparticles owing to the presence of hard SiC nano-reinforcements [21].

Kumar *et al* [22] also investigated the tribological behaviour of graphene nanoplatelet reinforced Mg–3Al nanocomposites produced using the microwave assisted powder metallurgy method followed by hot extrusion. The results of dry sliding and scratch testing experiments showed the best properties for the Mg–3Al–0.3GNP nanocomposite. In a similar study, Arab and Marashi [23] reported the formation of a protective lubricating layer in Mg/GNP nanocomposites which reduced the frictional coefficient and the intensity of adhesive and abrasive wear damage.

The sliding wear characteristics of Mg/CNT nanocomposites have been investigated by various researchers. In these studies, the addition of carbon nanotubes through different methods, such as stir casting, power metallurgy, and semi-solid mixing improved the wear resistance of Mg and Mg alloys [24–28]. Huang *et al* [25] tested the dry sliding wear response of stir cast AZ61/CNT nanocomposites. In this study, the experimental results (figure 6.8) revealed reduction in the wear loss and coefficient of friction. Analysis of the worn out surfaces revealed three dominant wear mechanisms, including abrasion, oxidation, and delamination. However, the extent of delamination was reduced for CNT nanocomposites when compared to the base alloy, AZ61. Powder metallurgy processed CNT reinforced AZ31 alloy nanocomposites were also tested for their wear properties [26, 27]. The results (see figures 6.9 and 6.10) showed reduced wear rates and frictional coefficients due to the self-lubrication effect of CNTs and the formation of a carbon film covering the surface.

In another study, the friction and wear performance of AZ31/(CNT+Al_2O_3) hybrid composites fabricated using the FSP method were investigated in comparison with AZ31 matrix alloy and composites reinforced with only nano-Al_2O_3 or CNT particles [28]. While all the composite formulations developed in this study displayed better wear resistance compared to the AZ31 base alloy, the composite containing only nano-Al_2O_3 particles exhibited a lower wear rate compared to the CNT containing composites, although the frictional coefficient of the nano-Al_2O_3 reinforced composites was larger compared to the matrix and the CNT containing

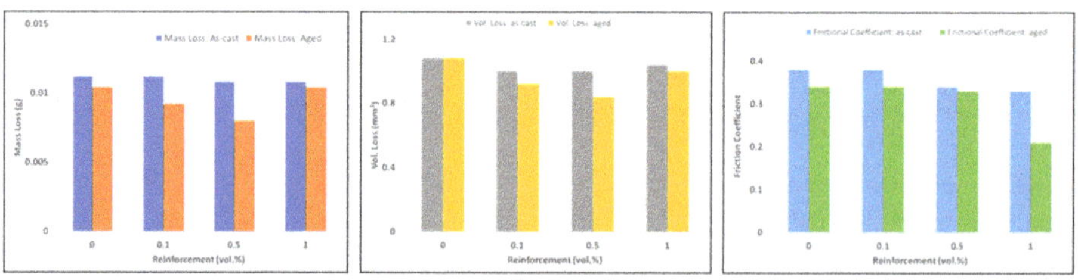

Figure 6.8. Variation of (a) mass loss, (b) volume loss, and (c) coefficient of friction with CNT concentration in AZ61/CNT nanocomposites. Based on data from [25].

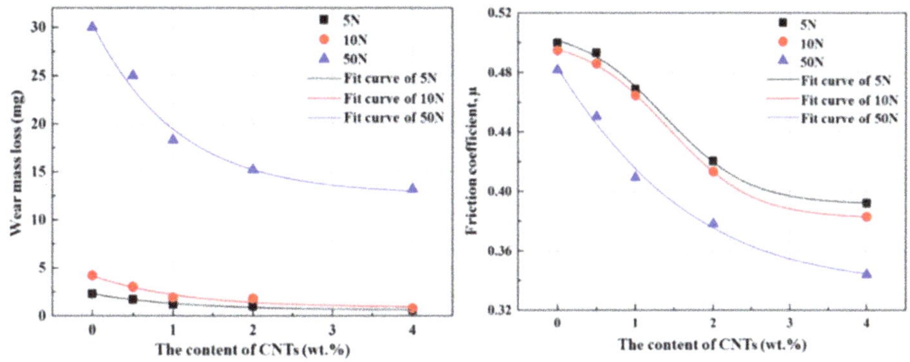

Figure 6.9. Wear rates and friction coefficients of Mg/CNT nanocomposites as reported by Zhou *et al.* Reprinted from [26], with permission from MDPI.

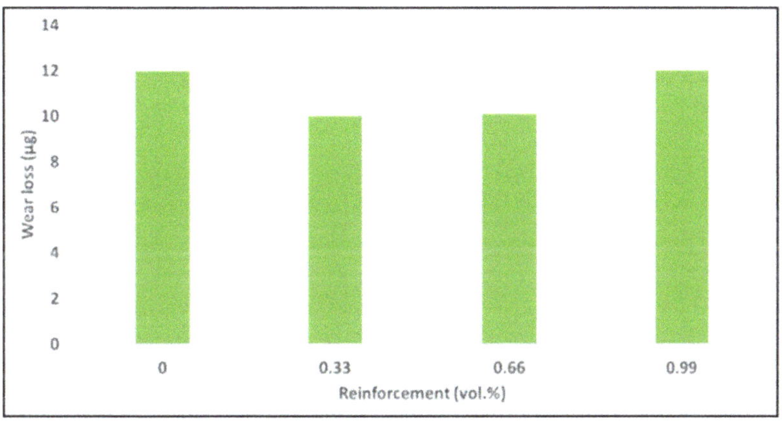

Figure 6.10. Wear rate of Mg/CNT nanocomposites as reported by Jayaraman *et al.* Based on data from [27].

composites (see figure 6.11). In the case of hybrid composites, the hybrid effect was reported. For loads larger than 1.95 MPa, the wear of AZ31/0.1Al_2O_3–0.2CNT was considerably lower and the wear mechanism was generally abrasive at lower loads. However, at normal loads greater than 1.3 MPa, the delamination of a mechanically mixed layer dominated the wear mechanism.

Kaviti *et al* [29] investigated the effects of boron nitride nanoparticles on the dry sliding wear properties of pure Mg. Dry sliding wear tests were conducted using a pin-on-disk set-up under different loads (5, 7, and 10 N) and sliding speeds (0.6, 0.9, and 1.2 m s^{-1}) for sliding distances of (500, 1000, and 1500 m). In this study, the authors reported an increase in the wear rate of pure Mg and Mg/BN nanocomposites due to increasing loads, sliding speeds, and sliding distances. With respect to reinforcement addition, the wear rates were found to reduce upon BN nanoparticle addition at all loads and sliding speeds for the lower sliding distance (500 m). However, in the case of a longer sliding distance (1500 m), the addition of nano-BN particles increased wear rates at larger loads (see figure 6.12). The authors

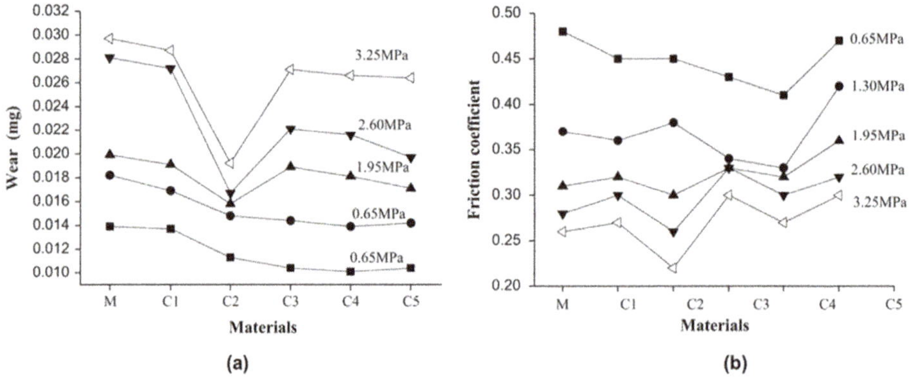

Figure 6.11. (a) Wear rates and (b) friction coefficients of FSP AZ31/CNT-Al_2O_3 hybrid composites. Reprinted from [28], Copyright 2019, with permission from Elsevier.

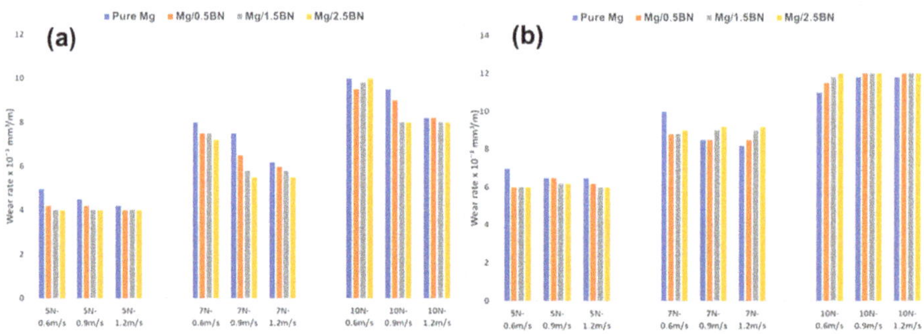

Figure 6.12. Wear rates of Mg/BN nanocomposites at different loads and sliding speeds for sliding distances (a) 500 m and (b) 1500 m. Reprinted from [29], Copyright 2018, with permission from Elsevier.

also reported reduced frictional coefficient (μ) values for Mg/BN nanocomposites at lower sliding speeds. However, at higher sliding speeds the Mg/BN nanocomposites exhibited a higher μ-value compared to pure Mg. From the results of micro-structural examination conducted on the worn out surfaces, the authors concluded that the wear mechanism in Mg/BN nanocomposites was dominated by abrasion and oxidation. While abrasion wear along with insignificant delamination was prominent at smaller test loads (5 N), the oxidation wear was dominant at larger test loads (10 N).

In another work, Kumar *et al* [30] compared the scratch induced wear between two different nano-reinforcements in the same Mg–0.4Ce matrix to understand the relative effect of ceramic reinforcements. While the nanocomposites exhibited better resistance to wear compared to pure Mg, the authors reported that the composite with ZnO nano-reinforcement was more wear resistant compared to the composites containing Y_2O_3 (see figure 6.13). Moreover, the ZnO-containing composite exhibited a reduced COF compared to the other composites. A significant drop in the wear rate in ZnO reinforced nanocomposite with an increase in the number of

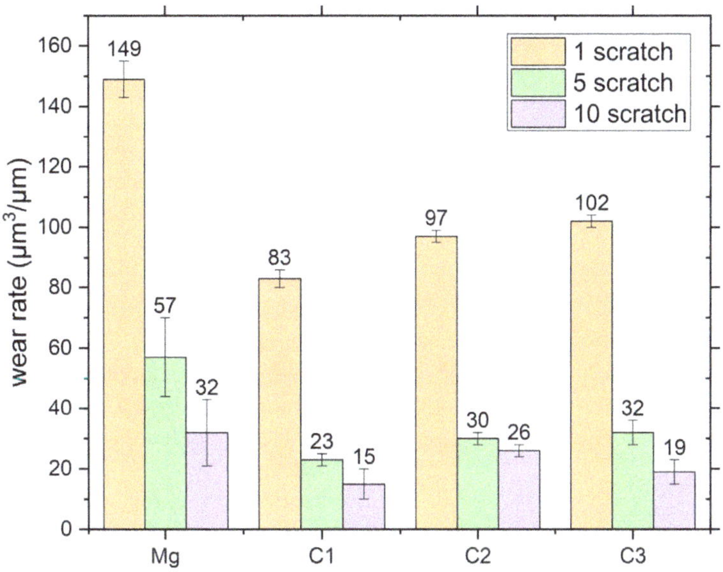

Figure 6.13. Wear rate for Mg, C1 (Mg–0.4Ce/1.5ZnO), C2 (Mg–0.4Ce/2Y_2O_3), and C3 (Mg–0.4Ce/2.84Y_2O_3) nanocomposites. Reproduced with permission from [30], © Taylor & Francis.

scratches was rationalized to be due to the better work-hardening capacity of Mg–ZnO composites. The authors suggested that micro-cutting was found to be dominant at lower penetration depths. At higher penetration depths, the wear mechanism transitions into spalling assisted by ploughing for pure Mg and ZnO reinforced composite and ploughing for Y_2O_3 composites.

6.5 Summary

Improving the wear resistance of soft magnesium alloys is one of the primary objectives of composite development as the incorporation of hard and strong reinforcing phases generally enhances the indentation resistance of the matrix material. In this context, the nano-lengthscale particle Mg composites with superior work-hardening capabilities are gaining increased attention. From the published literature, it can be understood that the actual wear response of Mg nanocomposites depends on the work-hardening/strengthening attributes of individual reinforcements and the efficiency of the processing methods. While most Mg nanocomposites undergo abrasion as the primary wear mechanism at low sliding speeds and test loads, oxidation and delamination type wear are also common at moderate load and sliding speed conditions. However, in most cases, the delamination wear was relatively insignificant, except for those composites containing agglomerated particles. On the other hand, adhesive type wear due to increased frictional heating and thermal softening was reported at high sliding speeds. In addition, the processing methods also play an important role in determining the wear mechanism as the poor interfacial bonding between the matrix and the reinforcement and the aggregation of nanoparticles are known to promote delamination and abrasive wear. Table 6.1

Table 6.1. Overview of wear mechanisms in Mg nanocomposites.

Matrix	Reinforcement	Processing method	Wear mechanism	Reference(s)
Mg, AZ31	Al_2O_3	DMD	Mostly abrasive wear, adhesive wear (at high speed), oxidation wear for larger sliding distances (>1500 m).	[8–10]
Mg	SiC	PM	Predominantly oxidation and delamination wear.	[19]
Mg	SiC	BM+SPS	Oxidation dominant.	[20]
Mg	GNP	BM+SPS	Abrasion and oxidation dominant.	[20]
Mg	GNP	FSP	Abrasive wear followed by oxidation and delamination. Adhesive wear decreased.	[23]
Mg	Y_2O_3 + Cu	PM	Predominantly abrasive, followed by delamination.	[15]
Mg	BN	PM	Mostly abrasion and insignificant delamination at low loads (5 N); oxidation dominated at larger loads (10 N).	[29]
AZ31	Al_2O_3	PM	Predominantly abrasive, followed by delamination.	[11]
AZ31	Al_2O_3	Casting	Abrasive wear dominant in AZ31, oxidation wear in nanocomposites.	[12]
AZ31	Al_2O_3	FSP	Adhesive wear in AZ31 before FSP which changed to abrasive in FSP processed nanocomposites.	[14]
AZ31	CNT	PM	Abrasion wear and delamination.	[26]
AZ31	CNT–Al_2O_3	FSP	Abrasion wear at low loads, but delamination at larger loads.	[28]
AZ61	CNT	Stir casting	Mostly abrasion and oxidation controlled. Reduced delamination.	[25]
AZ91	Al_2O_3	Melting and ultrasonic stirring	Abrasive and oxidation type wear.	[13]
AZ91–2Ca–0.3Sb	SiC	Squeeze cast	Mostly abrasive, followed by oxidation and insignificant delamination. Adhesion type wear in matrix material.	[21]
AE42	Saffil fibre + SiC	Squeeze cast	Abrasive wear and fracture/pull-out of saffil fibres.	[17, 18]

summarizes the wear mechanisms in Mg nanocomposites on the basis of matrix material, reinforcements, processing method, and test conditions.

References

[1] Kato K and Adachi K 2000 Wear mechanisms *Modern Tribology Handbook: Volume One: Principles of Tribology* (Boca Raton, FL: CRC Press)

[2] Kainer K U 2006 Basics of metal matrix composites *Combination of Materials for Light Metal Matrix Composites* (Weinheim: Wiley)

[3] Tyagi R and Davim J P 2015 Processing techniques and tribological behavior of composite materials *Processing Techniques and Tribological Behavior of Composite Materials* (Hershey, PA: IGI Global)

[4] Singh R A *et al* 2018 Tribological characteristics of magnesium nanocomposites *Mater. Today Proc.* **5** 16575–9

[5] Gupta M and Wong W L E 2015 Magnesium-based nanocomposites: lightweight materials of the future *Mater. Charact.* **105** 30–46

[6] 2017 Standard test method for wear testing with a pin-on-disk apparatus G99-17 *Annual Book of ASTM Standards* (Materials Park, OH: ASM International)

[7] Li X, Sosa M and Olofsson U 2014 A pin-on-disc study of the tribology characteristics of sintered versus standard steel gear materials *Wear* **340–341** 31–40

[8] Lim C Y H, Leo D K, Ang J J S and Gupta M 2005 Wear of magnesium composites reinforced with nano-sized alumina particulates *Wear* **259** 620–5

[9] Shanthi M, Nguyen Q B and Gupta M 2010 Sliding wear behaviour of calcium containing AZ31B/Al_2O_3 nanocomposites *Wear* **269** 473–9

[10] Srinivasan M *et al* 2012 Sliding wear behaviour of AZ31B magnesium alloy and nano-composite *Trans. Nonferrous Met. Soc. China (English Ed.)* **22** 60–5

[11] Kaviti R V P *et al* 2019 Improving the friction and wear characteristics of AZ31 alloy with the addition of Al_2O_3 nanoparticles *Mater. Res. Express* **6** 126505

[12] Habibnejad-Korayem M, Mahmudi R, Ghasemi H M and Poole W J 2010 Tribological behavior of pure Mg and AZ31 magnesium alloy strengthened by Al_2O_3 nano-particles *Wear* **268** 405–12

[13] Yadav S D, Bhingole P P, Chaudhari G P, Nath S K and Sommitsch C 2015 Hybrid processing of AZ91 magnesium alloy/nano-Al_2O_3 composites *Key Eng. Mater.* **651–653** 783–8

[14] Azizieh M *et al* 2018 Wear behavior of AZ31/Al_2O_3 magnesium matrix surface nano-composite fabricated via friction stir processing *J. Mater. Eng. Perform.* **27** 2010–7

[15] Hassan S F, Al-Qutub A M, Tun K S and Gupta M 2015 Study of wear mechanisms of a novel magnesium based hybrid nanocomposite *J. Tribol.* **137** 011601

[16] Fida Hassan S, Al-Qutub A M, Zabiullah S, Tun K S and Gupta M 2016 Effect of increasingly metallized hybrid reinforcement on the wear mechanisms of magnesium nano composite *Bull. Mater. Sci.* **39** 1101–7

[17] Mondal A K and Kumar S 2009 Dry sliding wear behaviour of magnesium alloy based hybrid composites in the longitudinal direction *Wear* **267** 458–66

[18] Mondal A K and Kumar S 2014 Dry sliding wear behaviour of magnesium alloy based hybrid composites in transverse direction *Mater. Sci. Forum* **783–786** 1530–5

[19] Zhang L, Luo X, Liu J, Leng Y and An L 2018 Dry sliding wear behavior of Mg-SiC nanocomposites with high volume fractions of reinforcement *Mater. Lett.* **228** 112–5

[20] Das A and Harimkar S P 2014 Effect of graphene nanoplate and silicon carbide nanoparticle reinforcement on mechanical and tribological properties of spark plasma sintered magnesium matrix composites *J. Mater. Sci. Technol.* **30** 1059–70

[21] Ganguly S, Sarkar S, Masanta M and Mondal A K 2019 Effect of SiC nanoparticles on the wear behaviour of squeeze-cast AZ91–2.0Ca–0.3Sb alloy *Mater. Sci. Technol.* **35** 1678–89

[22] Kumar P, Mallick A, Kujur M S, Tun K S and Gupta M 2019 Effects of graphene nanoplatelets on the tribological, mechanical, and thermal properties of Mg–3Al alloy nanocomposites *Int. J. Mater. Res.* **110** 534–42

[23] Arab M and Marashi S P H 2018 Graphene nanoplatelet (GNP)-incorporated AZ31 magnesium nanocomposite: microstructural, mechanical and tribological properties *Tribol. Lett.* **66** 156

[24] Selvamani S T, Premkumar S, Vigneshwar M, Hariprasath P and Palanikumar K 2017 Influence of carbon nano tubes on mechanical, metallurgical and tribological behavior of magnesium nanocomposites *J. Magnes. Alloy.* **5** 326–35

[25] Huang S J, Abbas A and Ballóková B 2019 Effect of CNT on microstructure, dry sliding wear and compressive mechanical properties of AZ61 magnesium alloy *J. Mater. Res. Technol.* **8** 4273–86

[26] Zhou M *et al* 2017 The effects of carbon nanotubes on the mechanical and wear properties of AZ31 alloy *Materials (Basel)* **10** 1385

[27] Jayaraman J, Kuppusamy R and Rao H 2016 Investigation on wear properties of AZ31-MWCNT nanocomposites fabricated through mechanical alloying and powder metallurgy *Sci. Eng. Compos. Mater.* **23** 61–6

[28] Lu D, Jiang Y and Zhou R 2013 Wear performance of nano-Al_2O_3 particles and CNTs reinforced magnesium matrix composites by friction stir processing *Wear* **305** 286–90

[29] Kaviti R V P, Jeyasimman D, Parande G, Gupta M and Narayanasamy R 2018 Investigation on dry sliding wear behavior of Mg/BN nanocomposites *J. Magnes. Alloy.* **6** 263–76

[30] Kumar A *et al* 2018 Investigations of wear response of pure Mg and Mg-0.4 Ce-Y_2O_3/ZnO nanocomposites using a single and repeated scratch tests *Tribol. Trans.* **61** 951–9

Chapter 7

The thermal response of magnesium nanocomposites: oxidation and ignition

Magnesium nanocomposites reinforced with a low volume fraction of nano-sized reinforcements offer an excellent combination of mechanical properties alongside low density. One of the major drawbacks restricting the application of Mg-based materials is their poor oxidation resistance due to their high surface reactivity, in particular at elevated temperatures, as Mg can also auto-ignite in the solid state when exposed to heat. Hence, understanding the thermal response of Mg nanocomposites in terms of oxidation and ignition is essential to evaluate their performance upon high temperature exposure during heat treatment, rolling, forming, welding, etc. Further, with the recent lifting of the ban on the use of magnesium in aircraft interiors, there is also a surge in research efforts to develop lightweight Mg materials with improved ignition resistance. In this regard, recent research works have emphasized the benefits of nanoscale ceramic reinforcements in improving the thermal response of Mg alloys and this chapter aims to summarize the published data pertaining to the oxidation and ignition characteristics of Mg nanocomposites. Special emphasis is also laid on the mechanisms by which the ignition temperature of Mg alloys can be improved.

7.1 Introduction

One of the primary concerns with magnesium alloys for application in the aerospace sector, despite their high specific strength and light weight, is their high surface reactivity. In general, the reaction between magnesium and oxygen leads to the formation of corrosion products such as $Mg(OH)_2$ and MgO. While a low temperature aqueous environment promotes the formation of $Mg(OH)_2$ due to the natural hydration of MgO film, $Mg(OH)_2$ decomposes at temperatures above 250 °C resulting in thicker MgO film [1]. Hence, the wet or aqueous corrosion characteristics of Mg alloys usually depend on the quality of the $Mg(OH)_2$ surface film, while

doi:10.1088/978-0-7503-3535-5ch7

the surface MgO film controls the extent of dry corrosion or oxidation loss at elevated temperatures. In addition to corrosion in wet and dry conditions, magnesium is also susceptible to ignition in the presence of a heat source. The melting temperature of Mg is 650 °C, but Mg auto-ignites at or below its melting temperature therefore sometimes igniting in the solid state. An example of such a magnesium fire was in Maywood, CA, USA, in 2016, when a metal-recycling plant caught fire and the presence of magnesium intensified the fire over time leading to significant damage [2]. Typically, a magnesium fire burns at a temperature exceeding 3000 °C (figure 7.1) [3]. Due to its tendency to ignite, magnesium usage was banned until 2014 by the Federal Aviation Administration (FAA) in aircrafts as it was considered detrimental to the safety of the aircraft in service. Initiatives to re-assess the ban by employing fire resistant magnesium alloys such as Elektron® 43 and, eventually, the lifting of the ban by the FAA renewed curiosity in exploring further options in designing magnesium alloys [4].

Conventionally, alloying additions are made to modify the surface film characteristics in order to improve the wet corrosion and oxidation resistance of magnesium [5, 6]. The modification of alloy chemistry and the formation of a passive surface film also help to improve the ignition temperature of Mg alloys [5]. Similar improvements are also expected from the addition of low volume fraction nano-reinforcement. Although the commonly used ceramic nanoparticles do not form secondary phases with the matrix, the presence of thermally stable nano-sized ceramic reinforcements is expected to increase the activation energy for improved surface stability and ignition temperature [7]. With the development of strong and tough magnesium nanocomposites, it is also imperative to check for their corrosion, oxidation, and ignition behaviour, particularly in the solid state, to render them useful in the applications. In this regard, several useful research attempts have been made in recent years and an attempt is made in this chapter to summarize the

Figure 7.1. AZ31 magnesium alloy burning at temperatures exceeding 3000 °C. Reprinted from [3], Copyright 2014, with permission from Elsevier.

effectiveness of nanoscale ceramic reinforcement addition on the oxidation (dry corrosion) and ignition behaviour of Mg alloys.

7.2 The oxidation behaviour of magnesium nanocomposites

Upon exposure to dry air at a high temperature, metals develop a layer of oxide on the exposed surface. Most metal oxides are stable compared to the base metals, and their protective nature can be assessed using the Pilling–Bedworth ratio (PBR), which is a measure of the surface passivation in dry air as given below [8, 9]:

$$\text{PBR} = \frac{\text{Volume of Oxide Produced}}{\text{Volume of Metal Consumed}} = \frac{A_o/\rho_o}{A_M/\rho_M} = \frac{A_o\rho_M}{A_M\rho_o}, \tag{7.1}$$

where A_o is the molecular weight of the oxide, A_M is the molecular weight of the metal, and ρ_M and ρ_o represent the density of the metal and oxide, respectively.

Table 7.1 lists the PBR values of some metals. While PBR values between 1 and 2 indicate the formation of a passivating oxide coating, PBR values greater than 2 imply cracking and spalling of a thicker oxide coating. On the other hand, a PBR less than 1 (as in the case of Mg) indicates the thin and porous nature of the oxide coating [8, 9]. Figure 7.2 shows the schematic of oxidation mechanism for all three cases.

Table 7.1. Pilling–Bedworth ratios for a few metal oxides [8, 9].

Metal oxide	Pilling–Bedworth ratio (PBR)
K_2O	0.5
Li_2O, Na_2O, CaO	0.6
BaO	0.7
MgO	0.8
CdO, CeO_2	1.2
Al_2O_3	1.3
Pb_3O_4	1.4
NiO	1.5
ZnO, BeO, PdO, ZrO_2	1.6
FeO, CuO	1.7
MnO	1.8
TiO_2	1.9
CoO, Cr_2O_3	2[a]
Fe_3O_4, Fe_2O_3, SiO_2	2.1[a]
Ta_2O_5	2.3
Nb_2O_5	2.7
V_2O_5	3.2
MoO_3, WO_3	3.4

[a] Although the PBR values of Cr_2O_3 and SiO_2 are greater than 2, the actual quantity of oxide formation will be less, which hinders large oxide volume accumulation and associated cracking/spalling.

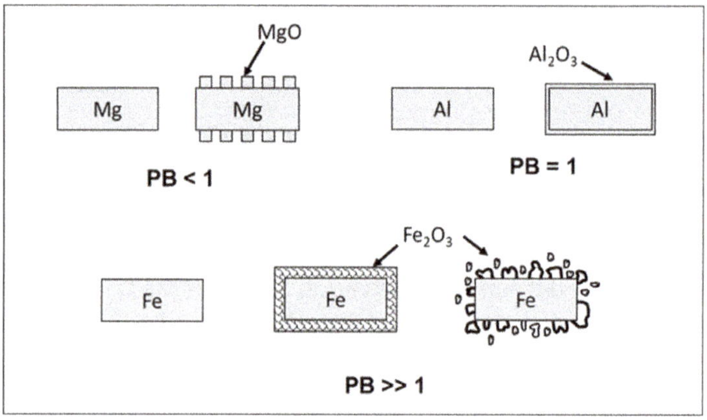

Figure 7.2. Schematic of oxidation for different PB ratios [10].

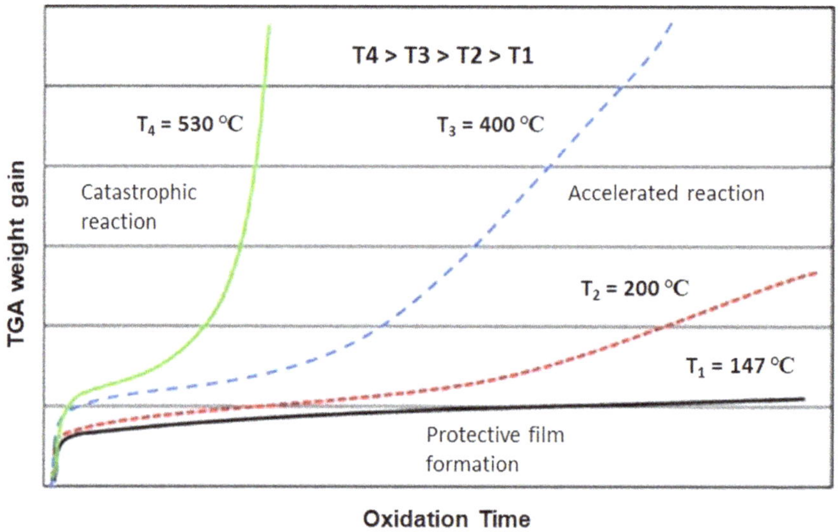

Figure 7.3. Oxidation kinetics of Mg at different temperatures. Reproduced from [11], with permission © Springer Nature.

The oxidation kinetics of Mg depends on the exposure time and temperature. While the formation of a compact oxide film and negligible weight gain offer reasonable protection at temperatures less than 200 °C, non-protective oxidation with linear kinetics occurs at temperatures around 400 °C. A further increase in temperature results in catastrophic oxidation (figure 7.3) [11].

Alloying additions affect the oxidation resistance of Mg. Most alloying additions increase the oxidation rate of Mg. Arrabal *et al* [6] proposed a mechanism (as shown in figure 7.4) to describe the oxidation behaviour of the AZ91 alloy. It involves the breakdown of loose oxide film due to the dissolution of the β-phase and porosities

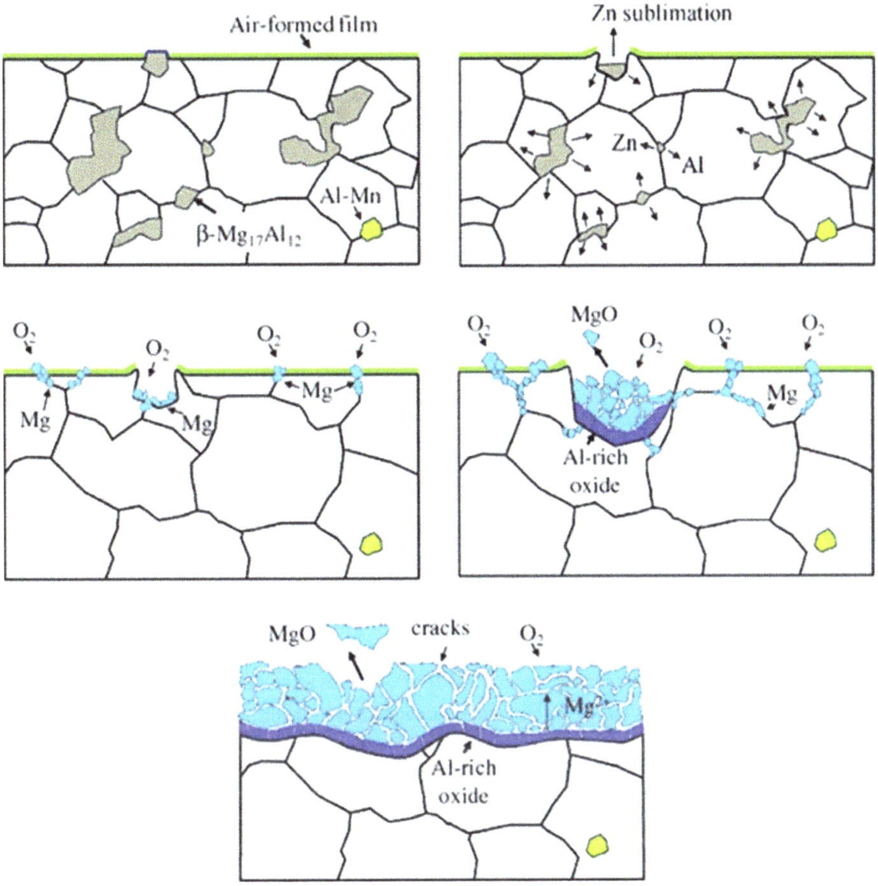

Figure 7.4. Oxidation mechanism in AZ91 alloy. Reproduced from [6], with permission © Springer Nature.

upon the sublimation of Zn when exposed to temperatures greater than 410 °C for a long time.

The addition of alloying elements such as beryllium, calcium, and rare earths improves the oxidation resistance of Mg alloys by forming stable and complex oxide films on the magnesium surface [6]. In an oxidizing environment, beryllium addition (up to 60 ppm) to Mg helps to delay the onset of non-protective oxidation although it does not affect the oxidation kinetics during the very early stages [12]. Calcium addition improves the oxidation resistance at high temperatures as it helps to convert the porous MgO oxide layer into a protective compact film of CaO–MgO. In the case of Al containing alloys, Ca addition results in Al_2Ca phases which improve the oxidation resistance of Mg alloys [13]. RE additions such as Ce or Y, on the other hand, modify the phase composition of native oxide (MgO) with RE oxide phases (Y_2O_3, Ce_2O_3, and CeO_2) [14]. In the case of Al the oxidation response typically depends on the concentration. Figure 7.5 shows the mechanism of

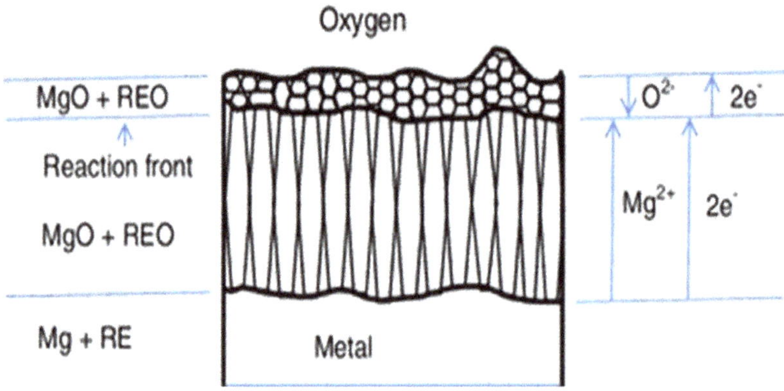

Figure 7.5. Mechanism of oxidation inhibition by RE additions. Reproduced from [11], with permission © Springer Nature.

oxidation inhibition in Mg alloys due to the addition of reactive elements with a high affinity to oxygen [11].

Published research suggest that the high temperature oxidation properties of Mg alloys can be controlled using thermally stable reinforcements [15]. For the first time, the role of nano-sized reinforcement on the oxidation characteristics of magnesium alloy was studied using thermogravimetric method under ambient atmospheric conditions by Nguyen *et al* [16]. In this study, the intrinsic influence of nano-alumina particulates (0.66 vol.%, 1.11 vol.%, and 1.50 vol.%) on the oxidation behaviour of magnesium alloy AZ31B was reported. Thermogravimetric (mass loss) measurements were carried out under ambient atmospheric conditions at temperatures ranging from 300 °C to 470 °C up to a time period of 7 h and the oxidation rates were measured using Wagner's expression [17] as shown below:

$$\text{Oxidation rate (in mils per year, mpy)} = \frac{534W}{DAT}, \tag{7.2}$$

where w is the mass loss in milligrams, D is the density of the specimen in g cm^{-3}, A is the area of the specimen in in.^{2}, and T is the exposure time in hours.

From the results (table 7.2 and figure 7.6) the authors reported a progressive increase in the oxidation resistance (a decrease in the oxidation rate) with an increase in the amount of reinforcement of AZ31B/Al_2O_3 nanocomposites. The reduction in the oxidation rate of alloy AZ31B as a result of nano-Al_2O_3 reinforcement was attributed to the uniform distribution of the nano-Al_2O_3 particulates, microstructure refinement, and fine second phase particles. The results of microstructural analysis (figure 7.6(b)) revealed a fairly protective oxide layer at temperatures below 450 °C. However, cracks and pores were found in the oxide layer at temperatures above 450 °C, which led to increased diffusion of air with the exposed surface, thus leading to linear oxidation kinetics and rampant oxidation. A further improvement in oxidation resistance was reported upon calcium addition to AZ31B/1.5Al_2O_3 nanocomposites (figures 7.7 and 7.8) [18].

Table 7.2. Oxidation rate of AZ31B and its nanocomposites [16].

	Oxidation rate at temperature (in mpy)				
Material	300 °C	350 °C	400 °C	450 °C	470 °C
AZ31B	51.9	74.6	89.2	165.5	753
AZ31B/0.66Al_2O_3	25.7	43.3	67.5	138.2	375.7
AZ31B/1.11Al_2O_3	13.6	27.7	50.3	117.2	307.7
AZ31B/1.50Al_2O_3	10.4	18.2	39.6	109.0	—
AZ31B/1.50Al_2O_3–1Ca	10.3	12.5	18.7	29.6	38.8
AZ31B/1.50Al_2O_3–2Ca	10.5	11.6	18.2	28.9	38.5
AZ31B/1.50Al_2O_3–3Ca	10.2	11.3	18.4	28.3	38.3

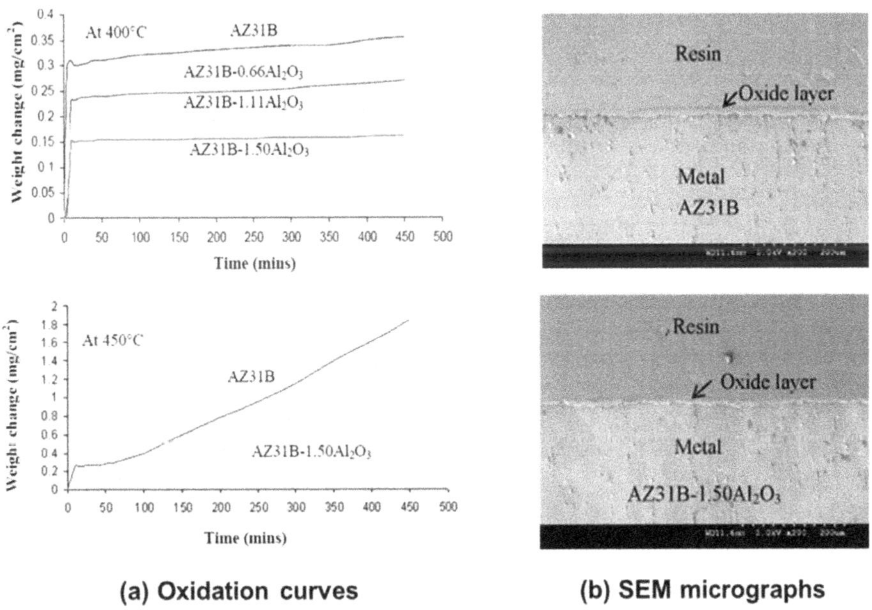

Figure 7.6. (a) Oxidation curves of AZ31/Al_2O_3 nanocomposites, and (b) cross-sectional microstructure showing the MgO oxide layer in AZ31B and AZ31/1.5Al_2O_3 nanocomposite. Reprinted from [16], Copyright 2009, with permission from Elsevier.

To summarize, the formation of porous MgO film on the surface of Mg upon exposure to air at elevated temperatures is not capable of isolating the base material from oxygen diffusion. This hence leads to non-protective accelerated oxidation depending on the temperature. In this regard, the use of nanoscale ceramic particles such as Al_2O_3 nanoparticles improves the oxidation resistance of Mg alloys by forming a reasonably protective surface oxide film.

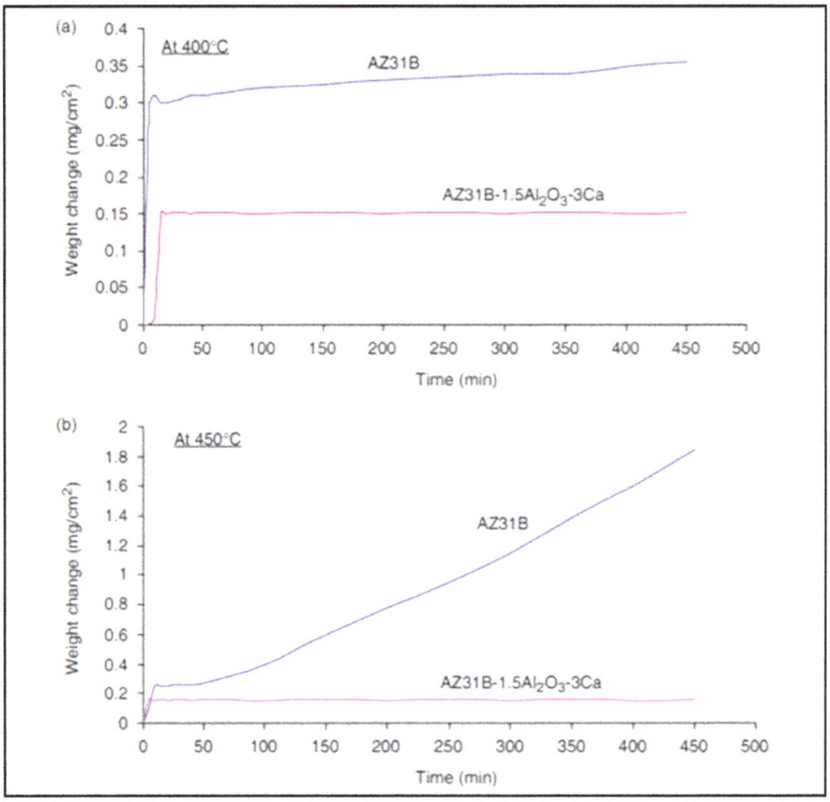

Figure 7.7. Oxidation kinetic curves of monolithic AZ31B and AZ31B/1.5Al_2O_3–3Ca at 400 °C and 450 °C. Adapted from [18].

7.3 Ignition behaviour of magnesium nanocomposites

Table 7.3 compares the ignition temperature and yield strength of Mg nanocomposites with some of the commonly used commercial and modified Mg alloys. With respect to ignition temperature, the addition of La_2O_3 in the micro- and nanoform seemed to be effective among the oxide reinforcements as it improved the ignition temperature of pure Mg by 40 °C [19] and AZ31 Mg alloy [20]. The addition of ZnO nanoparticles [21] improved the ignition temperature of Mg by only about 25 °C. The addition of silica nanoparticles also increased the ignition temperature by 20 °C [22]. Nano samarium oxide (Sm_2O_3) was reported to increase the ignition temperature to 650 °C from 590 °C in pure Mg [23]. Similarly, the addition of Fe_3O_4 increased the ignition temperature with the highest temperature of 635.2 °C recorded for the Mg/3Fe_3O_4 nanocomposite [24]. By far the highest ignition temperature noted with the addition of nano-reinforcements to pure Mg matrix was through the addition of CaO [25]. An ignition temperature of 755 °C was noted in Mg/1.5CaO, which is a considerably significant increase in ignition temperature compared to pure Mg (590 °C). Tekumalla *et al* [26], added nano-CaO to a Mg–Y matrix and observed the simultaneous existence of mechanical sturdiness and non-susceptibility to

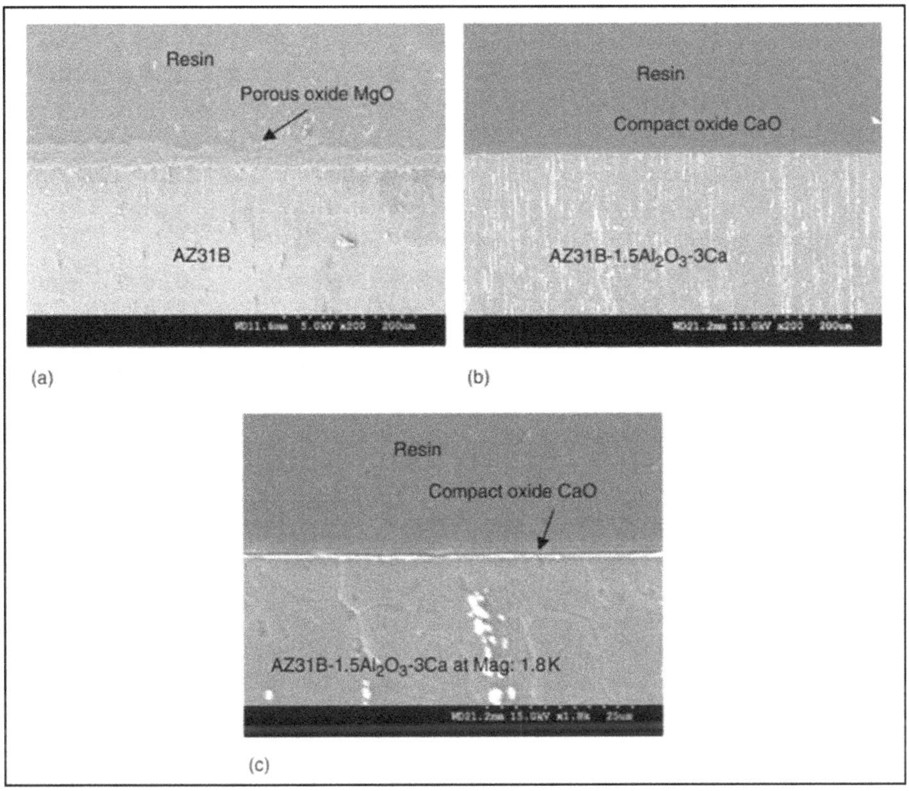

Figure 7.8. Cross-sectional images showing: (a) MgO porous oxide in the case of monolithic AZ31B, and (b) and (c) CaO thin and dense oxide in the case of AZ31B/1.5Al_2O_3–3Ca at 450 °C. Adapted from [18].

ignition. The Mg–1.8Y/1CaO nanocomposite exhibited an ignition temperature of 1045 °C, which is only marginally lower than the boiling point of magnesium (1091 °C). A holistic outline of the candidate materials in terms of relative density and ignition temperature is given in figure 7.9. Similarly to oxide nano-reinforcements, carbide nano-reinforcements also showed promise in improving the ignition resistance. SiC reinforced AM60B alloy, in addition to demonstrating superior mechanical properties, also had an enhanced ignition temperature to 603 °C for the composite from the 535 °C exhibited by AM60B alloy [27]. To understand the effectiveness of the carbide reinforcements clearly, nano boron carbide was added to an already FAA approved alloy, i.e. Elektron21, with the excellent ignition temperature of 741 °C. A steady increase was observed with the addition of boron carbide with an increased ignition temperature of 798 °C with the addition of 2.5 wt.% B_4C [28]. This reveals the effectiveness of nano-reinforcements, even in smaller amounts of 1%–3%, in significantly influencing the ignition temperature of the matrix.

From the published results so far, it can be clearly understood that the addition of nanoscale ceramic reinforcements generally improves the ignition temperature of Mg due to the following possible factors.

Table 7.3. Mechanical properties and ignition temperatures of selected Mg alloys and nanocomposites.

Materials	Composition (wt.%)	Yield strength (MPa)	Ignition temperature (°C)	Reference
Pure magnesium	Mg	87	581	Parende *et al* [22]
	Mg	130	590	Tekumalla *et al* [25]
Mg nanocomposites	AM60B/SiC	—	603	Nguyen and Lee [27]
	$Mg/5.9CeO_2$	111	607	Kujur *et al* [29]
	$Mg/2.7SiO_2$	114	611	Parende *et al* [22]
	Mg/1.5ZnO	131	613	Seetharaman *et al* [21]
	$Mg/1La_2O_3$	175	631	Tekumalla and Gupta [30]
	$Mg/3Fe_3O_4$	76	635.2	Johanes *et al* [24]
	$Mg/6.8Sm_2O_3$	128	650	Kujur *et al* [23]
	Mg/1.5CaO	125	755	Tekumalla *et al* [25]
	Mg–1.8Y/1CaO	266	1045	Tekumalla *et al* [26]
	$E21/2.5B_4C$	166	798	Tekumalla *et al* [28]
Commercial Mg alloys	ZK60A	275	499	
	ZK40A	255	500	
	AM60	130	525	
	AZ80A	330	542	
	AZ81A	275	543	
	ZK51A	275	552	
	AZ61	218	559	
	AZ63	200	573	
	AM50	230	585	
	AZ91	230	600	
	AZ31	175	628	Tekumalla and Gupta [19], Czerwinski [3, 31], Fassel *et al* [32], Liu *et al* [33]
	WE43[a]	250	644	
	E21[a]	141	741	
Modified commercial Mg alloys	ZA61–0.7Ca–1Y	210	680	
	AZ31–1Ca	190	690	
	AZ61–1Ca	204	710	
	ZA62–1Ca–1Y	176	740	
	AZ31–2Ca	255	750	
	AZ61–2Ca	230	760	
	AZ44–0.7Ca–0.6Y	140	770	
	GQ152K	309	935	

[a]FAA approved magnesium alloys [34].

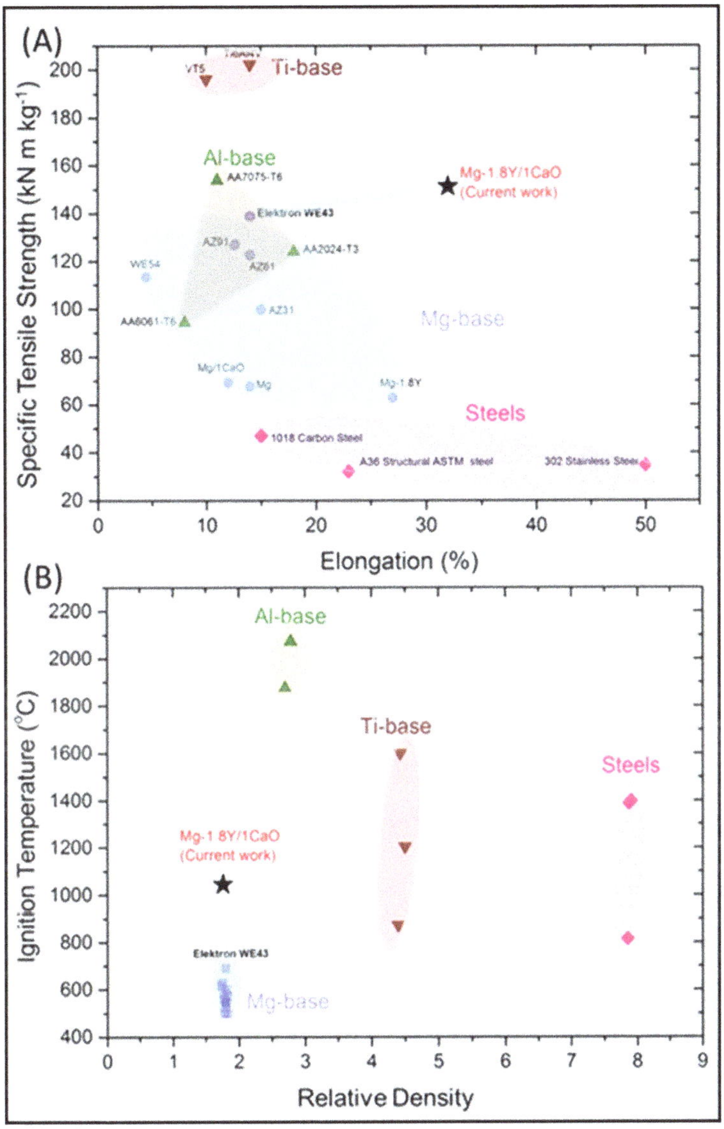

Figure 7.9. Excellent (a) mechanical properties and (b) ignition temperature of a low dense Mg–1.8Y/1CaO nanocomposite. Reproduced from [26], with permission © Springer Nature.

7.3.1 Thermal conductivity (*k*)

The ability of a material to conduct heat is given by its thermal conductivity. This plays a vital role in the determination of the ignition mechanism of a material. Heat transfer occurs at a lower rate in materials of low thermal conductivity than in materials of high thermal conductivity. For example, metals typically have high thermal conductivity and are very efficient at conducting heat, while the opposite is true for insulating materials such as ceramics. Hence, when a thermally stable and insulating ceramic reinforcement such as CaO or SiO_2 is added to a metal matrix it hinders heat transfer. The thermal conductivity of the nanocomposite is directly related to the amount of reinforcement added to the matrix. Hence, the theoretical thermal conductivity of the nanocomposite is computed based on the rule of mixtures following the equation

$$k_c = k_m \cdot V_m + k_r \cdot V_r, \tag{7.3}$$

where k is the thermal conductivity (W mK^{-1}), V is the volume fraction, and subscripts 'c', 'm', and 'r' refer to the composite, matrix, and reinforcement, respectively.

Figures 7.10(a) and (b) show the effective thermal conductivity of the nanocomposites (computed using the rule of mixtures given above) versus the ignition temperature of the nanocomposites. It was generally observed that the ignition temperature increased with a decrease in the thermal conductivity of the material. Further, a linear trend was observed with the progressive addition of reinforcement. This reveals that the nano-reinforcements in the Mg matrix acted as insulating sites, thereby delaying the onset of ignition in pure Mg. Note that the trend between the amount of reinforcement and the ignition temperature may not always be linear and much more work needs to be done to understand current behaviour.

In view of the ability of reinforcement to reduce the availability of a metallic matrix for ignition and the thermal conductivity of the material, the authors proposed a new parameter referred to as the ignition factor (thermal conductivity component). The IF_k can be defined as the ratio of the thermal conductivity of the composite to the thermal conductivity of the matrix in the overall available volume of the matrix material, V_m (by reducing the volume percent of reinforcement), and is given by

$$IF_k = V_m \times \frac{k_c}{k_m}, \tag{7.4}$$

where IF_k is the ignition factor (thermal conductivity component), V is the volume fraction, k is the thermal conductivity (W mK^{-1}), and subscripts 'c' and 'm' refer to the composite and matrix, respectively.

It was observed from figure 7.11(a) and (b) that the ignition factor follows a linear trend with the ignition temperature of magnesium nanocomposites.

7.3.2 Specific heat (C)

Apart from thermal conductivity, another influencing factor proposed was the specific heat of the material. The specific heat of a material is the amount of heat required to raise the temperature of 1 kg of material by 1 °C. It measures how much energy is stored in a heated object. Hence, it determines the likelihood of the material to ignite, i.e. the amount of energy the material can store before ignition of the material.

Like thermal conductivity, the specific heat of the nanocomposite is directly related to the amount of reinforcement added to the matrix. Hence, the theoretical

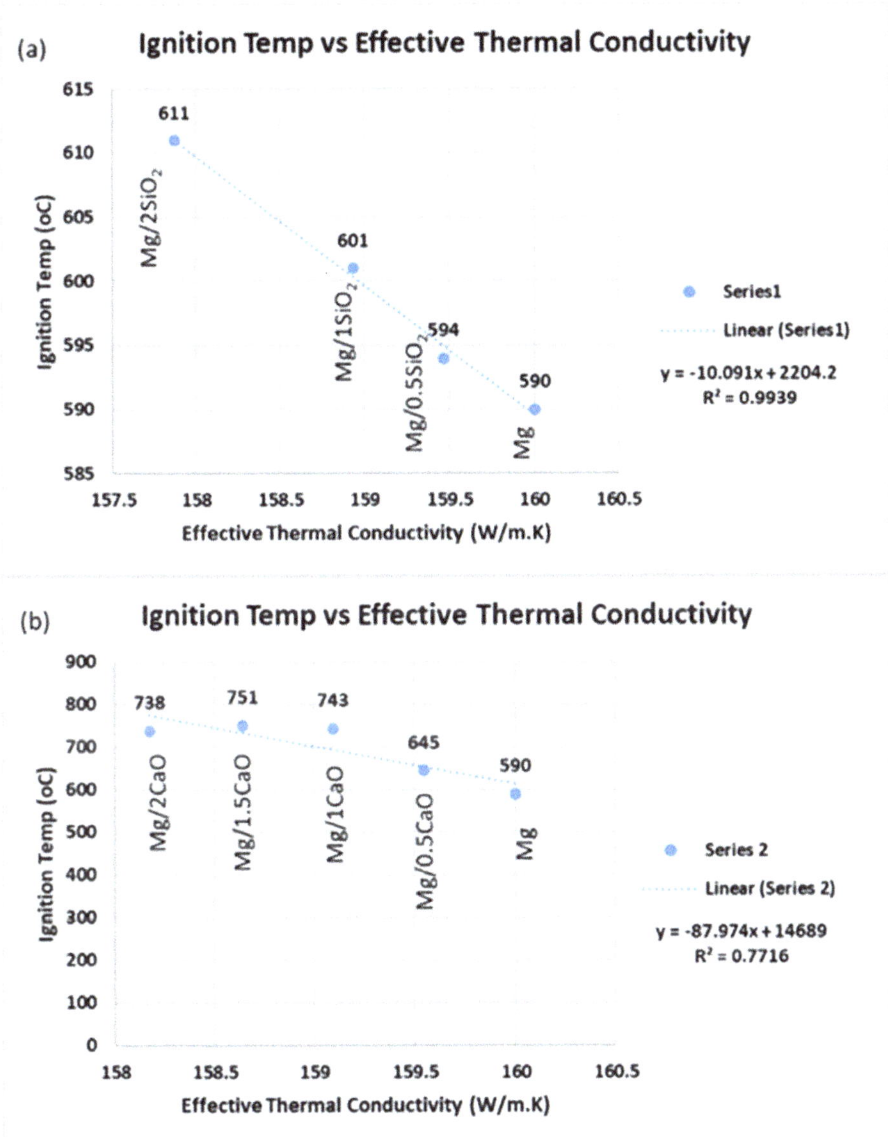

Figure 7.10. Ignition temperature versus effective thermal conductivity of (a) Mg/SiO_2 and (b) Mg/CaO nanocomposites. The theoretical thermal conductivity values of Mg, SiO_2, and CaO at 25 °C used in this work are 160 W mK^{-1} [35], 1.3 W mK^{-1} [36], and 13.807 W mK^{-1} [37], respectively.

specific heat of the nanocomposite is computed based on the rule of mixtures following the equation

$$C_c = C_m \cdot V_m + C_r \cdot V_r, \tag{7.5}$$

where C is the specific heat (J kg^{-1} K^{-1}), V is the volume fraction, and subscripts 'c', 'm', and 'r' refer to the composite, matrix, and reinforcement, respectively.

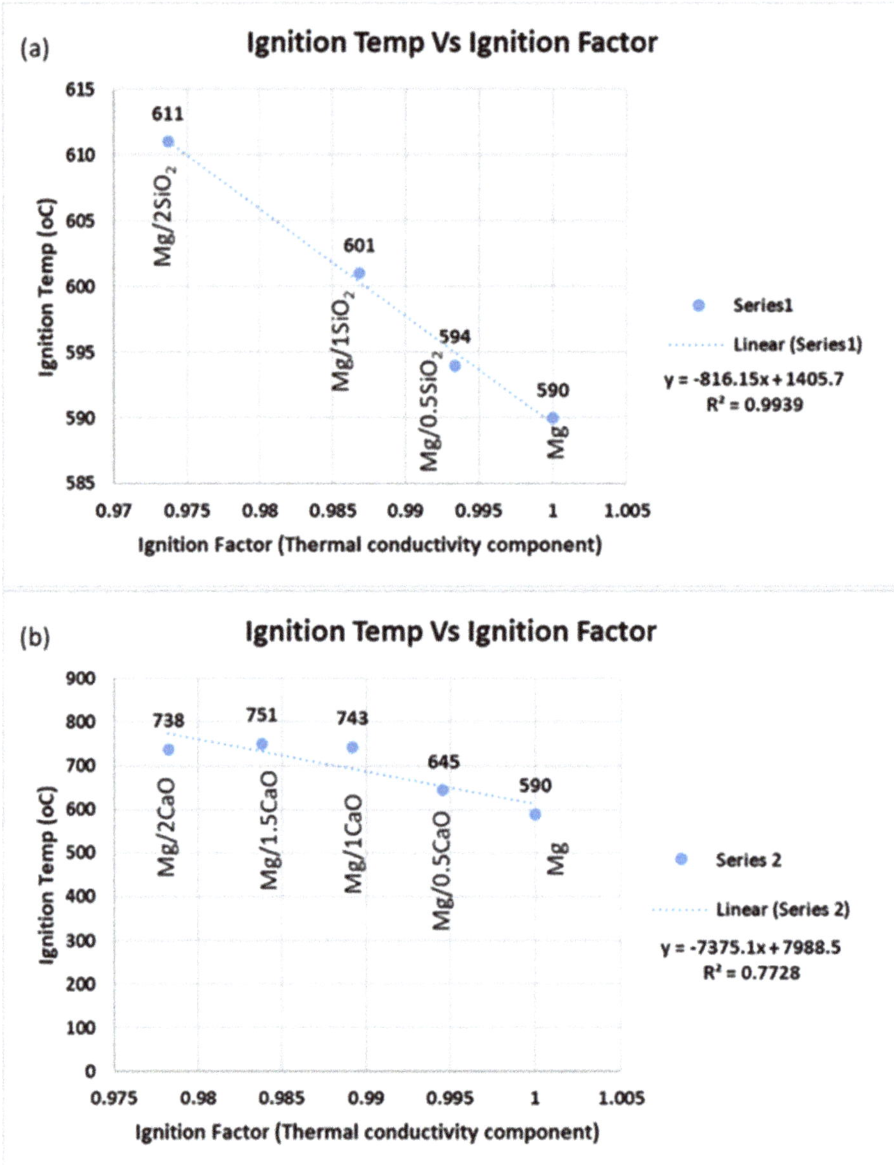

Figure 7.11. Ignition temperature versus ignition factor (thermal conductivity component) of (a) Mg/SiO_2 nanocomposites and (b) Mg/CaO nanocomposites.

Figure 7.12(a) and (b) show the effective specific heat of the nanocomposite (computed using the rule of mixtures) versus the ignition temperature of the nanocomposites. Based on the trend of the thermal conductivity of the material, it was generally observed that the ignition temperature increased with a decrease in the effective specific heat of the material. Further, a linear trend was observed with the progressive addition of reinforcement in this case as well. This indicates that the

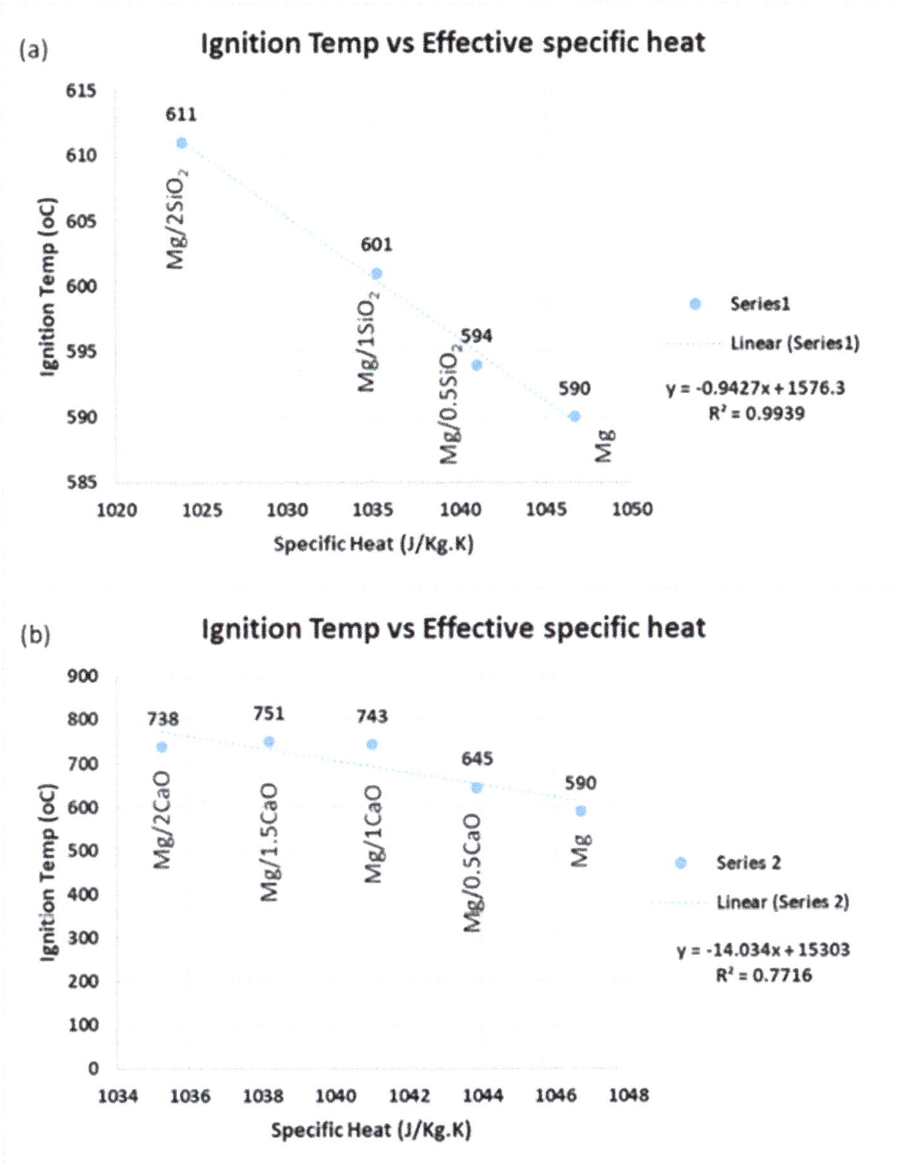

Figure 7.12. Ignition temperature versus effective specific heat of (a) Mg/SiO_2 nanocomposites and (b) Mg/CaO nanocomposites. The theoretical specific heat values of Mg, SiO_2, and CaO at 25 °C used in this work are 1046.7 J kg^{-1} K^{-1} [38], 680 J kg^{-1} K^{-1} [36], and 42.8 J kg^{-1} K^{-1} [39], respectively.

presence of nano-reinforcements in the Mg matrix led to lesser heat energy absorption of the material, thereby delaying the onset of ignition compared to pure Mg.

In view of the ability of reinforcement to reduce the availability of a metallic matrix for ignition and the specific heat of the material, the authors propose another measure, referred to as the ignition factor (specific heat) component (IFC). The IFC can be defined as the ratio of the effective specific heat of the composite to the specific heat of the matrix on the overall available volume of the matrix material, V_m (by reducing the volume percent of reinforcement), and is given by

$$\mathrm{IFC} = V_m \times \frac{C_c}{C_m}, \tag{7.6}$$

where IFC is the ignition factor (specific heat) component, V is the volume fraction, C is the specific heat (J kg^{-1} K^{-1}), and subscripts 'c' and 'm' refer to the composite and matrix, respectively.

It is observed from figures 7.13(a) and (b) that the ignition factor follows a linear trend with the ignition temperature of magnesium nanocomposites.

These factors are proposed to quantify the effect of reinforcements in the magnesium matrix. However, to achieve the desired 1000 °C ignition temperature, the addition of only reinforcements did not yield the desired results. With the addition of alloying elements and nanoparticles as reinforcements to the matrix, the governing mechanisms controlling the ignition behaviour of the material were seen to be completely different and are discussed in the following section.

7.3.3 The modification of the PB ratio through the synergistic effects of alloying and reinforcement additions

Although the insulating effect of the nanoparticles affected the thermal conductivity of the nanocomposite and reduced the heat absorption of the reinforcement, thus affecting the specific heat of the nanocomposite and aiding in enhancing the ignition temperature, it should be noted that this enhancement in ignition temperature is only to a limited extent. Another crucial factor for improving the resistance to ignition is the nature of the oxide layer formed on the metal surface, which is governed by the PBR, as mentioned earlier in section 7.2. Magnesium with a non-protective protective oxide layer (PBR ~08) ignites rapidly before reaching the melting temperature. Therefore, the modification of the surface oxide layer is essential to improve the ignition behaviour of magnesium [33]. For example, the addition of alloying elements with $1 < \mathrm{PBR} < 2$ such as Y [40] enhances the ignition temperature due to the formation of a protective and stable oxide layer adhering to the alloy surface. Similarly, in the case of Mg–1.8Y/1CaO nanocomposite, the formation of protective oxides along with the presence of thermally insulating reinforcement led to an ignition temperature beyond 1000 °C in the nanocomposite. A schematic of the oxidation mechanism in the Mg–1.8Y/1CaO nanocomposite is shown in figure 7.14.

7.4 Summary

Nanocomposites based on magnesium are seen to be promising in demonstrating an improvement in the thermal response in terms of oxidation and ignition characteristics. While it is established that the increase in oxidation resistance and ignition temperature is due to the addition of thermally insulating ceramic nanoparticles and the formation of a protective oxide layer, comprehensive studies demonstrating the

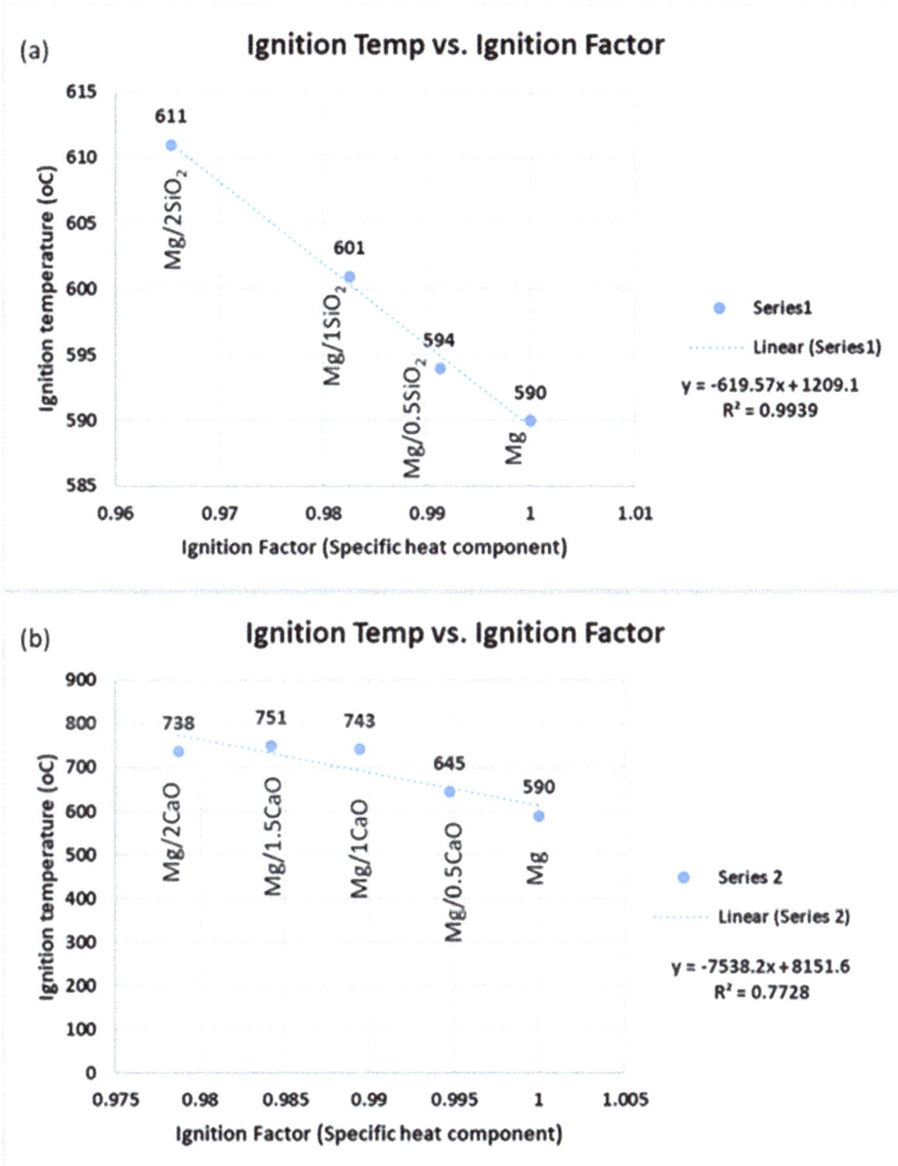

Figure 7.13. Ignition temperature versus ignition factor (specific heat component) of (a) Mg/SiO_2 nanocomposites and (b) Mg/CaO nanocomposites.

role of nano-sized reinforcements in improving the oxidation and ignition response of Mg nanocomposites are interesting to establish Mg nanocomposites' suitability for high temperature applications. However, more work is required to further understand the mechanisms that lead to enhancement in the ignition temperature beyond physical properties such as thermal conductivity and specific heat, as the

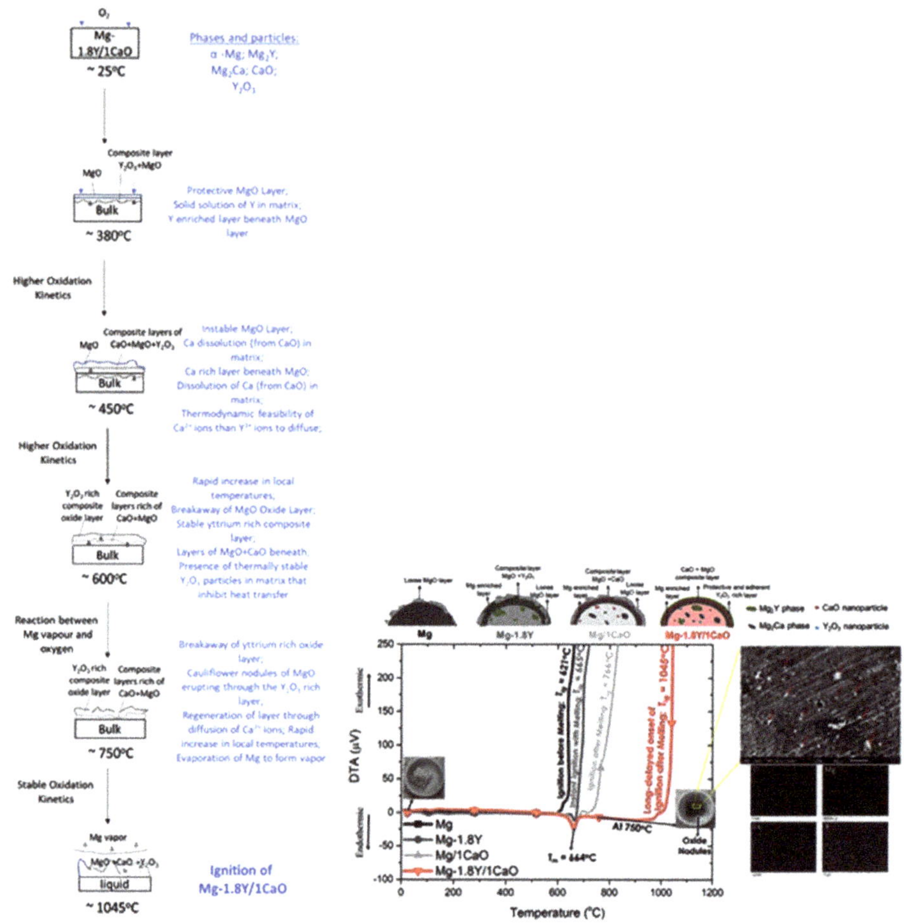

Figure 7.14. The mechanism of oxidation and ignition beyond 1000 °C in a Mg–1.8Y/1CaO nanocomposite. Reproduced from [26], with permission © Springer Nature.

presence of reinforcements also changes the microstructural aspects, which can be studied using some model alloy systems.

References

[1] Nie H, Schoenitz M and Dreizin E L 2016 Oxidation of magnesium: implication for aging and ignition *J. Phys. Chem.* C **120** 974–83

[2] City New Service 2016 Smell that smell? It's from burning magnesium in Maywood warehouse fire *Daily News* https://dailynews.com/2016/06/14/smell-that-smell-its-from-burning-magnesium-in-maywood-warehouse-fire/

[3] Czerwinski F 2014 Controlling the ignition and flammability of magnesium for aerospace applications *Corros. Sci.* **86** 1–16

[4] Summers B 2017 Designing an airline seat from scratch is not as easy as it looks *Skift* https://skift.com/2017/06/28/designing-an-airline-seat-from-scratch-is-not-as-easy-as-it-looks/

[5] Tan Q, Atrens A, Mo N and Zhang M X 2016 Oxidation of magnesium alloys at elevated temperatures in air: a review *Corros. Sci.* **112** 734–59
[6] Arrabal R, Pardo A, Merino M C, Mohedano M, Casajús P, Paucar K and Matykina E 2011 Oxidation behavior of AZ91D magnesium alloy containing Nd or Gd *Oxid. Met.* **76** 433
[7] Hosseini S G, Sheikhpour A, Keshavarz M H and Tavangar S 2016 The effect of metal oxide particle size on the thermal behavior and ignition kinetic of Mg–CuO thermite mixture *Thermochim. Acta* **626** 1–8
[8] Xu C and Gao W 2000 Pilling–Bedworth ratio for oxidation of alloys *Mater. Res. Innov.* **3** 231–5
[9] Pedeferri P 2018 High temperature corrosion *Corrosion Science and Engineering* (Cham: Springer)
[10] Shraddha n.d. Oxidation in metals: mechanism and kinetics, corrosion, metallury *Engineering Notes* https://engineeringenotes.com/metallurgy/oxidation/oxidation-in-metals-mechanism-kinetics-corrosion-metallurgy/43616 (Accessed: 20 May 2020)
[11] Czerwinski F 2012 Oxidation characteristics of magnesium alloys *JOM* **64** 1477–83
[12] Houska C 1988 Beryllium in aluminium and magnesium alloys *Met. Mater.* **4** 100–4
[13] Lee D B 2013 High temperature oxidation of AZ31+0.3wt.%Ca and AZ31+0.3wt.%CaO magnesium alloys *Corros. Sci.* **70** 243–51
[14] Bak S H and Lee D B 2009 Effect of Y and Y_2O_3 on oxidation of AZ91D Mg alloys between 400 °C and 500 °C *Trans. Nonferrous Met. Soc. China* **19** 871–4
[15] Ye H Z and Liu X Y 2004 Review of recent studies in magnesium matrix composites *J. Mater. Sci.* **39** 6153–71
[16] Nguyen Q B, Gupta M and Srivatsan T S 2009 On the role of nano-alumina particulate reinforcements in enhancing the oxidation resistance of magnesium alloy AZ31B *Mater. Sci. Eng.* A **500** 233–7
[17] Young D J 2016 *High Temperature Oxidation and Corrosion of Metals* (Amsterdam: Elsevier) 2nd edn
[18] Bau N Q 2009 Development of nano-composites based on magnesium alloys system AZ31B *PhD Thesis* National University of Singapore
[19] Tekumalla S and Gupta M 2017 An insight into ignition factors and mechanisms of magnesium based materials: a review *Mater. Des.* **113** 84–98
[20] Zhao S, Zhou H, Zhou T, Zhang Z, Lin P and Ren L 2013 The oxidation resistance and ignition temperature of AZ31 magnesium alloy with additions of La_2O_3 and La *Corros. Sci.* **67** 75–81
[21] Sankaranarayanan S, Pranav Nayak U, Sabat R K, Suwas S, Almajid A and Gupta M 2014 Nano-ZnO particle addition to monolithic magnesium for enhanced tensile and compressive response *J. Alloys Compd.* **615** 211–9
[22] Parande G, Manakari V, Meenashisundaram G K and Gupta M 2017 Enhancing the tensile and ignition response of monolithic magnesium by reinforcing with silica nanoparticulates *J. Mater. Res.* **32** 2169–78
[23] Kujur M S, Mallick A, Manakari V, Parande G, Tun K S and Gupta M 2017 Significantly enhancing the ignition/compression/damping response of monolithic magnesium by addition of Sm_2O_3 nanoparticles *Metals* **7** 357

[24] Johanes M, Tekumalla S and Gupta M 2019 Fe_3O_4 nanoparticle-reinforced magnesium nanocomposites processed via disintegrated melt deposition and turning-induced deformation techniques *Metals* **9** 1225
[25] Tekumalla S, Gupta M and Min K H 2018 Using CaO nanoparticles to improve mechanical and ignition response of magnesium *Curr. Nanomater.* **3** 44–51
[26] Tekumalla S, Nandigam Y, Bibhanshu N, Rajashekara S, Yang C, Suwas S and Gupta M 2018 A strong and deformable *in situ* magnesium nanocomposite igniting above 1000 °C *Sci. Rep.* **8** 7038
[27] Nguyen T D and Lee D B 2010 Oxidation of AM60B Mg alloys containing dispersed SiC particles in air at temperatures between 400 and 550 °C *Oxid. Met.* **73** 183–92
[28] Tekumalla S, Yuan N J, Haghshenas M and Gupta M 2019 Enhancing properties of aerospace alloy Elektron 21 using boron carbide nanoparticles as reinforcement *Appl. Sci.* **9** 5470
[29] Kujur M S, Manakari V, Parande G, Tun K S, Mallick A and Gupta M 2018 Enhancement of thermal, mechanical, ignition and damping response of magnesium using nano-ceria particles *Ceram. Int.* **44** 15035–43
[30] Sravya Tekumalla S S and Gupta M 2016 Enhancing tension-compression yield symmetry in magnesium by reinforcing with nano-La_2O_3 *Proc. iMagCon 2016, VIT University (Chennai, India)*
[31] Czerwinski F 2007 *Magnesium Injection Molding* (Berlin: Springer)
[32] Fassell W M, Gulbransen L B, Lewis J R and Hamilton J H 1951 Ignition temperatures of magnesium and magnesium alloys *JOM* **3** 522–8
[33] Liu M, Shih D S, Parish C and Atrens A 2012 The ignition temperature of Mg alloys WE43, AZ31 and AZ91 *Corros. Sci.* **54** 139–42
[34] Federal Aviation Administration 2014 Development of a laboratory scale flammability test for magnesium alloys used in aircraft seat construction *FAA report* DOT/FAA/TC-13/52 https://fire.tc.faa.gov/pdf/tc-13-52.pdf
[35] PeriodicTable.Com n.d. Thermal conductivity of the elements *PeriodicTable.Com* http://periodictable.com/Properties/A/ThermalConductivity.v.html
[36] AZO Materials n.d. Silicon dioxide *AZO Materials* https://azom.com/properties.aspx?ArticleID=1114
[37] Thermtest Instruments n.d. Materials thermal properties database *Thermtest Instruments* https://thermtest.com/materials-database#calcium-
[38] Engineers Edge n.d. Specific heat capacity of metals table chart *Engineers Edge* https://engineersedge.com/materials/specific_heat_capacity_of_metals_13259.htm
[39] NIST 2018 Calcium oxide *NIST Chemistry WebBook* webbook.nist.gov
[40] Tekumalla S, Yang C, Seetharaman S, Wong W L E, Goh C S, Shabadi R and Gupta M 2016 Enhancing overall static/dynamic/damping/ignition response of magnesium through the addition of lower amounts (<2%) of yttrium *J. Alloys Compd.* **689** 350–8

Chapter 8

Creep characteristics of magnesium-based nanocomposites

Owing to their low density, magnesium-based materials are highly promising for the structural light-weighting of automobile and aerospace components. Since these applications involve exposure to elevated temperatures, sufficient creep resistance is essential. However, it is well known that the creep resistance of magnesium-based materials is inferior at high temperatures. A promising approach to improve the creep resistance of Mg is to obstruct grain boundary sliding by using hard and strong reinforcements. In this regard, magnesium nanocomposites, particularly with thermally stable nanoparticles, are interesting and this chapter summarizes the responses of nanoparticle reinforced magnesium composites under different creep conditions (temperature and load).

8.1 Introduction

Creep refers to the time dependent thermally activated plastic deformation of a material under the application of a constant stress below the yield strength. Figure 8.1 shows a classical creep curve (deformation versus time) of a material depicting the three stages of creep. While the primary creep starts with a rapid creep rate that slows with time due to the strain hardening effects, the secondary creep has a relatively uniform creep rate due to the balance between strain hardening and thermal recovery. During tertiary creep, the creep rate accelerates due to necking and results in rupture or failure [1].

Figure 8.1 also highlights the dependence of the test temperature (T) and applied stress (σ). It shows that the increase in applied stress at a constant temperature or the increase in temperature at a constant stress increases the initial deformation and the secondary creep rate, and hence reduces the time to rupture. Since secondary creep covers a broad time span, the secondary creep rate ($\dot{\varepsilon}_s$) plays an important role in the evaluation of creep properties [2]. According to the Norton relationship, the general

doi:10.1088/978-0-7503-3535-5ch8

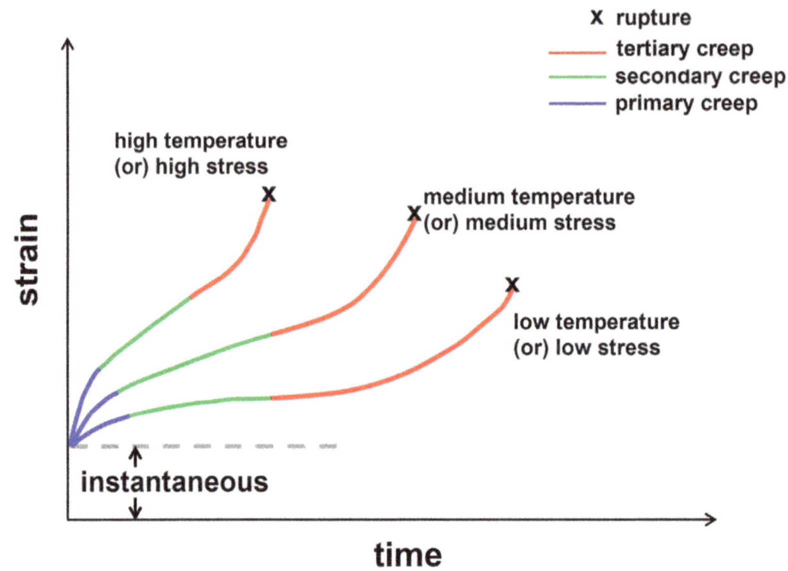

Figure 8.1. Typical creep curves showing different stages of creep [1].

dependence of creep rate on the temperature (T), applied stress (σ), and grain size (d) is

$$\dot{\varepsilon} = \frac{AD}{kT}\frac{\sigma^n}{d^p}. \tag{8.1}$$

In the above equation, A is the temperature dependent material constant, k is Boltzmann's constant, n is the stress exponent, p is the grain size exponent and D is the diffusion coefficient which is given by

$$D = D_o\exp\left(-\frac{Q}{RT}\right), \tag{8.2}$$

where D_o represents the coefficient of lattice diffusion, Q is the activation energy for creep, and R is the universal gas constant.

8.2 Mechanism of creep

Figure 8.2 shows the details of creep mechanisms at different temperature and stress levels. Based on the applied stresses and operating temperatures, the creep at elevated temperature can be either diffusion controlled or dislocation movement controlled. While the diffusion creep occurs at high temperatures and low stresses, the dislocation creep occurs at higher stresses.

In diffusion creep, the applied stress, being the driving force, controls the flow of atoms. While the diffusion of atoms occurs through the inside of grains at high temperature (Nabarro–Herring creep), the grain boundary diffusion is dominant at low temperatures (Coble creep). Since diffusion creep exhibits linear-viscous behaviour, the creep rate can be mathematically expressed in terms of the diffusion coefficient (D), applied stress (σ), and the grain size (d) as follows:

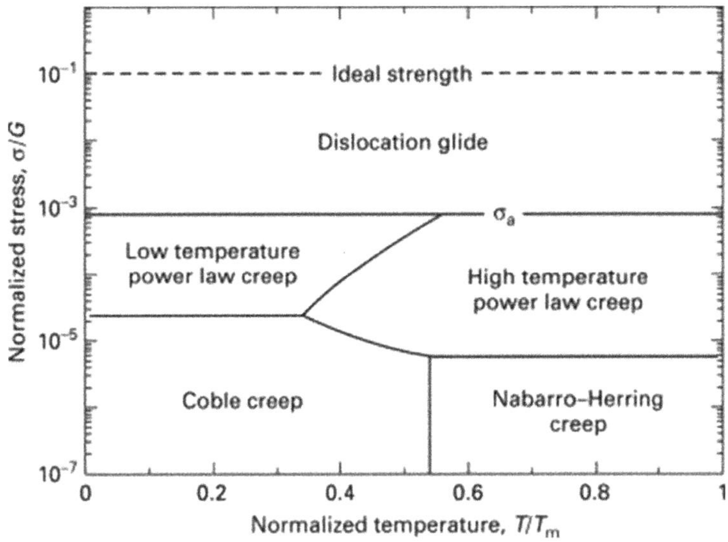

Figure 8.2. Creep mechanisms at different temperatures and stress levels. Reprinted from [3], Copyright 2008, with permission from Elsevier.

Table 8.1. Diffusion coefficients, and stress and grain size exponents for different types of creep [3].

Creep type	Diffusion coefficient (D)	Stress exponent (n)	Grain size exponent
Nabarro–Herring creep (volume diffusion)	D_o	1	2
Coble creep (grain boundary diffusion)	DGB	1	3
Grain boundary sliding	D_o	2	1
Dislocation creep (volume diffusion at high temperature)	D_o	3–5	0
Pipe diffusion-controlled dislocation creep (low temperature)	D_p	5–7	0

$$\dot{\varepsilon}_s = \frac{ADGb}{kT}\left(\frac{b}{d}\right)^p\left(\frac{\sigma}{G}\right)^n. \tag{8.3}$$

In the above equation G and b are the shear modulus and Burgers vector, respectively. Table 8.1 illustrates the values of diffusion coefficient and stress/grain size exponents for different creep mechanisms.

The minimum creep rate for Nabarro–Herring and Coble creep can be expressed as follows [3].

For Nabarro–Herring creep, $D = D_o$, coefficient for self-diffusion in a lattice, $p = 2$ and $n = 1$ in equation (8.3):

$$\dot{\varepsilon}_s = \frac{AD_oGb}{kT}\left(\frac{b}{d}\right)^2\left(\frac{\sigma}{G}\right)^1. \tag{8.4}$$

For Coble creep, $D = DGB$, coefficient for grain boundary diffusion, $p = 3$ and $n = 1$ in equation (8.3):

$$\dot{\varepsilon}_s = \frac{AD_{GB}Gb}{kT}\left(\frac{b}{d}\right)^3\left(\frac{\sigma}{G}\right)^1. \tag{8.5}$$

For grain boundary sliding (GBS) which is defined as the relative movement of two adjacent grains under an applied stress, the stress exponent value (n) is equal to 2, the grain size exponent is equal to 1, and diffusion coefficient (D) is equal to D_o:

$$\dot{\varepsilon}_s = \frac{AD_oGb}{kT}\left(\frac{b}{d}\right)^1\left(\frac{\sigma}{G}\right)^2. \tag{8.6}$$

At higher stresses, irrespective of the material type and grain size, dislocation creep occurs by the movement of dislocations and follows the power law ($°\varepsilon\alpha\sigma^n$). In general, there are two basic types of dislocation motion, (i) dislocation glide and (ii) dislocation climb, against obstacles such as grain boundaries, precipitates, other dislocations, and the solute atoms in solid solution. Therefore the rate of dislocation creep also depends on the number of dislocation sources and the distance between the dislocation pile-ups [3].

The following expression describes the minimum creep rate for dislocation creep [4]:

$$\dot{\varepsilon}_s = \frac{BD_s\sigma^{4.5}}{b^{0.5}N^{0.5}G^{3.5}kT}. \tag{8.7}$$

8.3 Evaluation of creep properties

In creep tests, the dimensional changes of a material are measured with time at different constant temperatures and constant (tension, compression, indentation, or impression) loads. Creep tests are usually conducted at low loads for a longer duration to predict the life expectancy of materials in long term applications that are strain limited. The test procedure involves heating the samples to a required temperature and measuring the gauge length at frequent intervals after applying predetermined loads without delay. Since temperature plays an important role in creep deformation, the testing is normally performed inside a chamber with precise and controlled heating/cooling and, typically, an average of 50 readings are taken for each temperature–stress combination.

8.4 Creep response of magnesium nanocomposites

Most magnesium alloys suffer from poor creep resistance due to the ease of GBS at temperatures above 150 °C [5, 6]. A promising approach to improve the creep resistance is to obstruct the GBS by using hard and strong reinforcements. While the effectiveness of micron-sized ceramic reinforcements has been extensively studied [7–11], only limited research has been conducted on the creep response of Mg composites reinforced with nanoscale reinforcements.

Ferkel and Mordike [12] investigated the creep behaviour of nano-SiC particle reinforced Mg nanocomposites. In the study the nanocomposites, prepared using powder mixing and ball milling methods, were tested for tension creep characteristics at temperatures between 200 °C and 300 °C under stresses between 35 and 45 MPa. From the experimental findings (figure 8.3), the nano-SiC particles were found to be beneficial in improving the creep resistance of Mg. Significant improvement comparable to that of the commercial creep resistant alloys such as WE43, WE54, and QE22 was realized for the ball-milled nanocomposite and the same was rightly attributed to the effectiveness of SiC nanoparticles in pinning the matrix grain boundaries in order to control the creep deformation.

In another study, Seetharaman *et al* [13] investigated the compression creep characteristics of ZnO nanoparticle containing Mg–5Sn–2Y alloy at 150 °C between 40 MPa and 80 MPa. From the reported creep data (figure 8.4(a)), it is clear that the Mg–5Y–2Y/1.5ZnO nanocomposite exhibits a much lower creep rate when compared to unreinforced Mg–5Sn–2Y. Although the addition of ZnO nanoparticles increased the resistance to deformation, the calculated values of stress exponent n

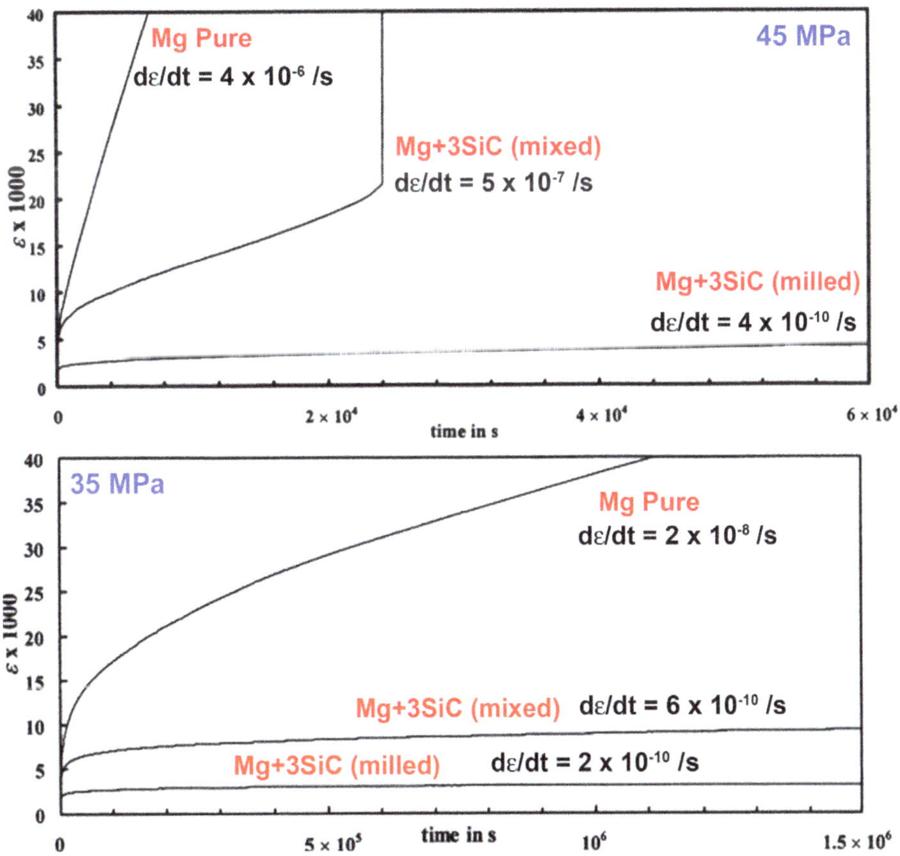

Figure 8.3. Creep curves of Mg/SiC nanocomposite at 200 °C at 35 MPa and 45 MPa. Reprinted from [12], Copyright 2001, with permission from Elsevier.

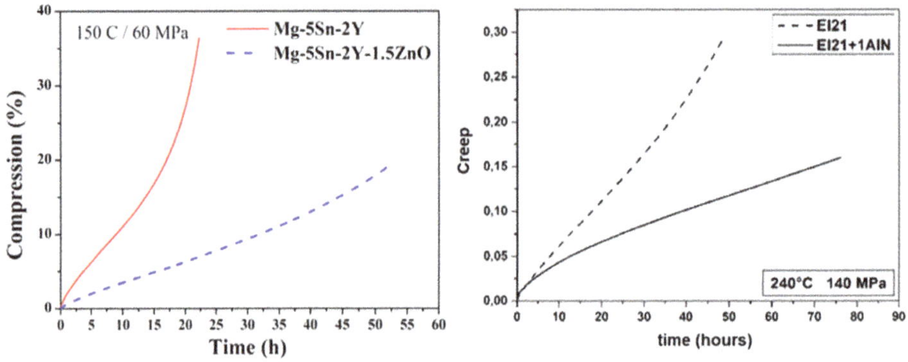

Figure 8.4. Creep curves of (a) Mg–5Sn–2Y/1.5ZnO nanocomposite at 150 °C, 60 MPa (adapted from [13]), and (b) El21/AlN nanocomposite at 240 °C, 140 MPa. Reprinted from [13, 14], Copyright 2015, 2016, with permission from Elsevier.

(6.9 for Mg–5Y–2Y and 7 for Mg–5Y–2Y/1.5ZnO) revealed no significant difference in the creep mechanism (pipe diffusion-controlled dislocation creep) between the nanocomposite and the unreinforced monolithic alloy. Similar improvement in creep resistance was also reported by Katsarou *et al* [14] for die cast Elektron 21 alloy reinforced with AlN nanoparticles (figure 8.4(b)). In this study, an ultrasonic assisted die casting method was used to fabricate the El21/AlN nanocomposites and creep tests were conducted at 240 °C under constant compression stresses between 70 and 200 MPa.

In a recent study, Yang *et al* [15] used the high shearing dispersion technique (HSDT) to disperse a nanoparticle mixture of AlN/Al (75% AlN and 25% metallic Al, produced by the electrical explosion of Al wires in argon and nitrogen atmosphere) in the Elektron 21 alloy. In this study, the high shearing speed (3000 rpm versus 500 rpm) during HSDT processing influenced the uniform distribution of AlN/Al NPs and the synergistic effects of evenly distributed Al and AlN NPs contributed towards a better creep resistance compared to traditional ultrasound assisted stirring. Overall, the El21/0.5%AlN/Al nanocomposite processed at 3000 rpm exhibited a better creep resistance, as shown in figure 8.5(a). The calculation of true creep stress exponents revealed the viscous glide of dislocation together with dislocation climbing as the dominant rate controlling mechanisms during creep deformation. The authors also investigated the effects of T6 heat treatment and reported a deterioration in the creep resistance of HSDT processed El21/1 wt.%AlN/Al nanocomposite due to the lesser amount of γ' and β' precipitates resulting from the formation of Al2RE precipitates (figure 8.5(b)) [16]. On the other hand, $Mg_{41}RE_5$ was found as the predominant intermetallic phase along with many nano-sized γ' and β' precipitates in the T6 heat-treated unreinforced El21 alloy. Based on the calculation of stress components, the authors reported no change in the creep mechanism due to heat treatment.

Indentation creep tests were conducted on ultrasonic stir cast AS41/Al_2O_3 nanocomposites at different temperatures and stresses ranging between 175 °C–250 °C and 109.2 MPa–140 MPa, respectively [17]. The results (figure 8.6) showed creep resistance

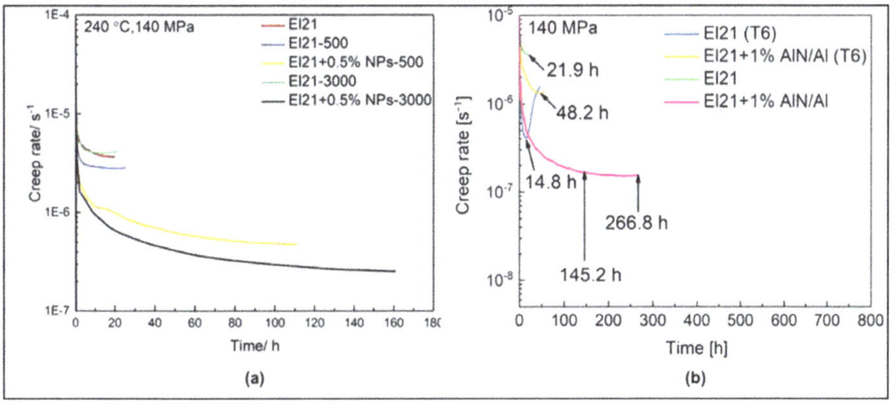

Figure 8.5. Creep rates at 240 °C for (a) non-sheared El21, El21–500, El21 + 0.5% NPs-500, El21–3000, and El21 + 0.5% NPs-3000, respectively (500 and 3000 denote the shearing speeds of 500 and 3000 rpm), and (b) HSDT processed El21+1%AlN nanocomposite in the as-cast and T6 heat-treated conditions. Reprinted from [15, 16], Copyright 2019, 2020, with permission from Elsevier.

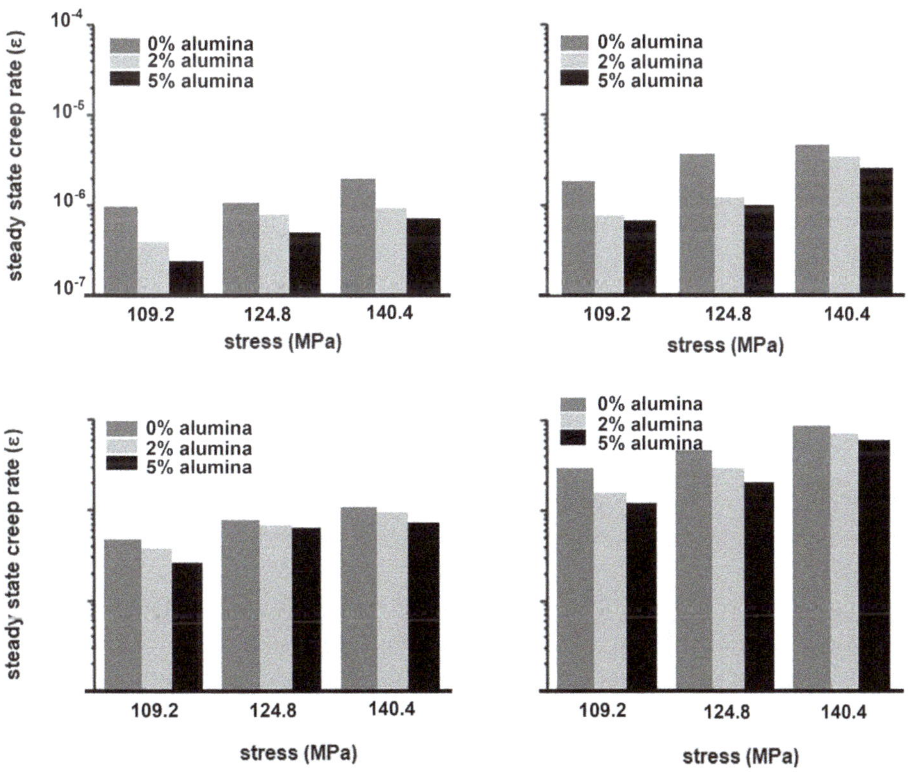

Figure 8.6. Creep rates of AS41/Al_2O_3 nanocomposites at (a) 175 °C, (b) 200 °C, (c) 225 °C, and (d) 250 °C. Reprinted from [17], Copyright 2014, with permission from Elsevier.

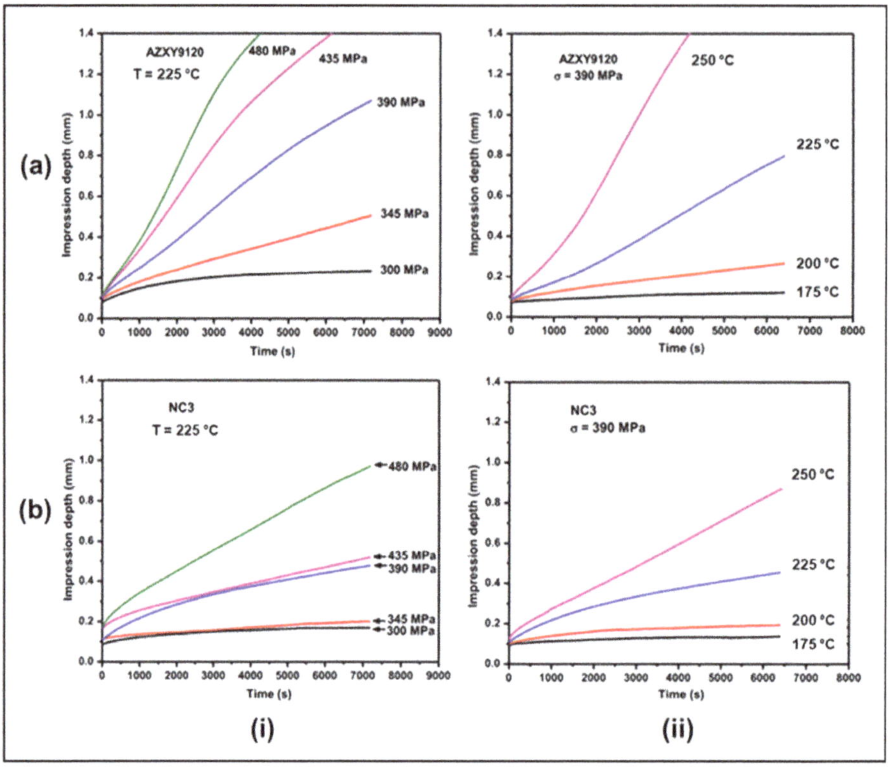

Figure 8.7. Effect of (i) applied stress and (ii) temperature on the impression creep behaviour of (a) AZ91 +2.0Ca+0.3Sb (AZXY9120) alloy and (b) AZ91+2.0Ca+0.3Sb/2.0SiC (NC3) nanocomposite. Reprinted from [18], Copyright 2018, with permission from Elsevier.

Table 8.2. Volume fraction of $Mg_{17}Al_{12}$ and Al_2Ca in AZ91–2Ca–0.3Sb/SiC nanocomposites as reported by [18].

Materials		$Mg_{17}Al_{12}$ phase (vol.%)	Al_2Ca phase (vol.%)
AZ91–2.0Ca–0.3Sb	AZXY91	9.45 ± 0.32	4.49 ± 0.19
AZ91–2.0Ca–0.3Sb/0.5SiC	NC1	8.18 ± 0.21	5.34 ± 0.24
AZ91–2.0Ca–0.3Sb/1.0SiC	NC2	6.78 ± 0.27	6.38 ± 0.27
AZ91–2.0Ca–0.3Sb/2.0SiC	NC3	5.43 ± 0.18	7.81 ± 0.12

improvement with increasing addition of nano-Al_2O_3 particles, which was attributed to the Orowan strengthening effects by the uniformly distributed nano-Al_2O_3 particles and the finer Mg_2Si precipitates. This study also confirmed 'pipe diffusion-controlled dislocation creep' as the dominant rate controlling deformation mechanism for AS41/ Al_2O_3 nanocomposites (figure 8.6).

Ganguly and Mondal [18] investigated the influence of SiC nanoparticle additions on the impression creep behaviour of squeeze cast AZ91–2Ca–0.3Sb alloy. The results (figure 8.7) of impression creep tests conducted under different temperature

Table 8.3. Summary of creep response for various magnesium nanocomposites.

Nanocomposite system	Processing method	Temperature (°C)	Stress (MPa)	True stress exponent	Rate controlling mechanism	Reference
Mg/3 vol.%SiC	Ball milling assisted powder metallurgy	200–300	35–45	10 at 200 °C 3 at 300 °C	Dislocation controlled	Ferkel and Mordike [12]
Mg–5Sn–2Y/1.5 wt.%ZnO	Disintegrated melt deposition followed by hot extrusion	150	40–80	7	Pipe diffusion-controlled dislocation creep	Sankaranarayanan *et al* [13]
El21/1 wt.%AlN	Ultrasonic assisted high pressure die casting	240	70–200	3.9	Dislocation climb and glide	Katsarou *et al* [14]
El21/0.5 wt.%AlN/Al	High shearing dispersion stirring	240	70–140	3.4–4.3	Viscous glide of dislocations and/or dislocation climbing	Yang *et al* [15]
El21/1 wt.%AlN/Al				3.4		Yang *et al* [16]
El21/1 wt.%AlN/Al (T6)				4.6		
AS41/Al_2O_3	Ultrasonic stir casting	175–250	109.2–140	3.55–6.5	Pipe diffusion-controlled dislocation creep	Kumar and Chaudhari [17]
AZ91–2Ca–0.3Sb/SiC	Squeeze casting	175–200	300–480	4.5–6.2	Dislocation climb controlled by pipe diffusion	Ganguly and Mondal [18]
Mg/BN	Microwave sintering assisted powder metallurgy	Room temperature	50 mN (non-destructive depth sensing method)	>3	Dislocation (power law) creep	Haghshenas *et al* [19]
Mg/Sm_2O_3	Microwave sintering assisted powder metallurgy	25–300		3–7	Dislocation (power law) creep	Haghshenas *et al* [20, 21]

and stress conditions (temperature range between 175 °C–250 °C and stresses between 300 MPa to 480 MPa for a time duration of 7200 s) showed superior creep resistance for the nanocomposites and the best creep resistance was reported for 2% SiC addition. From the microstructural analysis (table 8.2) it was found that the addition of SiC nanoparticles facilitated the formation of thermally stable Al_2Ca phases and controlled the dislocation motion to improve the creep resistance. The calculated values of stress exponents (4.5–6.2) suggested dislocation climb controlled by pipe diffusion as the rate controlling deformation mechanism for the AZ91–2Ca–0.3Sb alloy and its nano-SiC reinforced nanocomposites.

Haghshenas *et al* [19] used a two stage non-destructive depth sensing indentation method to assess the creep behaviour of Mg/BN nanocomposites at ambient (room) temperature. In this method, the specimens were initially loaded at different rates (between 0.05 and 50 mN s^{-1}) to a peak load of 50 mN and then held at a constant force for 400 s. From the load, displacement, and the time data, the creep resistance was recorded. While the authors confirmed better creep resistance due to the addition of BN nanoparticles, the stress exponent values (n) confirmed obstacle (particle or dislocation) limited glide as the dominant creep mechanism. Similar experiments were also performed on Mg/Sm_2O_3 nanocomposites [20]. While the addition of Sm_2O_3 nanoparticles strengthened the matrix by restricting the dislocation motion and grain boundary pinning, the room temperature creep was controlled by dislocation glide. However, under non-ambient conditions, dislocation climb was also reported [21].

8.5 Summary

The poor creep resistance of Mg and its alloys is one of the barriers for their applications in high temperature applications in different sectors. In this regard, the incorporation of thermally stable nanoparticles is being explored by various researchers to enhance the high temperature creep response of Mg alloys. Based on the published literature it can be clearly understood that the presence of nanoscale reinforcements with good thermal stability assists in impeding the movement of dislocations to impart high creep resistance. While the Mg nanocomposites exhibit superior creep resistance in general, the mechanism of creep depends on the composition of the alloy matrix, reinforcing phases, and the creep conditions, i.e. temperature and applied stress (as shown in table 8.3).

References

[1] Materion 2018 Stress-rupturing and creep rupturing *Technical Tidbits* issue 115 https://materion.de.com/-/media/files/alloy/newsletters/technical-tidbits/issue-no-115-stress-rupturing-and-creep-rupturing.pdf

[2] Hu H, Yu A, Li N and Allison J E 2003 Potential magnesium alloys for high temperature die cast automotive applications: a review *Mater. Manuf. Process.* **18** 687–717

[3] Maruyama K 2008 Fundamental aspects of creep deformation and deformation mechanism map *Creep-Resistant Steels* (Cambridge: Woodhorn) pp 265–78

[4] Yang H, Gavras S and Dieringa H 2020 *Creep characteristics of Metal Matrix Composites (Reference Module in Materials Science and Materials Engineering)* (New York: Elsevier)
[5] Kainer K U, Huang Y D, Dieringa H and Hort N 2010 Status of the development of creep resistant magnesium materials for automotive applications *Mater. Sci. Forum* **638–642** 73–80
[6] Bowles A L, Dieringa H, Blawert C, Hort N and Kainer K U 2005 Investigations in the magnesium–tin system *Mater. Sci. Forum* **488–489** 135–8
[7] Kumar S, Dieringa H and Kainer K U 2007 Thermal cycling behaviour of the magnesium alloy based hybrid composites in the transverse direction *Mater. Sci. Eng.* A **454–455** 367–70
[8] Kumar S, Dieringa H and Kainer K U 2005 Effect of particulate content on the thermal cycling behaviour of the magnesium alloy based hybrid composites *Composites* A **36** 321–5
[9] Svoboda M, Pahutová M, Kuchařová K, Sklenička V and Kainer K U 2007 Microstructure and creep behaviour of magnesium hybrid composites *Mater. Sci. Eng.* A **462** 220–4
[10] Dieringa H, Huang Y, Maier P, Hort N and Kainer K U 2005 Tensile and compressive creep behaviour of Al_2O_3 (Saffil®) short fiber reinforced magnesium alloy AE42 *Mater. Sci. Eng.* A **410–411** 85–8
[11] Arunachaleswaran A *et al* 2007 Creep behavior of AE42 based hybrid composites *Mater. Sci. Eng.* A **460–461** 268–76
[12] Ferkel H and Mordike B L 2001 Magnesium strengthened by SiC nanoparticles *Mater. Sci. Eng.* A **298** 193–9
[13] Seetharaman S, Dieringa H, Kainer K U and Gupta M 2015 Development and characterization of a new magnesium–tin–yttrium alloy and its nanocomposite *Proc. of Mg 2015–10th Int. Conf. on Magnesium Alloys and Their Applications (Jeju, Korea)*
[14] Katsarou L *et al* 2016 Microstructure, mechanical properties and creep of magnesium alloy Elektron21 reinforced with AlN nanoparticles by ultrasound-assisted stirring *Mater. Sci. Eng.* A **659** 84–92
[15] Yang H, Huang Y, Song B, Kainer K U and Dieringa H 2019 Enhancing the creep resistance of AlN/Al nanoparticles reinforced Mg-2.85Nd-0.92Gd-0.41Zr-0.29Zn alloy by a high shear dispersion technique *Mater. Sci. Eng.* A **755** 18–27
[16] Yang H *et al* 2020 Effects of heat treatment on the microstructural evolution and creep resistance of Elektron21 alloy and its nanocomposite *Mater. Sci. Eng.* A **789** 139669
[17] Kumar H and Chaudhari G P 2014 Creep behavior of AS41 alloy matrix nano-composites *Mater. Sci. Eng.* A **607** 435–44
[18] Ganguly S and Mondal A K 2018 Influence of SiC nanoparticles addition on microstructure and creep behavior of squeeze-cast AZ91-Ca-Sb magnesium alloy *Mater. Sci. Eng.* A **718** 377–89
[19] Haghshenas M, Islam R, Wang Y, Cheng Y T and Gupta M 2019 Depth sensing indentation of magnesium/boron nitride nanocomposites *J. Compos. Mater.* **53** 1751–63
[20] Haghshenas M, Song X, Hasannaeimi V, Mukherjee S and Gupta M 2020 Magnesium–samarium oxide nanocomposites: room-temperature depth-sensing nanoindentation response *Int. J. Light Mater. Manuf.* **3** 217–25
[21] Haghshenas M, Muhammad M, Hasannaeimi V, Mukherjee S and Gupta M 2019 Ambient and non-ambient temperature depth-sensing indentation of Mg-Sm_2O_3 nanocomposites *Int. J. Adv. Manuf. Technol.* **105** 2947–56

IOP Publishing

Magnesium-Based Nanocomposites
Advances and applications
Sankaranarayanan Seetharaman, Sravya Tekumalla and Manoj Gupta

Chapter 9

Machinability of magnesium-based nanocomposites

Magnesium-based nanocomposites with enhanced mechanical properties are envisaged for high performance applications. Unlike conventional microcomposites which require a large number of reinforcements to realize the strengthening benefits, only a small volume fraction of nano-sized reinforcements contribute to superior mechanical properties. In this regard, the mechanism of material removal in magnesium-based nanocomposites is expected to differ distinctively from that of the conventional microcomposites that suffer excessive tool wear due to the large number of abrasive particle reinforcements. Hence, the addition of low volume fraction nano-reinforcement is promising, and therefore it is essential to identify the highly productive and cost-effective machining methods applicable to Mg-based nanocomposites. In this context, some of the recent studies assessed the machinability attributes of Mg nanocomposites. This chapter summarizes the published research findings on the machinability of Mg nanocomposites.

9.1 Introduction

Machining is one of the most suitable manufacturing techniques for producing near net shape components. In general, the cutting conditions and tool geometry together with the materials' properties affect the machining characteristics of any material (see figure 9.1). Saravan and Natarajan [1] reported the significant influence of cutting speed on the surface roughness and the cutting force of AZ91D alloy-based composite reinforced with 5 vol.% graphite particles (size: 50 μm). According to Davies *et al* [2], the chip morphology is also primarily influenced by the cutting speed, while the changes in tool rake angle and reinforcement particle size/volume fraction showed little influence. Pramanik *et al* [3] also recommended higher cutting speeds and lower feed rates to obtain a better surface finish. However, the enhanced friction resulted in excessive tool wear at a lower feed rate. In this regard,

doi:10.1088/978-0-7503-3535-5ch9

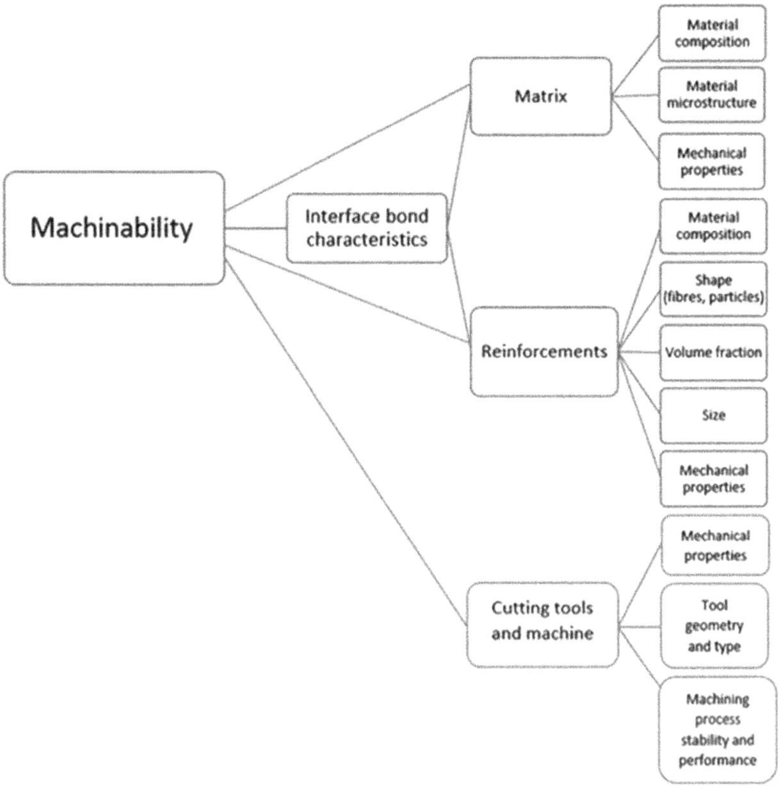

Figure 9.1. Factors affecting machinability. Reproduced from [4, 5], with permission © Springer Nature.

unconventional machining processes without physical contact between the cutting tool and the component are interesting [4]. However, they have limitations such as excessive heat generation in laser machining, resulting in poor surface finish and poor conductivity of ceramic reinforcements affecting the material removal rate, thus causing severe electrode wear in electrical discharge machining (EDM) [4].

As discussed in previous chapters, the mechanical properties of Mg are improved significantly through the addition of hard ceramic or metallic nanoparticles [6]. Unlike micron-sized particle reinforcements which are added in large amounts, the addition of very low volume fractions of nano-sized reinforcements contributes to comparable strengths without a significant effect on the ductility. Therefore, the material removal mechanism of magnesium-based nanocomposites is expected to differ distinctly from that of the conventional microcomposites. While a large amount of highly abrasive ceramic particles leads to larger cutting forces and excessive tool wear in micron-sized particle reinforced magnesium composites (figure 9.2(a)) [7], the addition of a low volume fraction of nano-reinforcement seems promising. The reduction in particle size and volume fraction can decrease the cutting tool wear, as reported in the case of 5–15 vol.% SiC particle (particle size: 15–38 μm) reinforced Mg–Al/SiC composites (figure 9.2(b)) [8, 9]. Hence, it is widely perceived that the mechanism of material removal in nanocomposites differs

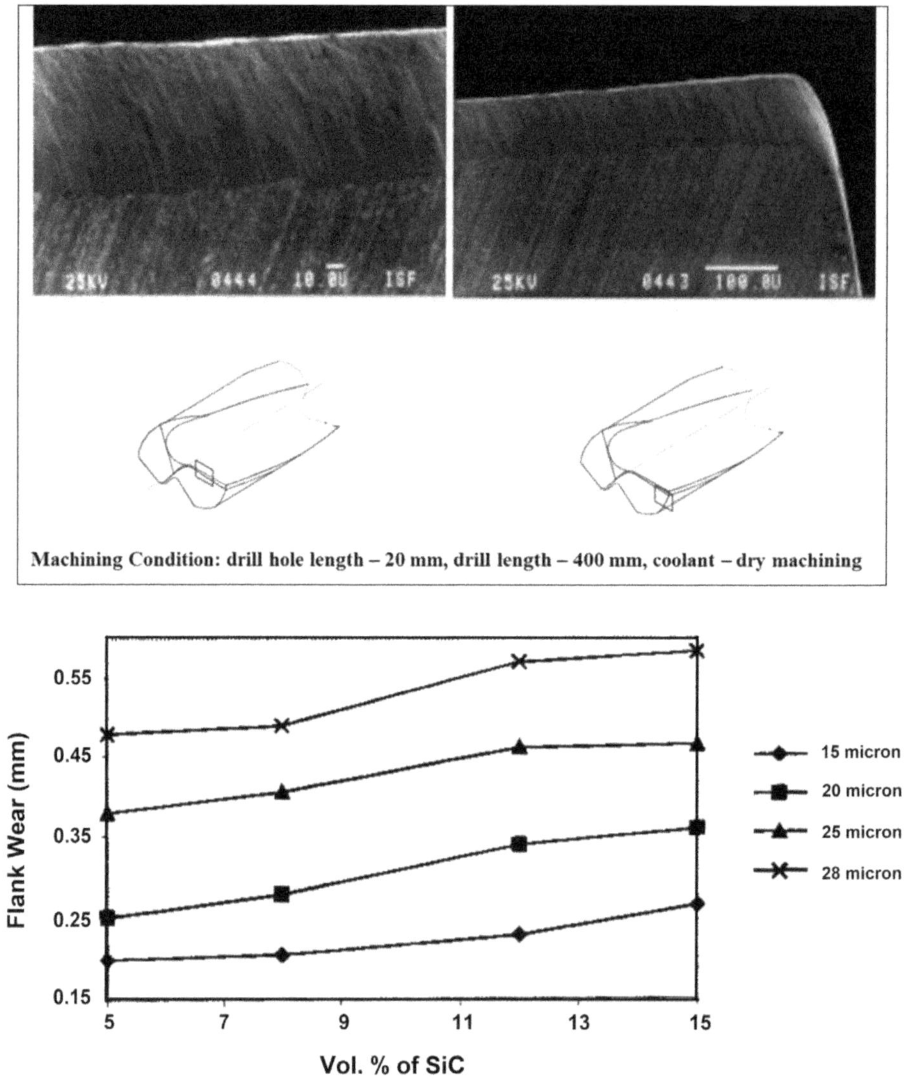

Figure 9.2. (Top) Worn cutting edge after machining AZ91/20 vol.%SiC composite and (bottom) the effect of SiC particle size and volume fraction on flank wear. (Top) Reproduced with permission from [9], © John Wiley & Sons. (Bottom) Reproduced with permission from [8], © Trans Tech Publications.

distinctly from that of the conventional microcomposites, and the published literature suggests that the machining parameters must be fine-tuned to suit the variation in reinforcement particles [10, 11]. For example, Sahin and Sur [12] recommended higher cutting speeds for fine particle reinforced composites while lower cutting speeds are suggested for the machining of composites reinforced with coarse particles based on the milling experiments conducted on Al/SiC nano-composites. Teng *et al* [13] reported that the nano-sized particles remained intact without fracture and produced continuous chips during the machining of SiC nanoparticle reinforced Al composites, while the micron-sized particles easily break and form discontinuous chips. In this context, several investigations in recent years have focussed on the machining characteristics of nanoparticle reinforced magnesium composites and the key attributes are summarized in this chapter.

9.2 Machinability of magnesium-based nanocomposites

For the first time, Ponappa *et al* [14] investigated the effects of EDM process parameters on the quality of high aspect ratio drill holes (diameter—0.5 mm, height—12 mm) in microwave sintered Mg/Al_2O_3 nanocomposites. A Taguchi orthogonal L9 array design involving process parameters such as pulse-on time (T_{on}), pulse-off time (T_{off}), voltage gap (S_v), and servo speed (SEN) was used (see table 9.1).

The machinability was assessed in terms of the effectiveness ratio (η) of response variables: (i) surface roughness and (ii) taper of the drilled hole calculated using the following equation:

$$\eta = -10 \log[(1/N) * \sum(Y_i^2)) \text{ for } i = 1 \text{ to } N, \tag{9.1}$$

where Y_i is the value of surface roughness or taper for a specific test, n is the number of tests, and N is the total number of data points.

From the calculated values, the authors identified the optimum set of process parameters, as shown in figure 9.3. Using statistical variance analysis, the authors also confirmed the significant role of pulse-on time and servo speed in maintaining

Table 9.1. Design of experiments based on the Taguchi orthogonal L9 array in [14]. Reproduced with permission, © Springer.

	Process parameters*			
Exp no.	T_{on}	T_{off}	S_v	SEN
1	1	1	1	1
2	1	2	2	2
3	1	3	3	3
4	2	1	2	3
5	2	2	3	1
6	2	3	1	2
7	3	1	3	2
8	3	2	1	3
9	3	3	2	1

***Details of process parameters**

Factor	**Unit**	**Level 1**	**Level 2**	**Level 3**
T_{on}	μs	2	3	4
T_{off}	μs	3.5	4.5	5.5
S_v	V	2	3	4
SEN	mm min^{-1}	24	50	82

Materials	Level	Parameters: T_{on} (μs) (A)	Parameters: T_{off} (μs) (B)	Parameters: S_v (V) (C)	Parameters: SEN (mm/min) (D)	Optimum Parameters	Predicted	Experimental
Effectiveness Ratio for Taper							**Taper**	
$Mg/0.8Al_2O_3$	Level 1	**22.7263**	**22.4377**	20.8314	21.3546	**A1B1C2D2** (T_{on} - 2 μs, T_{off} - 3 μs S_v - 3 V, SEN - 50 mm/min)	0.0565	0.0612
	Level 2	21.8782	20.7545	**21.4258**	**21.489**			
	Level 3	18.5933	20.0055	20.9405	20.3541			
$Mg/1.2Al_2O_3$	Level 1	24.3263	22.4511	23.1284	**27.6759**	**A2B2C3D1** (T_{on} - 3 μs, T_{off} - 4.5 μs S_v - 4 V, SEN - 24 mm/min)	0.04509	0.0461
	Level 2	**25.0161**	**23.226**	22.3781	21.9075			
	Level 3	19.4995	23.1648	**23.3353**	19.2584			
Effectiveness Ratio for Surface Roughness							**Surface Roughness**	
$Mg/0.8Al_2O_3$	Level 1	**-11.748**	-14.352	-13.6058	**-11.201**	**A1B3C2D1** (T_{on} - 2 μs, T_{off} - 5.5 μs S_v - 3 V, SEN - 24 mm/min)	1.6632	3.4090
	Level 2	-13.1828	-14.7577	**-13.2104**	-14.5109			
	Level 3	-16.1208	**-11.9418**	-14.2354	-15.3396			
$Mg/1.2Al_2O_3$	Level 1	**-13.6064**	-14.379	-14.4378	**-11.8313**	**A1B2C3D1** (T_{on} - 2 μs, T_{off} - 4.5 μs S_v - 4 V, SEN - 24 mm/min)	3.553	4.6977
	Level 2	-13.9075	**-14.1121**	-14.3767	-15.2328			
	Level 3	-15.4757	-14.4985	**-14.1751**	-15.9256			

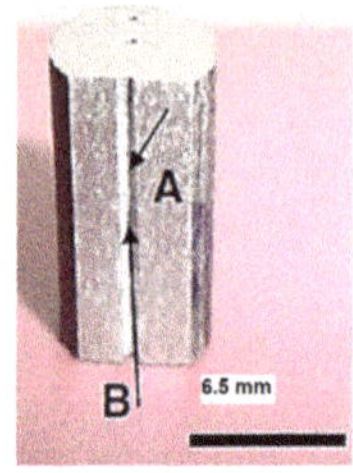

Figure 9.3. Predicted and experimental values of taper and surface roughness for optimized process parameters. Reproduced from [14], with permission © Springer Nature.

the dimensional tolerance of the high aspect ratio drilled hole. The reduced heat flux due to the increased time interval between consecutive sparks and the faster heat removal by dielectric fluid resulted in a reduced taper and improved surface finish. These observations were similar to those published on micron-sized particle reinforced Mg composites [15, 16]. Further, the experiments (see figure 9.3) conducted using optimized process parameters also confirmed acceptable values of taper and surface finish for both $Mg/0.8Al_2O_3$ and $Mg/1.2Al_2O_3$ nanocomposites. The authors also reported that abnormal arcing and random spark discharges were avoided in nanocomposites when compared to microcomposites which normally display bands and craters.

Li *et al* [17–20] performed micromachining experiments to understand the machinability characteristics of 5–15 vol.% nano-SiC particle reinforced Mg nano-composites. Micromachining is a mechanical material removal process used to machine high accuracy 3D components with dimensions or feature sizes ranging from several microns to several millimetres [21, 22]. Micro-manufacturing methods attract more attention from both industries and academia to meet the increasing demand for high precision and miniature components with superior mechanical properties [23]. In this study, the machinability characteristics were assessed in terms of the cutting forces, machined surface finish quality, and surface roughness. The cutting force profile for Mg-MMCs was distinctly different for each tool path when compared to pure Mg. While the cutting force increased with increasing spindle speed, feed rate, and nanoparticle volume fraction, the largest increasing slope occurred at the transition from 5 to 10 vol.% nano-SiC particles due to the rapid change in the mechanical properties of the material (figure 9.4). Similarly, the cutting force was remarkably high when the spindle speed was fixed at 60 000 rpm for the composite reinforced with 15 vol.% SiC nanoparticles.

Jian *et al* [20] also compared the specific cutting energy trends from the micro-machining experiments conducted for pure Mg and 10 vol.% Mg/nano-SiC. The variation in specific cutting energy versus the uncut chip thickness are identified as different regions, (i) elastic recovery, (ii) ploughing, and (iii) shearing zones, as marked in figure 9.5. It highlights the need for additional energy to machine the

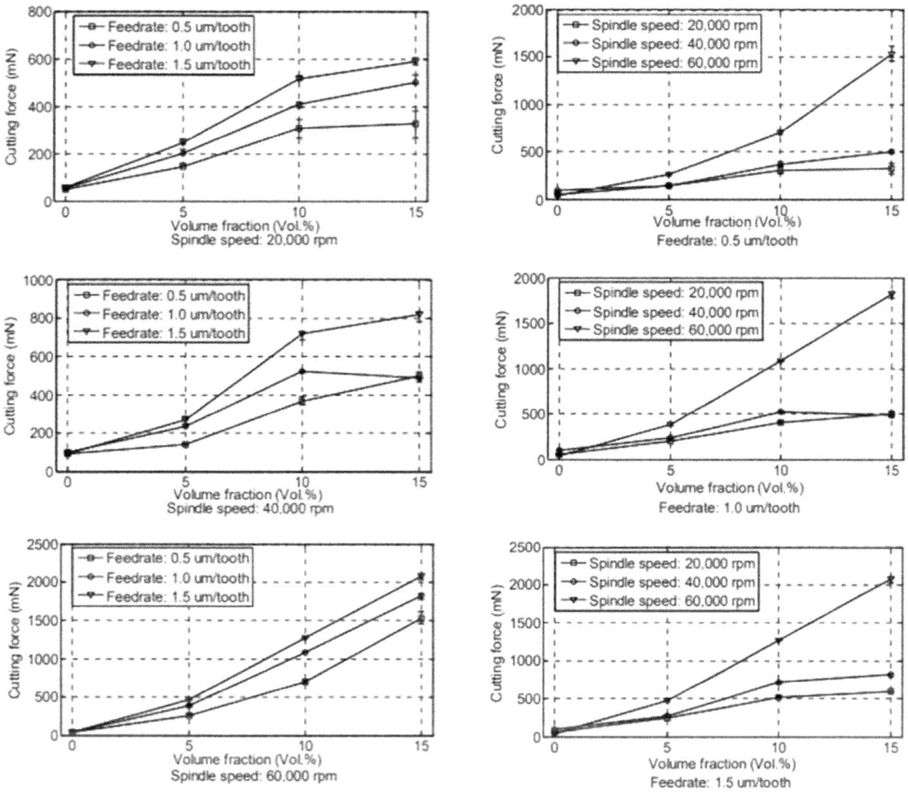

Figure 9.4. Variation in cutting force with respect to reinforcement volume fraction at specific spindle speed and feed rate. Reproduced with permission from [17, 18], © ASME.

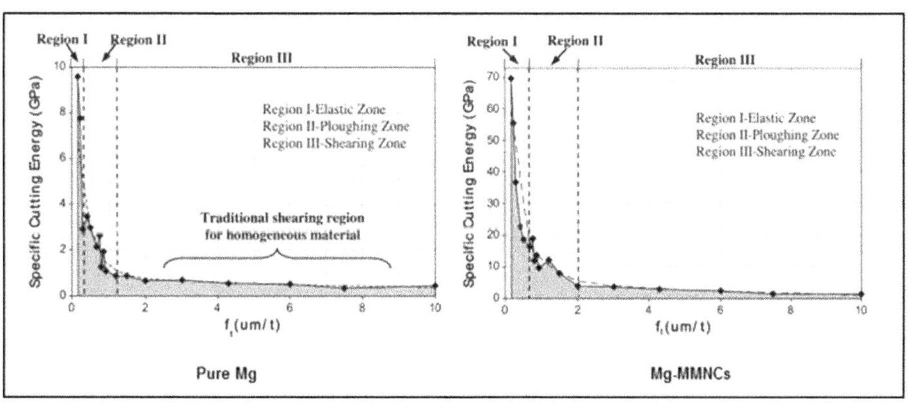

Figure 9.5. (a) Specific cutting energy versus nominal feed per tooth for pure Mg and Mg-MMCs. Reproduced with permission from [20], © ASME.

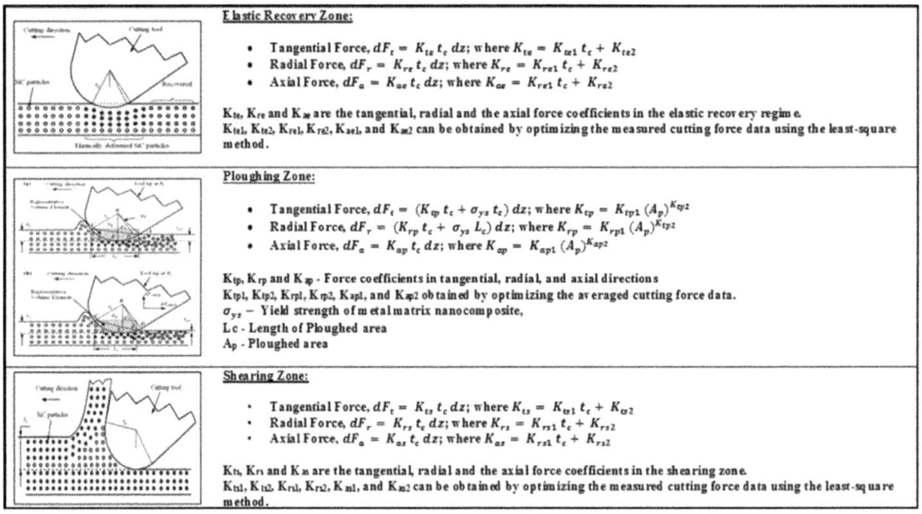

Figure 9.6. Cutting force model for elastic recovery, ploughing, and shearing mechanism. Reproduced with permission from [20]. Copyright 2013 ASME.

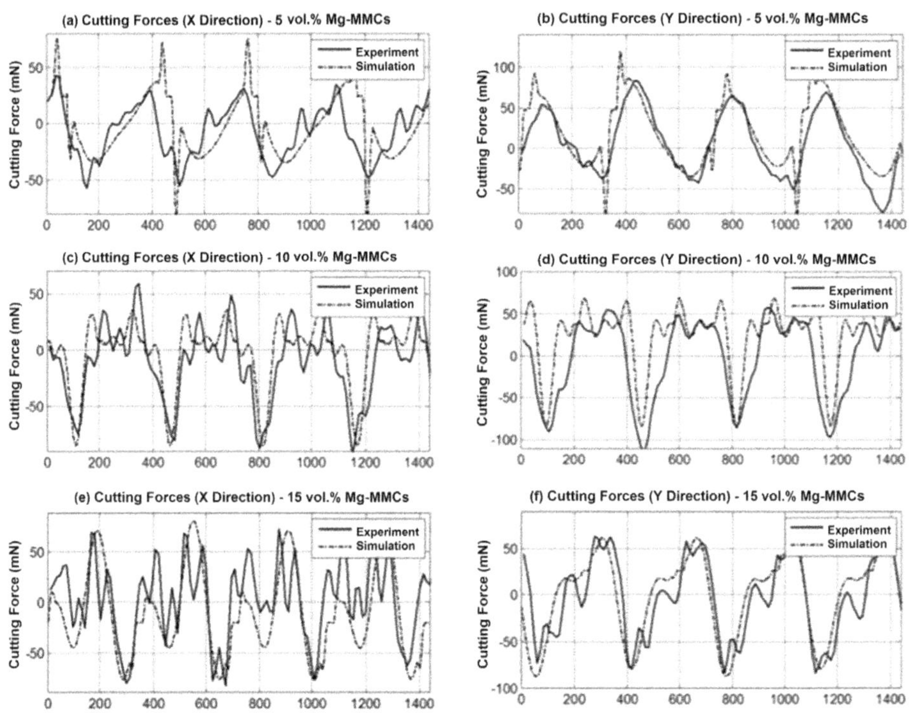

Figure 9.7. Comparison of experimental and simulated cutting forces for Mg nanocomposites as reported by Liu *et al.* Reproduced with permission from [20], © ASME.

nanocomposites when compared to Mg and the same was attributed to the particle strengthening effects. While the peak value for cutting pure Mg-MMCs in the elastic zone is around 10 GPa, it increased to 70 MPa for cutting 10 vol.% Mg/SiC. Further, the elastic zone and the ploughing zone are also wider for Mg-MMCs than pure Mg. Based on the experimental results, the authors also proposed a theoretical model to analyze the micro-machinability of the SiC nanoparticle reinforced Mg-MMCs. Since the cutting mechanism varies between the regions, the components of cutting force such as tangential force (F_t), radial force (F_r), and axial force (F_a) were modelled as shown in figure 9.6.

The model proposed in this work was also experimentally validated for different parameters that are representative of three different cutting conditions, i.e. elastic recovery, ploughing, and shearing. The comparison between experimental and simulated cutting forces (as shown in figure 9.7) confirmed the effectiveness of the

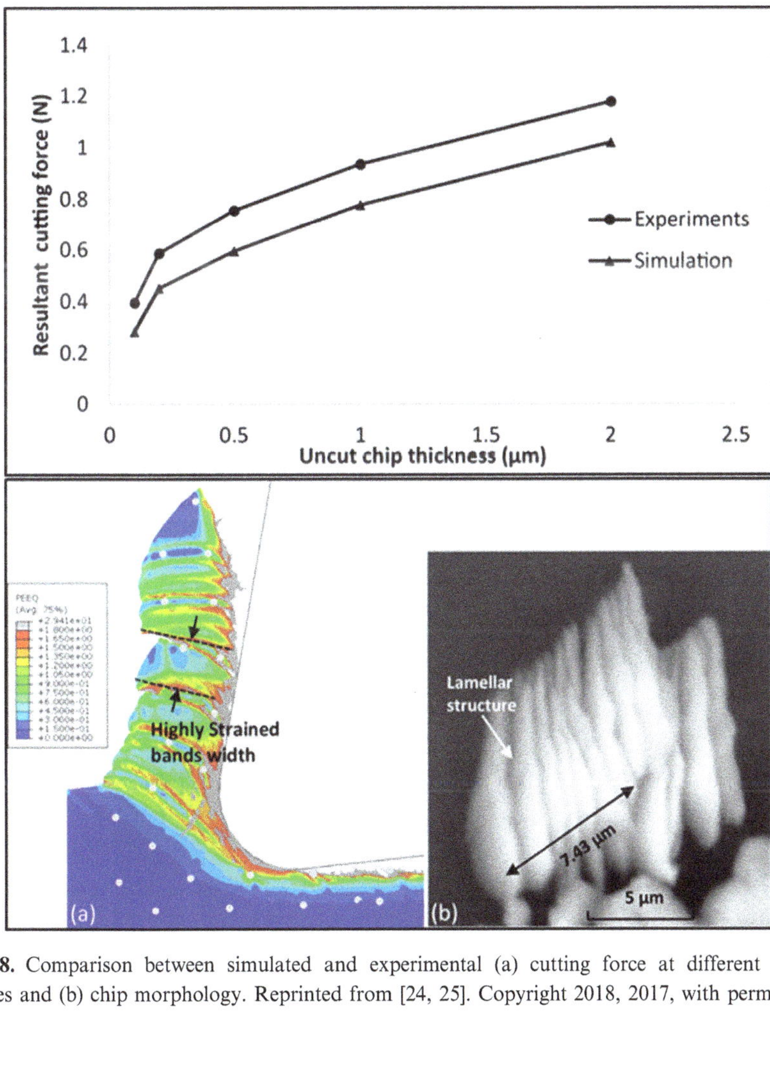

Figure 9.8. Comparison between simulated and experimental (a) cutting force at different uncut chip thicknesses and (b) chip morphology. Reprinted from [24, 25]. Copyright 2018, 2017, with permission from Elsevier.

model in terms of major cutting force characteristics in nanoparticle reinforced Mg-MMCs.

Teng *et al* [24, 25] also performed a simulation study to understand the micro-machining of SiC nanoparticle reinforced Mg nanocomposites. A two-dimensional micromechanical finite element model was developed based on the cutting-edge radius using ABAQUS-Explicit. The simulation results were analysed in terms of stress/strain distribution in order to understand chip formation, tool–particle interaction, and the cutting force (figure 9.8). The investigation of tool–particle interaction provided a better understanding of premature tool wear due to stress concentration on the particle and ploughing between the particles and the flank face. The machined chips showed highly strained bands in the vicinity of particles and intact particles without fracture. The comparison of simulated results with experimental cutting force and chip morphology showed satisfactory correlation (figure 9.8).

The micro-machinability of Mg-based nanocomposites reinforced with 1.98 vol. % nano-sized Ti or TiB_2 were tested by a micro-end milling process using UT coated carbide end mills [26–28]. The cutting force during the machining process was measured and the machined surface was analysed for surface morphology and roughness. From the experimental results it was deduced that the spindle speed and

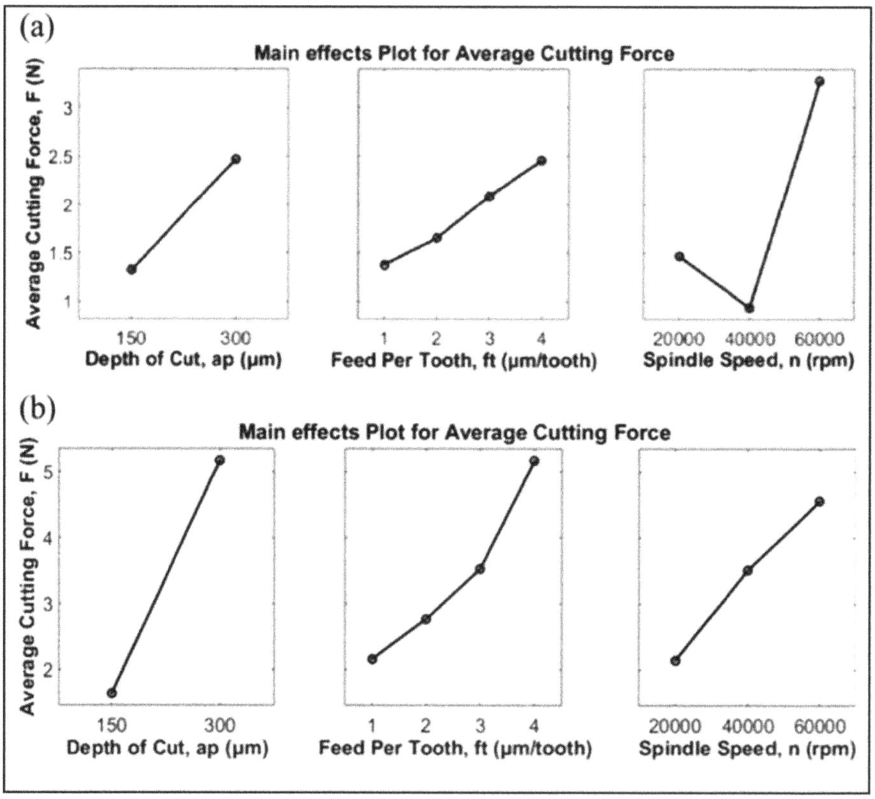

Figure 9.9. Variation of cutting force for (a) Mg/TiB_2 and (b) Mg/Ti MMCs with respect to depth of cut, feed per tooth, and spindle speed. Reproduced from [28], with permission © Springer Nature.

depth of cut have a significant influence on surface roughness (figure 9.9). Among the nanocomposites, the Mg/Ti nanocomposite derived a larger cutting force when compared to that of Mg/TiB_2. However, Mg/TiB_2 exhibited better machinability in terms of surface morphology.

Teng *et al* [29] also investigated the micro-machinability of DMD processed Mg/ZnO and Mg/BN nanocomposites in comparison to pure Mg. From the experimental results (figure 9.10), it was found that the cutting force for machining the nanocomposites was relatively smaller when compared to pure Mg and the same was attributed to the improved deformation characteristics due to nano-reinforcement addition. An increase in depth of cut and feed increased the cutting force requirements for all the materials. Among the nanocomposites, the cutting force requirement for Mg/2.5ZnO was much larger. Analysis of the machined surface showed evidence of elastic recovery in Mg/2.5BN due to increased ductility. Similarly, the chip morphology showed distinct differences. Mg/2.5BN

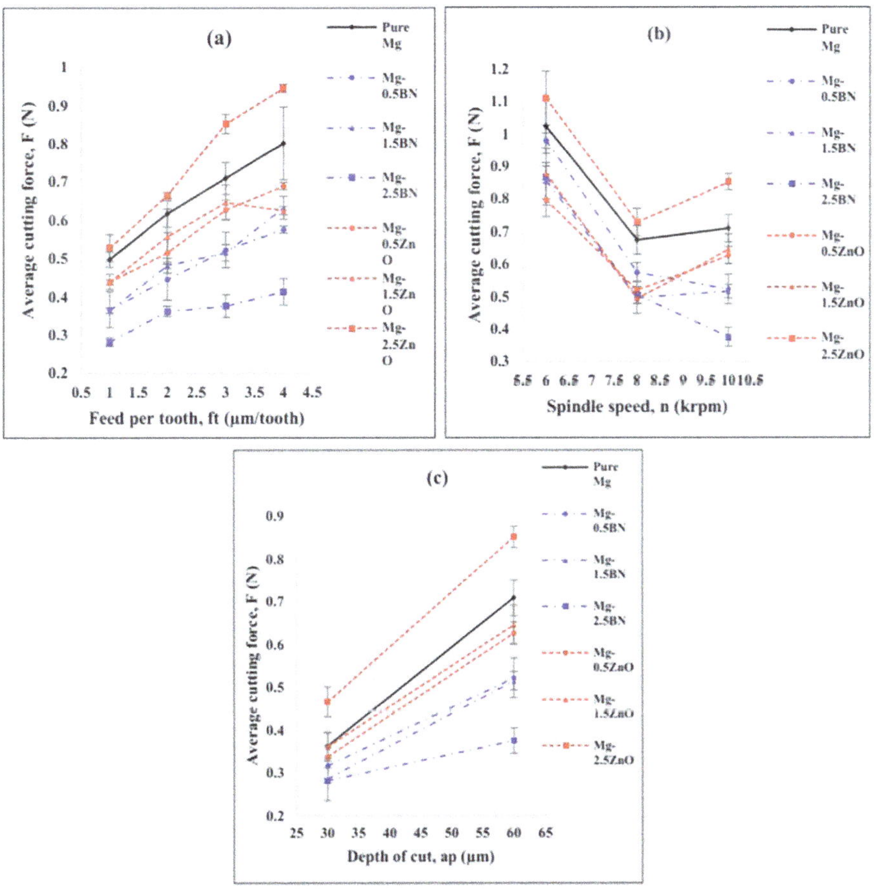

Figure 9.10. The effect of cutting parameters on the average cutting forces for nano-BN and nano-ZnO reinforced Mg nanocomposites. Reproduced from [29], with permission © Springer Nature.

demonstrated tightly curled and short chips compared to pure Mg and Mg/2.5ZnO due to its reduced compressive ductility (figure 9.11).

Sun *et al* [30] investigated the micro-drilling characteristics of Mg-based composites reinforced with SiO_2 nanoparticles. In this study, the reinforcement volume fraction and the drilling parameters influenced the chip morphology, as shown in figure 9.12. Because of the improved ductility and lower hardness, the chip formation was easier in pure Mg and the chips are normally long and continuous at a low feed rate. However, the presence of hard reinforcements acted as chippers to result in short spiral type chips in Mg/SiO_2 nanocomposites. At high feed rates and spindle speeds the thermal softening and easy breaking of chips due to the increased friction and heat generation resulted in short spiral chips in pure Mg, but the nanocomposites exhibited discontinuous and segmented chips. The feed rate and spindle speed also affected the dimensional accuracy (figure 9.13). While the increase in feed rate increased the cutting force and heat generation to result in increased surface roughness, increased spindle speeds reduced the surface roughness slightly. With respect to burr height, while an increase in spindle speed and feed rate resulted in increased burr height, an increase in reinforcement volume fraction decreased the same. The authors also reported that the cutting force profile consisted of three

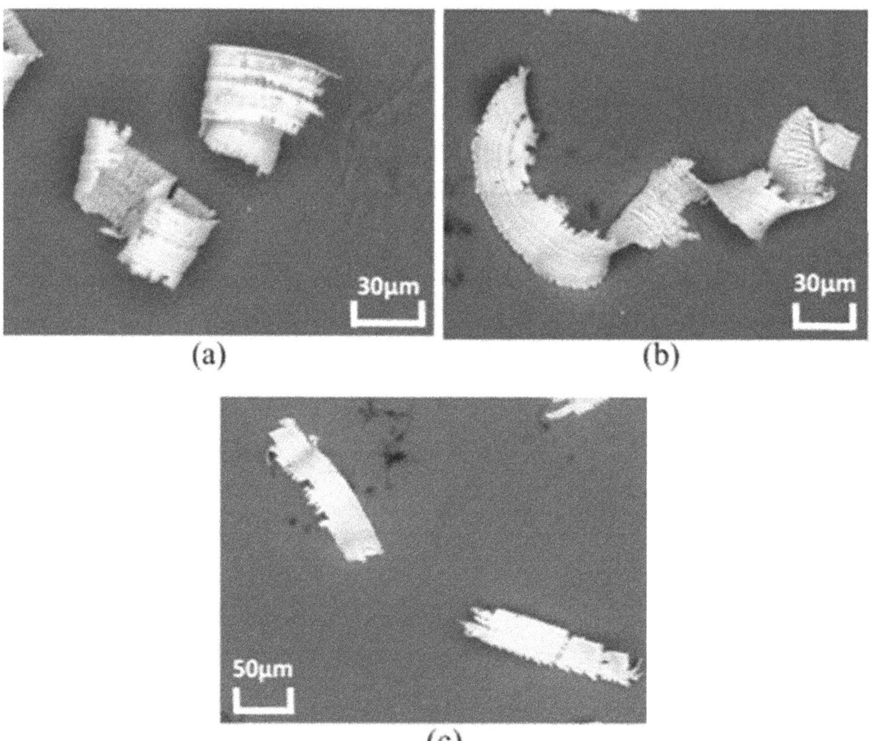

Figure 9.11. Surface morphology of chips of (a) Mg/2.5BN, (b) pure Mg, and (c) Mg/2.5ZnO at feed per tooth –0.5 mm, depth of cut –30 μm, and spindle speed –6000 rpm. Reproduced from [29], with permission © Springer Nature.

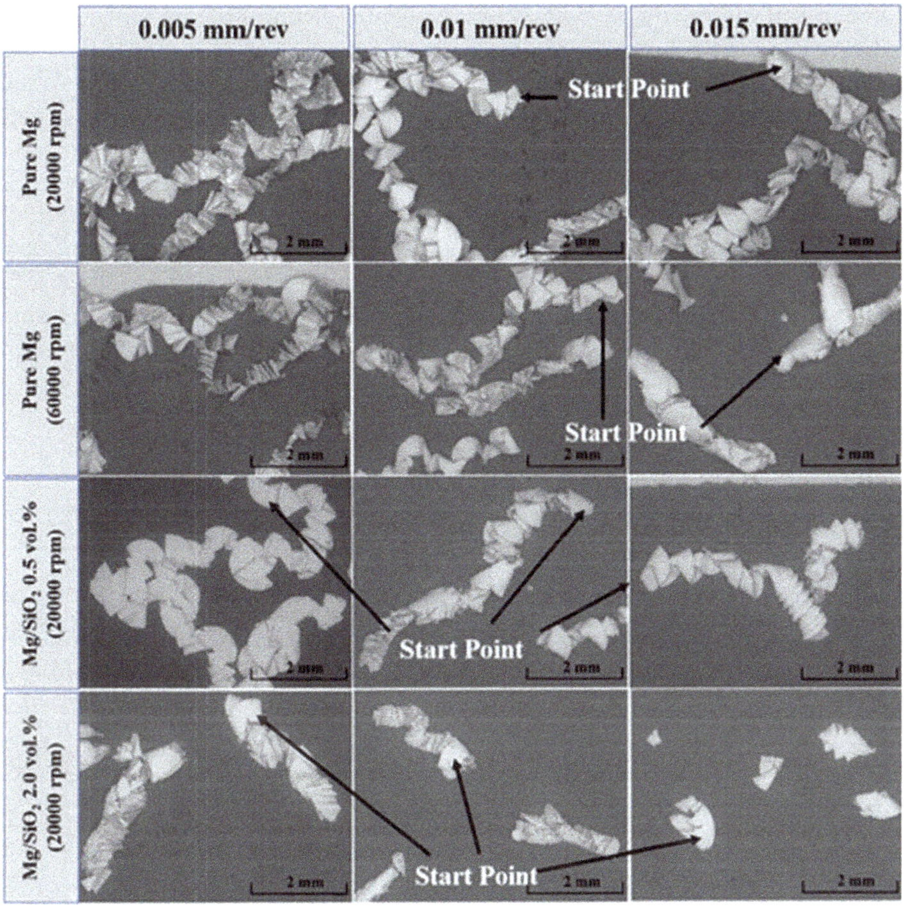

Figure 9.12. Chip morphology in Mg/SiO$_2$ nanocomposites. From [30], reproduced with the author's permission.

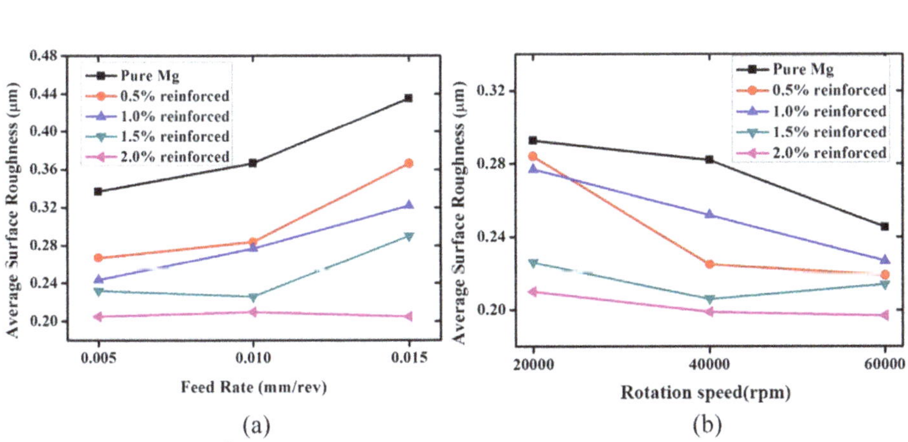

Figure 9.13. The effect of feed rate and spindle speed on the surface roughness of Mg/SiO$_2$ nanocomposites. From [30], reproduced with the author's permission.

distinct zones with increasing values of uncut chip thickness (figure 9.14). For a chip thickness less than 0.3 μm, the cutting mechanism involved ploughing off by elastic deformation without chip formation. In this zone, the cutting force was found to increase with an increasing amount of SiO_2 nanoparticles and the largest cutting force was recorded for 1% SiO_2. In the transition zone with chip thickness between 0.7 and 1.7 μm, the plastic and elastic deformation occurred simultaneously to cause a significant reduction in the cutting force. When the chip thickness was greater than 1.7 μm, continuous chips were formed through shearing under plastic deformation. Similar to cutting force, the surface roughness profile also exhibited three regions with respect to the uncut chip thickness. Based on the cutting force and surface roughness profiles, the minimum cutting thickness of Mg/SiO_2 nanocomposite was reported as 1.1 μm while it was 1.7 μm for pure Mg. Similar observations were also reported for HA nanoparticle reinforced Mg composites (figure 9.15) [31].

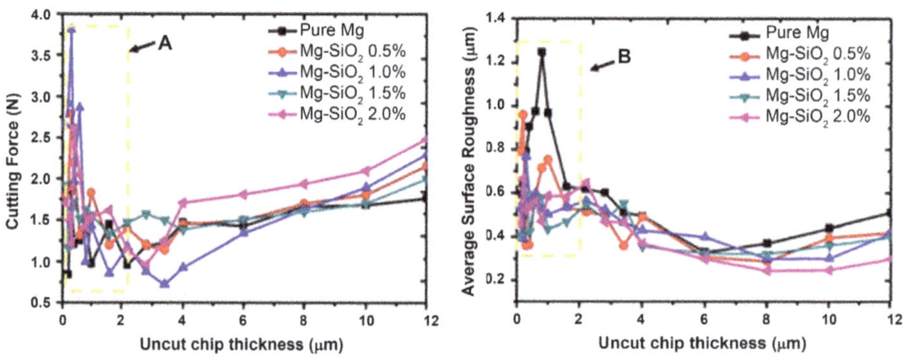

Figure 9.14. Cutting force and surface roughness profiles of Mg/SiO_2 nanocomposites with respect to uncut chip thickness. From [30], reproduced with the author's permission.

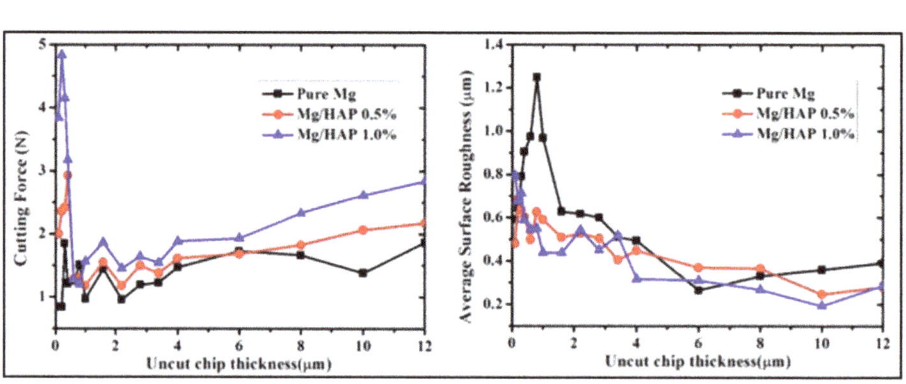

Figure 9.15. Cutting force and surface roughness profiles of Mg/HA nanocomposites with respect to uncut chip thickness. From [31], © IOP Publishing. Reproduced with permission. All rights reserved.

9.3 Summary

Magnesium-based nanocomposites with comparable/enhanced mechanical properties are promising replacements for aluminium alloys to achieve structural weight reduction. While the mechanical properties of Mg-based nanocomposites have been investigated extensively, it is equally important to understand the mechanism of material removal during the machining processes, as it is one of the most suitable manufacturing techniques for producing near net-shaped components. However, only very few studies have explored the machinability attributes of Mg-based nanocomposites. In conventional machining processes such as turning, milling, drilling, and grinding, or unconventional machining methods such as laser machining and EDM, the quality of machined parts generally depends on the material's properties and the process variables. In this regard, the presence of hard and strong particle reinforcements modifies the cutting force requirement and the chip morphology to cause increased tool wear at a larger feed rate. However, the cutting tool wear rates and the surface roughness of parts machined from Mg nanocomposites are significantly smaller when compared to the conventional microcomposites. Based on the published literature, the use of a low feed rate and high cutting speed are recommended for the machining of Mg nanocomposites in order to achieve a better surface finish. The unconventional machining processes such as EDM also displayed improvements in terms of dimensional accuracy and surface finish when compared to microcomposites.

References

[1] Saravan M and Natarajan N 2015 Study of cutting forces in machining of magnesium composite by response surface methodology *Carbon Sci. Technol.* **7** 36–58

[2] Davis B, Dabrow D, Ju L, Li A, Xu C and Huang Y 2017 Study of chip morphology and chip formation mechanism during machining of magnesium-based metal matrix composites *J. Manuf. Sci. Eng. Trans. ASME* **139** 091008

[3] Pramanik A, Basak A K, Dong Y, Shankar S and Littlefair G 2018 Milling of nanoparticles reinforced Al-based metal matrix composites *J. Compos. Sci.* **2** 13

[4] Li J and Laghari R A 2019 A review on machining and optimization of particle-reinforced metal matrix composites *Int. J. Adv. Manuf. Technol.* **100** 2929–43

[5] Markopoulos A P, Pressas I S, Papantoniou I G, Karkalos N E and Davim J P 2015 Machining and machining modeling of metal matrix composites—a review *Modern Manufacturing Engineering* (Berlin: Springer) pp 99–141

[6] Clyne T W and Withers P J 1995 *An Introduction to Metal Matrix Composites* (Cambridge: Cambridge University Press)

[7] Weinert K and König W 1993 A consideration of tool wear mechanism when machining metal matrix composites (MMC) *CIRP Ann. Manuf. Technol.* **42** 95–8

[8] Li X P and Lu L 2003 Study of the reinforcement percentage of Mg-Al-SiC MMC in relation to the mechanical properties and machinability *Mater. Sci. Forum* **437–438** 185–8

[9] Weinert K and Lange M 2001 Machining of magnesium matrix composites *Adv. Eng. Mater.* **3** 975–9

[10] Zhang Z and Chen D L 2008 Contribution of Orowan strengthening effect in particulate-reinforced metal matrix nanocomposites *Mater. Sci. Eng.* A **483–484** 148–15

[11] Malaki M, Xu W, Kasar A K, Menezes P L, Dieringa H, Varma R S and Gupta M 2019 Advanced metal matrix nanocomposites *Metals* **9** 330

[12] Sahin Y and Sur G 2004 The effect of Al_2O_3, TiN and Ti (C,N) based CVD coatings on tool wear in machining metal matrix composites *Surf. Coat. Technol.* **179** 349–55

[13] Teng X, Chen W, Huo D, Shyha I and Lin C 2018 Comparison of cutting mechanism when machining micro- and nano-particles reinforced SiC/Al metal matrix composites *Compos. Struct.* **203** 636–47

[14] Ponappa K, Aravindan S, Rao P V, Ramkumar J and Gupta M 2010 The effect of process parameters on machining of magnesium nano alumina composites through EDM *Int. J. Adv. Manuf. Technol.* **46** 1035–42

[15] Ranjith R, Tamilselvam P, Prakash T and Chinnasamy C 2019 Examinations concerning the electric discharge machining of AZ91/5B4CP composites utilizing distinctive electrode materials *Mater. Manuf. Process.* **34** 1120–8

[16] Arunkumar L and Raghunath B K 2013 Electro discharge machining characteristics of Mg/SiCP metal matrix composites by powder metallurgy (P/M) techniques *Int. J. Eng. Technol.* **5** 4335–8

[17] Li J, Liu J, Liu J, Ji Y and Xu C 2013 Experimental investigation on the machinability of SiC nano-particles reinforced magnesium nanocomposites during micro-milling processes *Int. J. Manuf. Res.* **8** 64–84

[18] Li J, Liu J and Xu C 2010 Machinability study of SiC nano-particles reinforced magnesium nanocomposites during micro-milling processes *ASME 2010 Int. Manuf. Sci. Eng. Conf.*

[19] Liu J, Li J, Ji Y and Xu C 2011 Investigation on the effect of SiC nanoparticles on cutting forces for micro-milling, magnesium matrix composites *ASME 2011 Int. Manuf. Sci. Eng. Conf.*

[20] Jian L, Juan L and Chengyin X 2013 Cutting force prediction on micromilling magnesium metal matrix composites with nanoreinforcements *J. Micro Nano-Manuf.* **1** 0110101

[21] Bhattacharyya B 2015 Electrochemical micromachining for nanofabrication *MEMS and Nanotechnology* (New York: Elsevier)

[22] Taniguchi N 1983 Current status in, and future trends of, ultraprecision machining and ultrafine materials processing *CIRP Ann. Manuf. Technol.* **32** 573–82

[23] Hasan M, Zhao J and Jiang Z 2019 Micromanufacturing of composite materials: a review *Int. J. Extrem. Manuf.* **1** 012004

[24] Teng X, Huo D, Chen W, Wong E, Zheng L and Shyha I 2018 Finite element modelling on cutting mechanism of nano Mg/SiC metal matrix composites considering cutting edge radius *J. Manuf. Process.* **32** 116–26

[25] Teng X, Huo D and Chen W 2017 An investigation into finite element modelling of micro machining of nano Mg/SiC metal matrix composites *Proc. 17th Int. Conf. Eur. Soc. Precis. Eng. Nanotechnology, EUSPEN 2017*

[26] Teng X, Huo D, Wong W L E and Gupta M 2015 Experiment based investigation into micro machinability of Mg based metal matrix composites (MMCs) with nano-sized reinforcements *2015 21st Int. Conf. Autom. Comput. Autom. Comput. Manuf. New Econ. Growth, ICAC 2015*

[27] Teng X, Huo D, Shyha I, Chen W and Wong E 2018 An experimental study on tool wear behaviour in micro milling of nano Mg/Ti metal matrix composites *Int. J. Adv. Manuf. Technol.* **96** 2127–40

[28] Teng X, Huo D, Wong E, Meenashisundaram G and Gupta M 2016 Micro-machinability of nanoparticle-reinforced Mg-based MMCs: an experimental investigation *Int. J. Adv. Manuf. Technol.* **87** 2165–78

[29] Teng X, Huo D, Wong W L E, Sankaranarayanan S and Gupta M 2017 Machinability investigation in micro-milling of Mg based MMCs with nano-sized particles, ed K Solanki et al *Magnesium Technology 2017 (Minerals, Metals & Materials Series)* (Cham: Springer) pp 61–9
[30] Sun F, Huo D, Fu G, Teng X, Kannan S and Zhang H 2020 Micro-drilling of Mg-based MMCs reinforced with SiO_2 nanoparticles: an experimental approach *Proc. Inst. Mech. Eng.* B **234** 1473–85
[31] Sun F Z, Fu G Y and Shen J R 2019 Experimental investigation on micro-drilling properties of Mg-based MMCs reinforced with HAP nanoparticles *IOP Conf. Ser.: Mater. Sci. Eng.* **657** 012047

IOP Publishing

Magnesium-Based Nanocomposites
Advances and applications
Sankaranarayanan Seetharaman, Sravya Tekumalla and Manoj Gupta

Chapter 10

Wet corrosion and biocompatibility of magnesium nanocomposites

Magnesium composites reinforced with hard and strong ceramic reinforcements are excellent candidates for lightweight applications. A judicious selection of matrix and reinforcement helps in achieving the desired properties in terms of improved strength, ductility, wear resistance, fatigue life, and creep resistance. However, the micro-galvanic corrosion due to the larger difference in the corrosion potential of Mg matrix and ceramic reinforcements results in higher corrosion rates. In this regard, reducing the reinforcement volume fraction is perceived to be beneficial. Hence, low volume fraction nano-sized reinforcements have been explored recently. As the biocorrosion characteristics of Mg make it ideal for biodegradable implants, recent studies on Mg nanocomposites also focus on the biodegradation and cell toxicity. This chapter summarizes the outcomes from various research works that are focussed on the effects of nano-sized reinforcements on the saltwater corrosion and biocompatibility of Mg and Mg alloys.

10.1 Introduction

Magnesium has low electrochemical potential (see figure 10.1). The difference in electrode potential between the Mg alloy matrix and the heavy metal impurities/second phase particles/reinforcements results in galvanic corrosion. Further, the poor stability of surface oxide films also contributes towards the rampant material loss upon exposure to saltwater solution [1].

As the heavy metal impurities (e.g. Fe, Ni, Cu) in Mg alloys are more noble compared to Mg, they serve as cathodes to cause severe galvanic corrosion. Similarly, pitting corrosion occurs when Mg alloys are exposed to chloride ions and the resultant corrosion pits are normally found adjacent to the second phase particles. The grain boundaries where precipitation and segregation of second phase particles often occur are also the preferred sites for intergranular corrosion.

doi:10.1088/978-0-7503-3535-5ch10

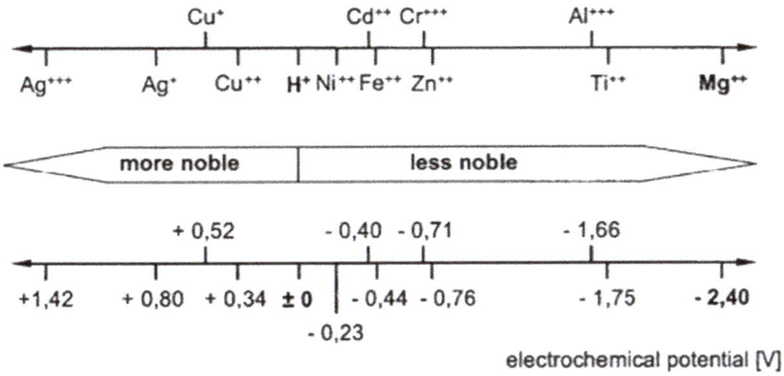

Figure 10.1. Standard reduction potentials of different metals. Reproduced from [2], with permission © Springer Nature.

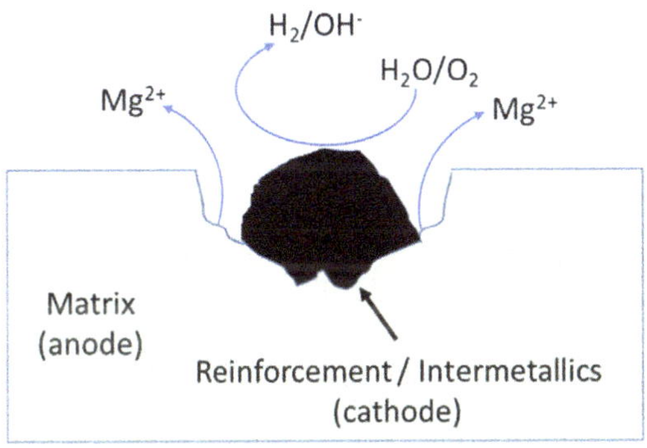

Figure 10.2. Schematic of corrosion in Mg alloys and composites. Reprinted from [3], Copyright 2017, with permission from Elsevier.

Hence, the corrosion mechanism in Mg materials depends on several factors including the chemical composition of the base material and reinforcement, microstructural aspects, the type of corrosion medium, film characteristics, etc. Exposure to a corrosive medium can also affect the crack propagation and fatigue response. For example, the fatigue strength and fatigue life of Mg alloys are significantly affected when exposed to sodium chloride. Fatigue cracks are known to initiate from surface/subsurface defects, inclusions, and corrosion pits.

10.2 The mechanism in magnesium and magnesium alloys

The corrosion of magnesium in aqueous solution (as shown in figure 10.2) can be explained using the following electrochemical reactions. While hydrogen gas evolution and protective layer formation reduces the rate of corrosion reaction, the breakdown of weak protective layers allows faster degradation.

Anodic reaction: $Mg \rightarrow Mg^{2+} + 2e^-$.

Cathodic reaction: $2H_2O + 2e^- \rightarrow H_2 + 2OH^-$.

Product formation: $Mg^{2+} + 2OH^- \rightarrow Mg(OH)_2$.

Similarly, in a physiological environment, the presence of dissolved oxygen, proteins, amino acids, chloride, and hydroxide ions affects the degradation behaviour of Mg-based materials and the corrosion mechanism (schematic as shown in figure 10.3) is explained below.

Anodic reaction: $Mg \rightarrow Mg^{2+} + 2e^-$.

Cathodic reaction: $Mg + 2H_2O \rightarrow Mg(OH)_2 + H_2$.

Product formation: $Mg(OH)_2 + 2Cl^- \rightarrow MgCl_2$.

Product formation: $Mg^{2+} + 2Cl^- \rightarrow MgCl_2$.

Magnesium, being naturally biodegradable (i.e. it degrades on its own in body fluids) is highly suitable for making temporary implants. Compared to other implant materials, the properties of Mg closely match those of human bone (table 10.1). However, its faster corrosion rate in body fluids necessitates the need for controlled degradation. There is also a risk of toxicity due to the release of ions such as Al and Zn. Hence, the biocorrosion and cytotoxicity of Mg-based materials has attracted huge research interest in the development of next-generation biomedical implants.

The effects of different alloying elements on the corrosion resistance and biocompatibility of Mg are summarized in table 10.2. While reinforcements, in general, have detrimental effects on the corrosion resistance of Mg alloys, nano-lengthscale reinforcements are believed to improve the surface chemical activity and

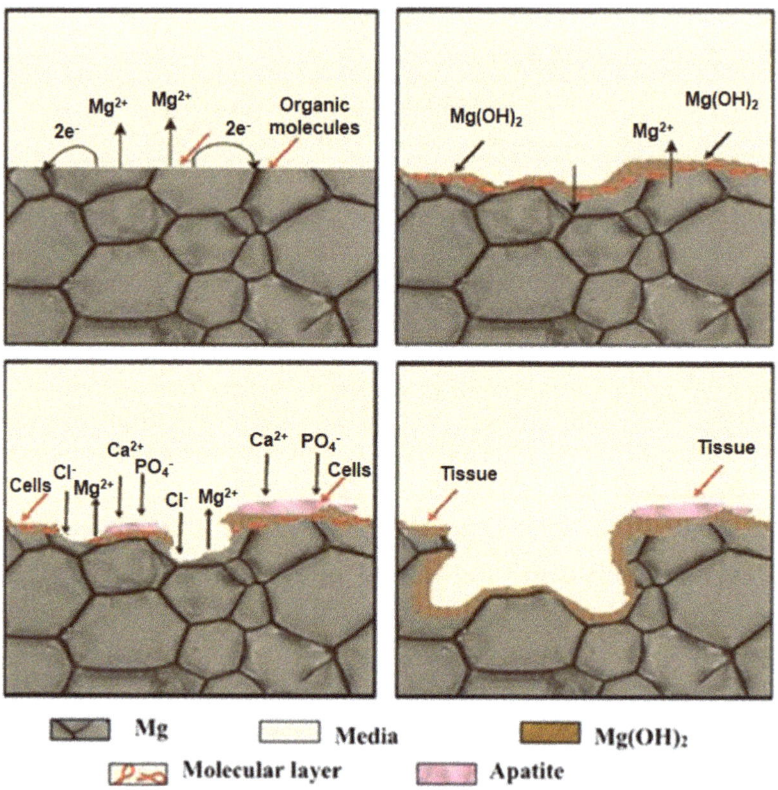

Figure 10.3. Degradation mechanism of Mg *in vivo*. From [4], © KeAi.

Table 10.1. General properties of implant materials [5–8].

Material	Density (g cc^{-1})	Elastic modulus (GPa)	Compressive yield strength (MPa)	Tensile yield strength (MPa)	Ultimate tensile strength (MPa)	Elongation (%)
Cortical bone	1.6–2.0	5–23	164–240	105–115	35–283	1.07–2.10
Cancellous bone	1.0–1.4	0.01–1.5	—		1.5–38	
Ti_6Al_4V	4.5	114	—	760–880	830–1025	12
Synthetic HA	3.05–3.15	70–120	100–900		40–200	
DL-PLA	—	1.9–2.4	—		29–35	
Cast magnesium	1.74	41		21	90	13
Die cast AZ91D	1.81	45	160	150	230	3
Extruded AZ31	1.78	45	60–70	125–135	235	7
Extruded WE43 T5	1.84	44	—	148	280	18
Stainless steel	7.7	189–205	170–310		480–620	30–40

promote the formation of a protective surface film [3]. However, the actual passivation and cell adhesion benefits will depend on the nature of individual reinforcement and the processing methods. Table 10.3 summarizes the characteristics of commonly used biomedical ceramics. From the available literature, it can be understood that the corrosion research on Mg-based nanocomposites is still in its infancy. While some of the earlier studies focussed on the saltwater corrosion resistance, several recent studies report the biocorrosion behaviour of Mg nanocomposites.

10.3 Evaluation of corrosion resistance

The corrosion characteristics of Mg nanocomposites in an aqueous medium can be evaluated using the immersion, salt spray, potentio-dynamic polarization, and electrochemical impedance spectroscopy (EIS) methods.

In immersion and salt spray tests, the samples are exposed to an aerated salt solution or a corrosive medium (as listed in table 10.4) for a specific duration and mass measurements are carried out to calculate the corrosion rate. The corrosion rates (in mm yr^{-1}) are calculated from the measured mass loss values using the following relation [14]:

$$\text{Average corrosion rate: } P_h \text{ (in mm yr}^{-1}) = 2.10\,\Delta W, \tag{10.1}$$

where ΔW is the mass loss (in mg cm^{-2} day^{-1}).

Table 10.2. The effects of different alloying elements on the corrosion resistance and biocompatibility of Mg [8–11].

Alloying element	Corrosion resistance	Biocompatibility	Presence in human body	Presence in blood serum
Aluminium	A network distribution of $Mg_{17}Al_{12}$ in Mg alloys enhances corrosion resistance; however, the coupled micro-galvanic effects between the $Mg_{17}Al_{12}$ phase and Mg matrix at the same time increase the corrosion rate.	Toxic. May cause Alzheimer's disease and affect osteoclast viability.	300 mg	2.1–4.8 $\mu g\ L^{-1}$
Manganese	Enhances corrosion resistance by reducing impurities with a small quantity of Mn addition.	Helps in bone formation and protects against infection.	12 mg	0.8 $\mu g\ L^{-1}$
Zinc	Inhibits the harmful effects of Fe and Ni impurities on the corrosion; enhances the corrosion resistance of Mg alloys at a content below 5%.	Essential for immune system. Higher concentrations of Zn exhibit neurotoxicity.	2 g	46 $\mu mol\ L^{-1}$
Zirconium	Small amounts of Zr addition (less than 2%) enhances the corrosion resistance; otherwise significantly deteriorates the corrosion resistance.	Toxic if used in higher concentrations.	250 mg	—
Rare earth elements	Increase corrosion resistance in general through protective passive film formation.	Many rare earth elements exhibit anticarcinogenic properties. Accumulation in bone and liver. However, toxic when used in large quantities.	—	—
Yttrium	Influence on corrosion resistance under debate; depends on the composition of Mg alloys; reduces the corrosion resistance of binary Mg–Y alloys with a concentration >2%.	Accumulation in bone and liver. Affect cell viability if concentration greater than 1.6 $\mu m\ L^{-1}$.	—	47 μg

(*Continued*)

Table 10.2. (*Continued*)

Alloying element	Corrosion resistance	Biocompatibility	Presence in human body	Presence in blood serum
Lithium	Enhances corrosion resistance at a concentration below 9% in pure Mg; accelerates the corrosion rate significantly with higher Li addition.	Assists in bone formation. Reduces kidney function and leads to central nervous system disorder.		2–4 ng g^{-1}
Calcium	Excessive addition of Ca in pure Mg deteriorates corrosion resistance; Ca concentration in Mg alloys should be less than 1%.	Provides strength to bones to support the soft tissues.	1100 g	0.919–0.993 mg L^{-1}
Strontium	The influence on corrosion depends on the volume fraction of $Mg_{17}Sr_2$ phase in Mg alloys. Optimal content less 2%.	Prevents bone resorption.	0.3 g	0.17 mg
Silicon	Corrosion potential of Mg_2Si similar to pure Mg.	Cross linking agent of connective tissue base membrane structures. Necessary for growth as bone calcification.	18 mg	—
Silver	Worsens the corrosion resistance although minor additions are proven to be beneficial.	Antibacterial effect.	—	11–26 mg L^{-1}
Tin	Adversely affects the corrosion resistance of Mg.	Carcinogenic. When used in larger amounts leads to tin accumulation in liver.	3 mg	<0.1 μg L^{-1}
Copper	Detrimental.	Allergen. Trace element in cells. Cytotoxic.	200 mg	74–131 μmol L^{-1}
Nickel	Detrimental.	Carcinogenic and toxic. Strong allergen which induces metal sensitivity.	10 mg	0.05–0.23 μg L^{-1}
Iron	Detrimental.	Essential for life. Metabolically regulated and stored.	4–5 g	5.0–17.6 g L^{-1}

Table 10.3. Characteristics of biomedical ceramic reinforcements [12, 13].

Material	Advantages	Limitations	Applications
Alumina (Al_2O_3)	Biocompatible and bioinert. High hardness, strength, and abrasion resistance. Stress shielding. Permeability and inflammation response.	Low fracture toughness and high surface roughness.	Load-bearing applications. Used in femoral head, porous coatings for femoral stems. Bone screws, plates, and knee prosthetics.
Zirconia (ZrO_2)	Fracture toughness and flexural strength better than alumina. Low Young's modulus. Chemically inert and biocompatible. Non-toxic.	Femoral head made of 100% zirconia reportedly failed within short time after implantation. So, not actively used in orthopaedics.	Used in restorative dentistry, femoral heads, artificial knees, bone screws, and plates.
Yttria-stabilized tetragonal zirconia polycrystals (Y-TZP)	Yttria has high biocompatibility.	Low-temperature degradation. Poor bioactivity and osteo-generation.	Mostly used in dental applications.
Silicon nitride (Si_3N_4)	Improves the wear resistance. Facilitates bone-cell adherence.		Promotes bone fusion in spinal surgery. Coatings for prosthetic hips and knee joints.
Silica (SiO_2)	High biocompatibility. Enhance bone mineral density. Has stimulatory effects on osteoblasts. Better cytocompatibility and osteo-conductivity.	Inhibits osteoclasts *in vitro*.	Used in artificial bone and dental implants.
Zinc oxide (ZnO)	Biosafe and biocompatible. Cytotoxicity depends on the concentration and exposure time. Higher cytotoxicity compared to alumina and TiO_2.	Higher concentrations are highly cytotoxic.	Used in medical disinfection, sunscreens, and cosmetics. ZnO nanowires are used to develop potentiometric electrochemical sensors.
Titania (TiO_2)	Good mechanical properties and chemical stability. Improves bioactivity. Least cytotoxic compared to other metal oxides.	—	Coatings for dental implants. Improve osseointegration of titanium implants.

(Continued)

Table 10.3. (*Continued*)

Material	Advantages	Limitations	Applications
Iron oxide (Fe_3O_4)	Biocompatible and non-toxic. Superior magnetic characteristics and serves as a promising nanocarrier in the human body.	—	For targeted drug delivery applications.
TiB_2, TiC, and TiN	Improves the wear resistance. No effect on cytotoxicity. Non-irritant and biocompatible.	—	Load-bearing applications. Mostly used as a coating material.
Carbon nanotubes	Improves the hardness, wear resistance, and chemical inertness, and provides a low friction surface. Mainly used as coatings and a reinforcement material.	Adverse cell response and cell detachment from surrounding tissues due to carbon release.	Used as coatings and reinforcement materials. Used in hard tissue implants, scaffolds, microcatheters, as a substrate for neuronal growth disorders, heart valves, percutaneous devices, and dental implants.
Hydroxyapatite (HA) ($Ca_{10}(PO_4)_6(OH)_2$)	Chemical similarity with natural bone. Bioresorbable, bioactive, and biocompatible. Increased cell adhesion, growth, and proliferation.	Poor mechanical properties. Unpredicted osteointegration.	Used as coatings for implants for femoral knee, femoral hip, tibial components, acetabular cup.
Fluorapatite (FA) $Ca_{10}(PO_4)_6F_2$	Lower dissolution and better cell adherence than HAp. Improved osteo-conductivity and bone formation compared to HAp.	Poor mechanical properties.	Used in coatings and dental/bone implants.
β-tricalcium phosphate ($Ca_3(PO_4)_2$)	Gives excellent biological response. Stable in body fluid and dry/moist air.	Poor mechanical properties.	Used in surface coatings on metal implants for bone repair and augmentation.

Table 10.4. Chemical composition of different physiological corrosion media that are applicable for Mg nanocomposites [16].

		Composition	
S. no.	Chemical	Reagent/ions	Amount (in mM lit^{-1})
1	Hank's balanced salt solution (HBSS)	NaCl	136.75
		KCl	5.37
		$CaCl_2$	1.85
		$NaHCO_3$	3.15
		K_2HPO_4	0.42
		NaH_2PO_4	0.65
		$MgSO_{4.7}H_2O$	0.81
		Glucose	5.55
2	Phosphate-buffered saline (PBS) solution	NaCl	137
		KCl	2.7
		Phosphates	10
3	Ringer solution	NaCl	7.2 g
		KCl	0.37 g
		$CaCl_2$	0.17 g
4	Simulated body fluid (SBF)	Na^+	142
		K^+	5
		Ca^{2+}	2.5
		Mg^{2+}	1.5
		Cl^-	148.8
		HPO_4^{2+}	1
		HCO_3^-	4.2
		SO_4^{2+}	0.5
5	Modified-simulated body fluid (m-SBF)	Na^+	122.5
		K^+	5
		Ca^{2+}	2.5
		Mg^{2+}	1.5
		Cl^-	103
		HPO_4^{2+}	1
		HCO^{3-}	10
		SO_4^{2+}	0.5
6	Dulbecco's modified Eagle's medium (DMEM)	NaCl	109.4
		KCl	5.37
		$CaCl_2$	1.8
		$NaHCO_3$	44.05
		NaH_2PO_4	1.04
		Mg_2SO_4	0.81
		Glucose	5.55
		Amino acid	11.01

(*Continued*)

Table 10.4. (*Continued*)

S. no.	Chemical	Composition	
		Reagent/ions	Amount (in mM lit^{-1})
7	DMEM + 10% foetal bovine serum (FBS)	NaCl	109.4
		KCl	5.37
		$CaCl_2$	1.8
		$NaHCO_3$	44.05
		NaH_2PO_4	1.04
		Mg_2SO_4	0.81
		Glucose	5.55
		Amino acid	11.01
		FBS	10%
		Penicillin	1%

The rate of hydrogen gas evolution during immersion tests is also used to evaluate the corrosion rate in terms of mass loss as shown below [15]:

$$\text{Mass loss, } \Delta W \text{ (in mg cm}^{-2}\text{ day}^{-1}) = 1.085\, V_h, \tag{10.2}$$

where V_h is the hydrogen gas evolution in mL cm^{-2} day^{-1}.

Although the actual corrosion rates cannot be calculated, the proper monitoring of pH values helps in understanding the corrosion mechanism as it is known that the release of hydroxide ions (OH–) during corrosion increases the pH concentration.

Potentio-dynamic polarization involves a three-electrode system consisting of (i) a platinum electrode as the auxiliary electrode, (ii) a saturated calomel electrode as the standard electrode, and (iii) the test sample as the working electrode. Based on the experimental results, a Tafel plot (as shown in figure 10.4) is used to calculate the electrode potential (E_{corr}) and current density (I_{corr}). While low corrosion current density (I_{corr}) represents a low corrosion rate or high corrosion resistance of materials, the loss of chemical reactivity indicates passivation or protective film formation on the material's surface. From the current density (I_{corr}, in μA cm^{-2}), the average corrosion rate (in mm yr^{-1}) can be calculated as follows [17]:

$$\text{Average corrosion rate, } P_i = \frac{0.13 I_{corr}}{\rho} M, \tag{10.3}$$

where M represents the equivalent mass in g and ρ is the density in g cm^{-3}.

EIS is a non-destructive test used to measure the cell impedance of a surface subjected to frequency dependent AC polarization [19]. The results are usually represented using a Nyquist plot where the polarization resistance can be obtained from the semicircle diameter. With its major advantage being its ability to detect individual layers on the surface of Mg, this method can be used effectively to understand the levels of protection offered by various coatings.

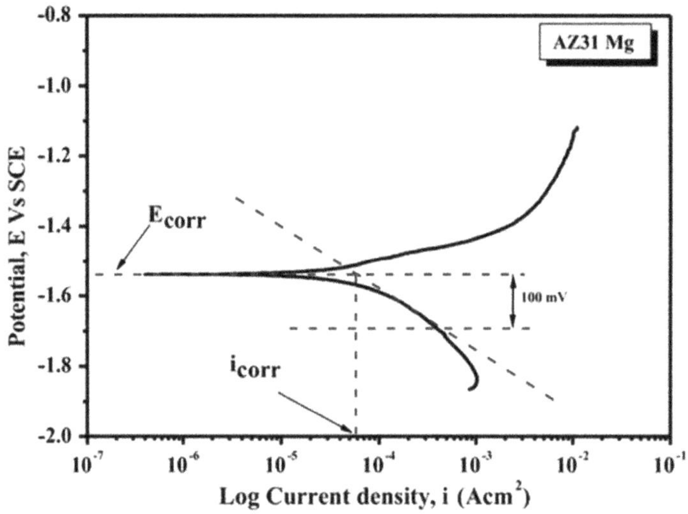

Figure 10.4. Tafel plot used in the calculation of corrosion potential and current density. Reproduced from [18], with permission © Springer Nature.

10.4 Wet corrosion of magnesium nanocomposites

Kukreja *et al* [20] investigated the corrosion characteristics of DMD processed AZ31B/Al_2O_3 nanocomposites in a freely aerated 3.5 wt.% NaCl solution. In this study, the average corrosion rate was calculated using the current density obtained from the Tafel extrapolation of polarization curves. From the results, it was observed that the corrosion current density was less for the AZ31B/Al_2O_3 nanocomposite and the average corrosion rate for the AZ31B/Al_2O_3 nanocomposite (9 mm yr^{-1}) was approximately one-third that for monolithic AZ31B alloy (27 mm yr^{-1}).

The corrosion resistance of DMD processed AZ31/Al_2O_3 nanocomposites was tested using hydrogen evolution, weight loss and electrochemical methods in simulated body fluid (SBF) [18]. The results showed significant reduction in the weight loss and hydrogen evolution due to Al_2O_3 nanoparticles after 336 h of exposure in SBF (figure 10.5). Electrochemical analysis in SBF showed the shifting of the open circuit potential towards the nobler direction due to nano-Al_2O_3 addition (figure 10.6).

A similar trend was also observed in the results of EIS measurements. The Nyquist plots (figure 10.7) obtained from EIS results showed the existence of two capacitive arcs in the high and medium frequency regions with an inductive loop in the low-frequency region which represents three different interfacial reactions [18]. The first capacitive arc in the high-frequency region is accompanied by electrochemical charge transfer and double layer/oxide film behaviour [18]. The second capacitive arc in the medium frequency region is ascribed to mass transference relaxation because of the dispersion of magnesium ions through the corrosion film [18]. In addition, the inductive loop in the low-frequency region is accompanied by the relaxation of surface coverage owing to the attachment of magnesium intermediate species [18].

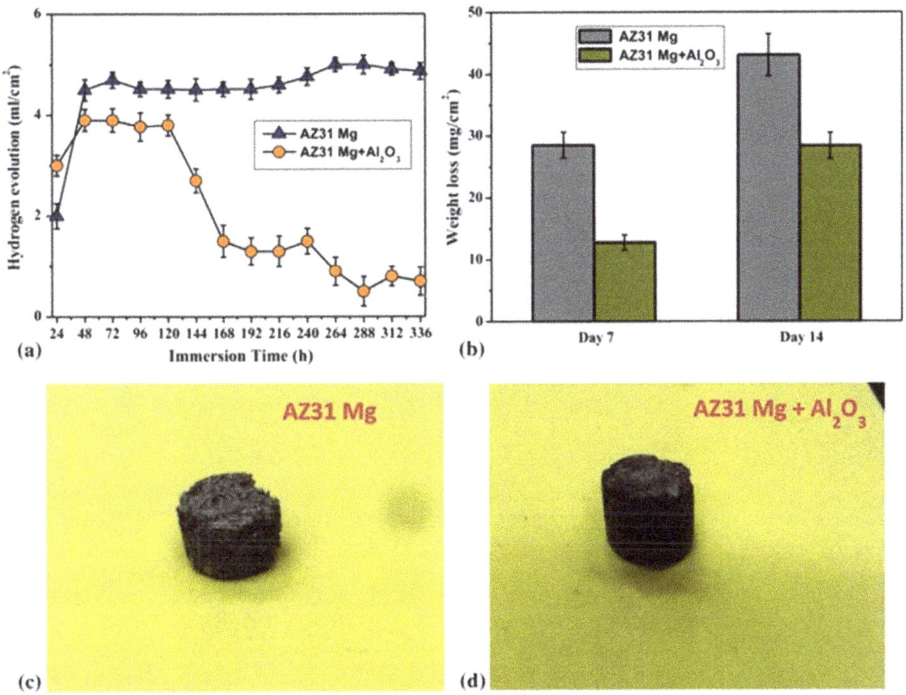

Figure 10.5. Results of immersion tests conducted on AZ31/Al_2O_3 nanocomposites. Reproduced from [18], with permission © Springer Nature.

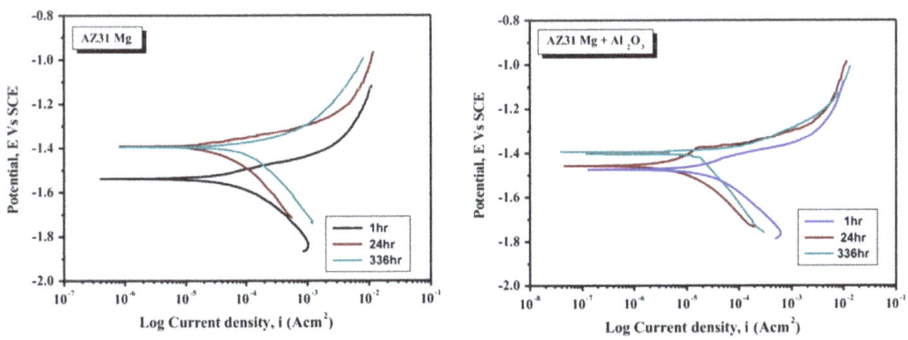

Figure 10.6. Potentio-dynamic polarization curves of AZ31/Al_2O_3 nanocomposites. Reproduced from [18], with permission © Springer Nature.

Aung *et al* [21] used the electrochemical, immersion, and hydrogen evolution methods to evaluate the corrosion properties of Mg/CNT nanocomposites in 3.5 wt. % NaCl solution. From the polarization curves, the corrosion potential and current density values were calculated (as shown in table 10.5). The corrosion rates were also directly calculated from the hydrogen gas evolution, as shown in figure 10.8. The results indicated good agreement among all testing methods, which showed an

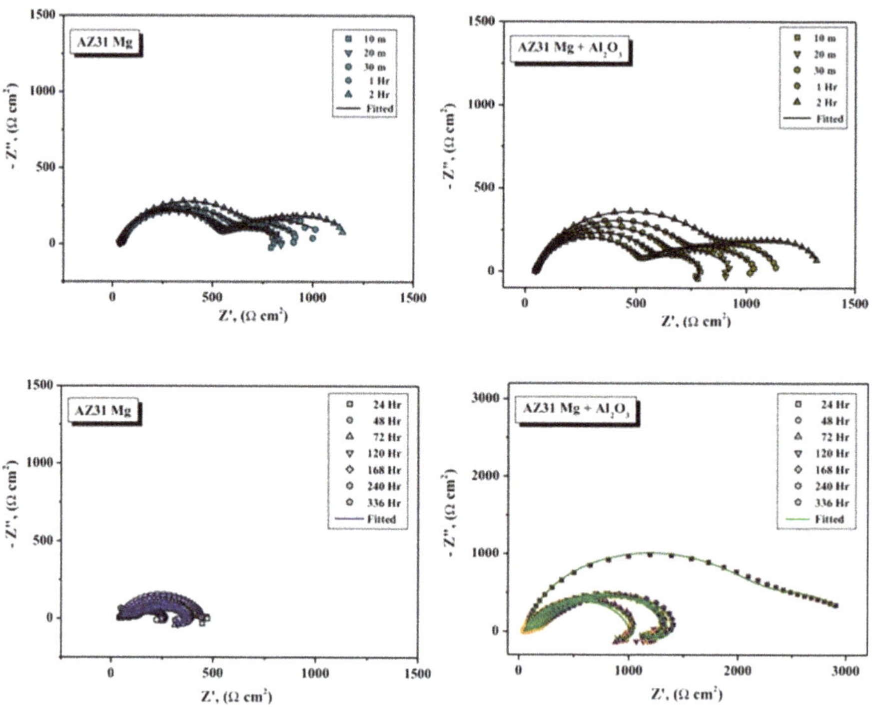

Figure 10.7. Nyquist plots of AZ31/Al_2O_3 nanocomposites. Reproduced from [18], with permission © Springer Nature.

Table 10.5. E_{corr} and I_{corr} values of Mg/CNT nanocomposites as reported by [21].

Sample	Corrosion potential, E_{corr} (V)	Current density, I_{corr} (mA cm^{-2})
Mg	−1.59	0.027
Mg/0.3CNT	−1.57	0.056
Mg/1.3CNT	−1.50	0.572

increase in the corrosion rate for CNT reinforced Mg nanocomposites due to the micro-galvanic action between the cathodic CNTs and the anodic Mg matrix. On the other hand, Kumar *et al* [22] reported an improvement in the corrosion resistance of DMD processed AZ31 alloy in SBF medium due to CNT reinforcement (figure 10.9).

Graphene nanoplatelet (GNP) (G-5 and G-15) reinforced Mg nanocomposites were produced using a powder metallurgy method and assessed for degradation behaviour in Hanks' solution (HBSS) [23]. The results of hydrogen evolution and polarization studies showed improved corrosion resistance for Mg–GNP composites (table 10.6). Compared to pure Mg, Mg–GNP composites reinforced with 0.1 wt.%

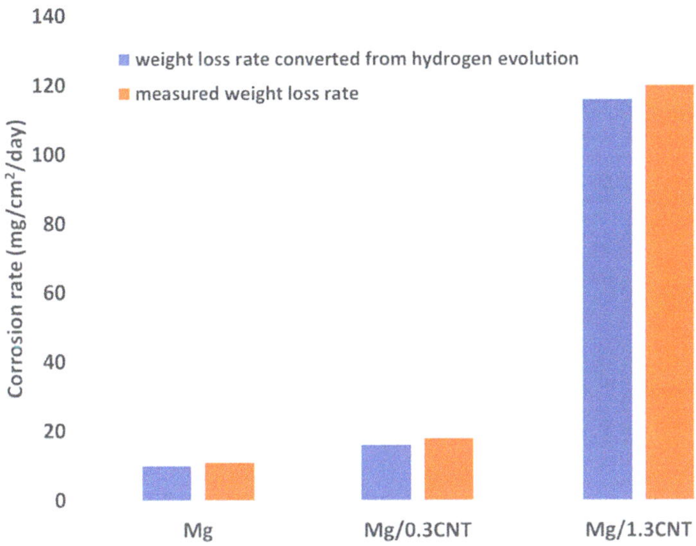

Figure 10.8. Corrosion rate of Mg/CNT nanocomposites. Data from [21].

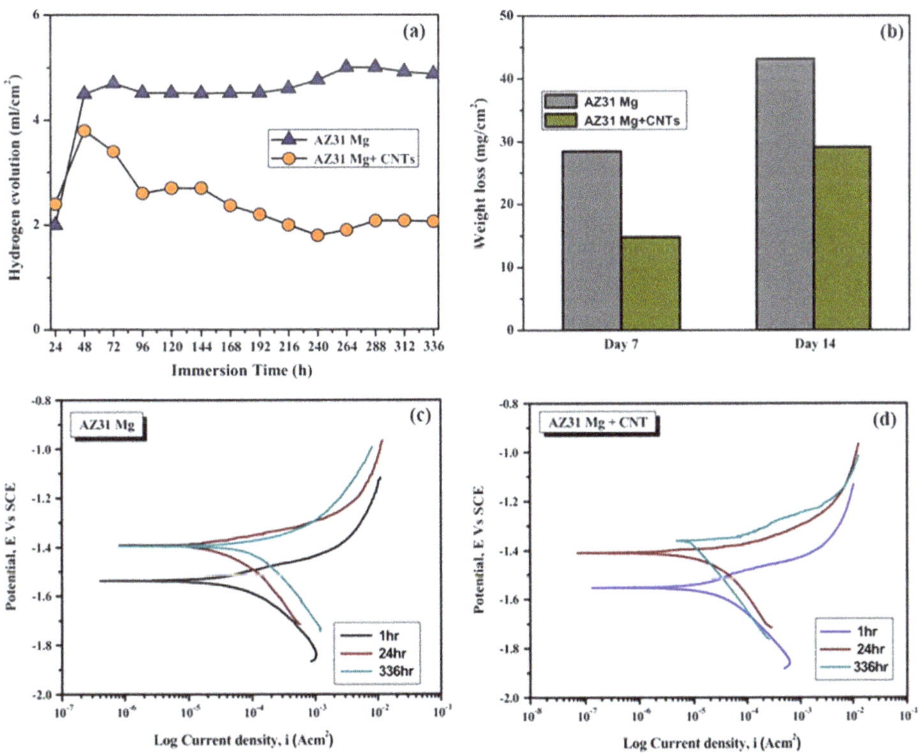

Figure 10.9. Corrosion properties of AZ31/CNT nanocomposites. Reproduced from [22], with permission © Springer Nature.

Table 10.6. Corrosion characteristics of Mg/GNPs composites as reported in [23].

	Hydrogen gas evolution		Polarization	
Material	Vol. of H_2 (V_h) mL cm^{-2} day^{-1}	Corrosion rate mm yr^{-1}	Corrosion current (I_{corr}) mA cm^{-2}	Corrosion rate (P_i) mm yr^{-1}
Pure Mg	32.52±1.58	74.11±3.61	0.81±0.02	25.02±1.79
		G-15 GNP		
Mg/0.10GNP	15.16±0.41	34.57±0.94	0.42±0.02	13.28±0.27
Mg/0.20GNP	26.93±1.08	61.38±2.48	0.58±0.03	18.50±0.83
Mg/0.30GNP	30.41±1.39	69.32±3.16	0.35±0.04	11.00±1.08
		G-5 GNP		
Mg/0.10GNP	29.55±2.61	67.35±5.95	0.50±0.05	15.13±0.91
Mg/0.20GNP	31.54±2.97	71.88±6.77	0.69±0.03	21.45±1.91
Mg/0.30GNP	34.93±1.82	79.62±4.17	0.91±0.03	28.74±3.06

Table 10.7. The effect of GNP on the electrochemical properties of Mg–0.3Sr–0.3Ca alloy as reported in [24].

Materials	Corrosion potential, E_{corr} (V)	Current density, I_{corr} (μA cm^{-2})	Corrosion rate (mm yr^{-1})
Mg–0.3Sr–0.3Ca	−1.878	7.486	0.171
Mg–0.3Sr–0.3Ca/0.2GNP	−1.823	6.730	0.153

G-15 and G-5 GNPs showed the best corrosion resistance due to the *in situ* formation of MgH_2 that resulted in slower H_2 production rates and lower I_{corr} values.

Ramezanzade *et al* [24] used the electrochemical test to evaluate the corrosion behaviour of GNP reinforced stir cast Mg–0.3Sr–0.3Ca alloy in SBF. From the Tafel plots, the corrosion rate was calculated using the E_{corr} and I_{corr} values, which showed a reduction in the corrosion rate of the nanocomposite due to GNP reinforcement (table 10.7).

Using the powder metallurgy method, magnesium composites with different weights of nano-HA (0%, 2%, 4%, and 8%) were fabricated [25]. The evaluation of biodegradation properties in phosphate-buffer saline showed a 35%–40% reduction in the degradation rate due to nano-HA addition (figure 10.10). Similarly, the results of pH measurement of the solution showed an increase in the pH values which was affected by the HA content and duration, which was due to the formation of a $Mg(OH)_2$ layer with stronger corrosion resistance (figure 10.10). Similar improvement in corrosion resistance due to nano-HA was also reported for Mg/HA nanocomposites developed using microwave sintering assisted powder metallurgy (figure 10.11) [26].

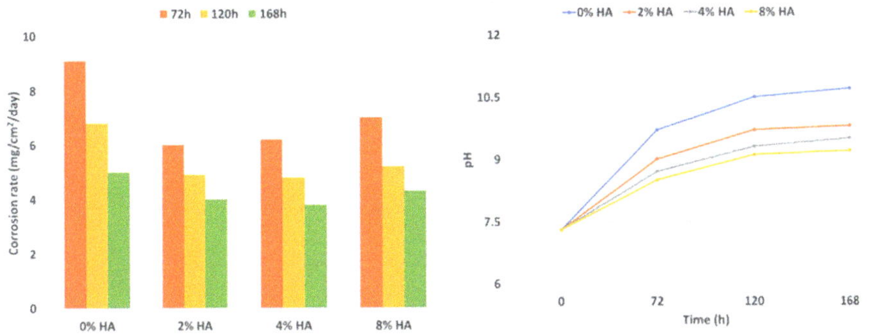

Figure 10.10. Corrosion rate and pH values of PM processed Mg/HA nanocomposites. Data from [25].

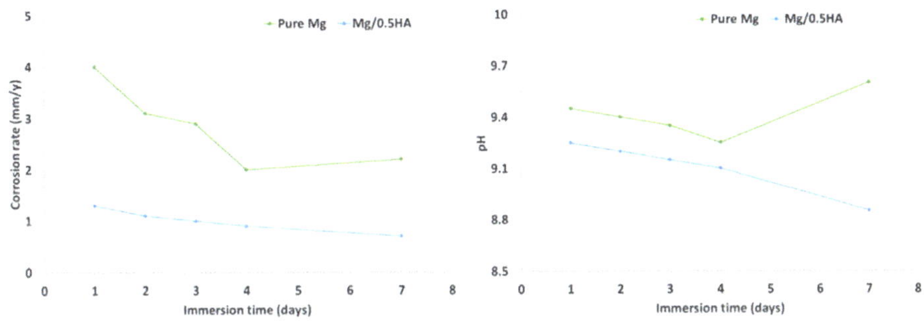

Figure 10.11. Results of corrosion studies conducted on microwave sintering processed Mg/HA nanocomposites. Data from [26].

Jaiswal *et al* [27] investigated the corrosion behaviour of powder metallurgy processed Mg–Zn/HA nanocomposites in SBF using potentio-dynamic polarization and immersion methods. From the Tafel plots, it was observed that the addition of HA decreased the corrosion current density (I_{corr}) and shifted the corrosion potential (E_{corr}) towards the positive side (table 10.8). The addition of HA up to 5% increased the corrosion resistance by ~40%. Similarly, the results of immersion tests performed for up to 8 weeks showed an increasing rate of apatite layer formation due to increasing HA content (see figures 10.12(a) and (b)). While the rate of apatite formation was initially high, it decreased after 3 days and was almost constant after 28 days of immersion. Similarly, the measured Mg^{2+} ion concentration was reduced upon HA addition, which is also indicative of reduced corrosion in Mg/HA nanocomposites (figure 10.12(c)). The best corrosion resistance was reported for Mg–3Zn/5HA due to the uniform distribution of HA particles that facilitate a lower galvanic corrosion rate and high resistance to localized pitting in longer duration immersion testing.

Nano-sized HA reinforced Mg–Zn–Si alloy-based nanocomposites were also developed using the DMD method and studied for their corrosion properties using

Table 10.8. Electrochemical corrosion of Mg–3Zn/HA nanocomposites from [27].

Samples	Corrosion potential, E_{corr} (V)	Current density, I_{corr} (μA cm^{-2})	Corrosion rate (mm yr^{-1})
Mg–3Zn	−1.78 ± 0.025	962.41 ± 2.16	2.0579
Mg–3Zn/2HA	−1.76 ± 0.060	777.02 ± 3.77	1.6366
Mg–3Zn/5HA	−1.65 ± 0.020	571.97 ± 4.17	1.1746
Mg–3Zn/10HA	−1.69 ± 0.022	682.16 ± 4.97	1.3577

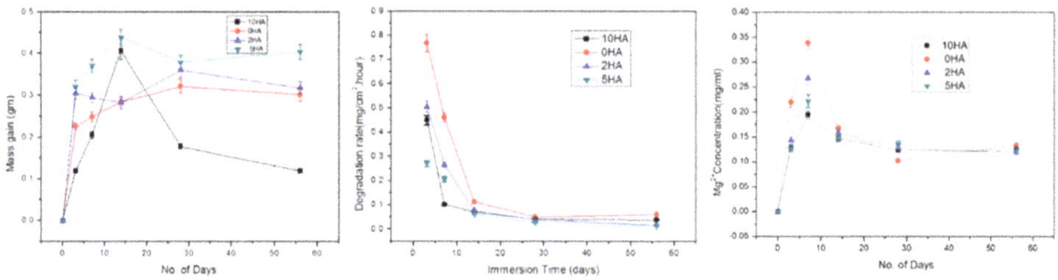

Figure 10.12. (a) Mass loss, (b) hydrogen evolution, and (c) Mg^{2+} concentration of Mg–3Zn/HA nanocomposites. Reprinted from [27], Copyright 2018, with permission from Elsevier.

immersion tests in HBSS [28]. While the corrosion rate was high for all the samples at the end of day 1, a declining rate was reported after day 4 until the end of the cycle at 21 days (figure 10.13). The pH readings also increased drastically at the beginning of the corrosion cycle and attained dynamic stabilization thereafter.

Mg/HA–MgO nanocomposites containing different amounts of nano-HA and nano-MgO particles were developed using the powder metallurgy method and investigated for electrochemical and immersion characteristics in SBF to understand the effect of nano-MgO modification on the corrosion behaviour of Mg/HA nanocomposites [29]. From the polarization curves, the corrosion current density (I_{corr}), corrosion potential (E_{corr}), pitting corrosion potential (E_{pit}), and corresponding corrosion rate (P_i) were calculated, as shown in table 10.9. The polarization resistance was also calculated from the current density (I_{corr}) and anodic/cathodic slopes (β_a, β_c) as follows:

$$R_p = \frac{\beta_a \beta_c}{2.3(\beta_a + \beta_c)} I_{corr}. \tag{10.4}$$

The results showed an increase in the corrosion resistance due to nano-MgO modification. While the addition of 5% MgO alongside a decrease in HA content to 20% resulted in no significant change in current density, the electrochemical potential was found to shift towards a more noble region, indicating an improvement in the corrosion resistance. Similarly, the cathodic slope also decreased which is an indication of decreased hydrogen evolution rate. The corrosion resistance was further improved when MgO and HA were changed to 10% and 12.5%, respectively.

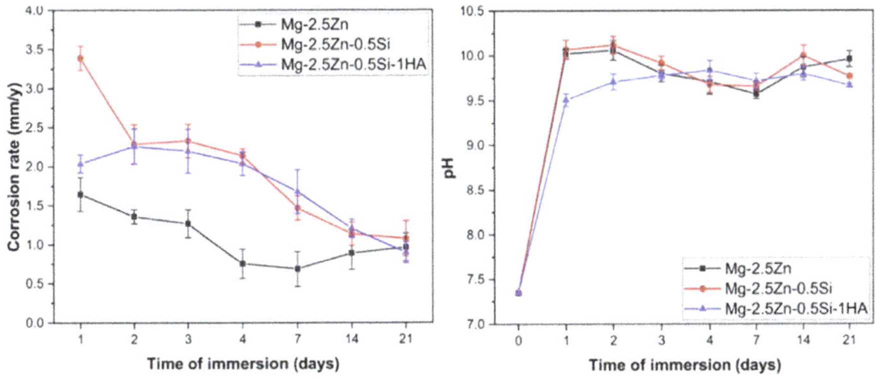

Figure 10.13. Corrosion rates and pH values of nano-HA reinforced Mg–Zn–Si alloys. Reprinted from [28], Copyright 2020, with permission from Elsevier.

Table 10.9. Electrochemical test results of Mg/HA–MgO nanocomposites in SBF as reported by [29].

Composition	Designation	Corrosion potential, E_{corr} mV	Current density, I_{corr} μA cm^{-2}	Corrosion rate, $_{Pi}$ mm yr^{-1}	Cathodic slope, β_c	Anodic slope, β_a	Polarization resistance, R_P kΩ cm^2
Mg/27.5HA	MHM0	−1487.3	187.4	4.28	411.2	149.6	0.25
Mg/20HA–5MgO	MHM5	−1434.6	203.6	4.65	344.9	235.3	0.29
Mg/12.5HA–10MgO	MHM10	−1258.2	46.8	1.06	200.8	391.4	1.23
Mg/5HA–15MgO	MHM15	−1293.8	85.2	1.94	288.5	330.7	0.78

However, the further increase in MgO or reduction in HA did not improve the corrosion resistance. Results from immersion and hydrogen evolution experiments also corroborated well with the electrochemical measurements (figure 10.14).

The effects of friction stir-processing and nano-HA addition on the corrosion resistance of cast Mg and AZ31 alloy were studied using immersion and electrochemical tests in supersaturated SBF medium [30, 31]. The results of immersion testing showed reduced weight loss due to FSP and nano-HA addition (figure 10.15). Similarly, the increase in pH values was also less for FSP processed and nano-HA added pure Mg and AZ31 (figure 10.15). The results of polarization experiments (table 10.10) also showed an improvement in the corrosion resistance due to FSP and nano-HA addition, as indicated by the shifting of corrosion potential towards a more positive side and reduced current density. A similar trend was also reported for the test samples immersed in SBF for 72 h.

Johanes *et al* [32] studied the weight loss of Mg/Fe_3O_4 nanocomposites in HBSS solution (at 37 °C) processed by DMD and turning induced deformation (TID)

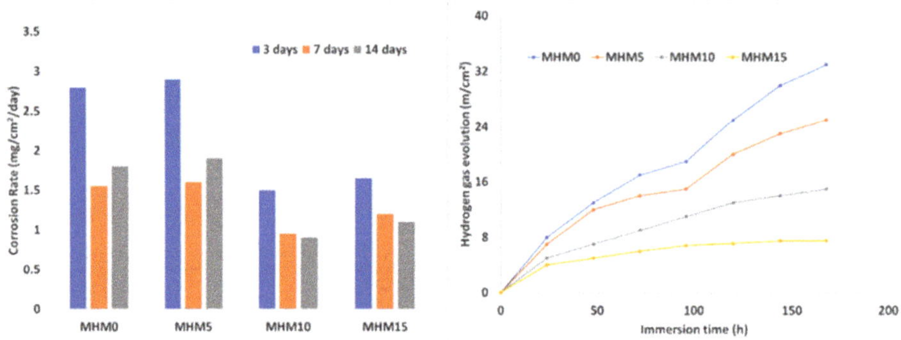

Figure 10.14. Results of weight loss and hydrogen evolution studies. Data from [29].

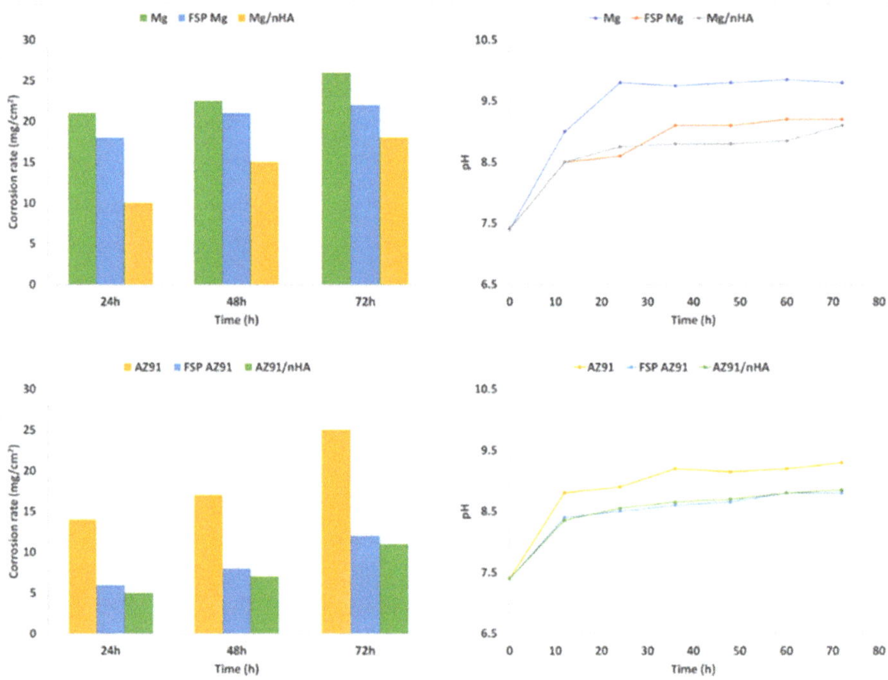

Figure 10.15. Corrosion rates and pH values of Mg and AZ91 alloy nanocomposites containing HA nanoparticles. Data from [30, 31].

methods. The authors observed the highest biocorrosion resistance in $Mg/2Fe_3O_4$ (DMD) nanocomposite among all the materials (figure 10.16). Further, the authors highlight the importance of porosity in controlling the corrosion resistance. The materials processed by TID generally showed higher porosity and hence an increasingly higher weight change due to immersion with higher Fe_3O_4 content. This led to the complete dissolution of TID materials and failure to survive immersion tests past 8 h.

Table 10.10. E_{corr} and I_{corr} values for nano-HA reinforced Mg and AZ91 nanocomposites (before and after biomineralization in SBF for 72 h) as reported in [30, 31].

		Corrosion potential, E_{corr} (V)		Current density, I_{corr} (mA cm^{-2})	
Sample	Condition	Before	After	Before	After
Mg	As-cast	−1.88	−1.46	8.71	0.10
	FSP	−1.61	−1.38	2.79	0.10
Mg/nHA	FSP	−1.57	−1.23	0.67	0.02
AZ31	As-cast	−1.71	−1.69	15.81	0.41
	FSP	−1.51	−1.63	4.67	0.32
AZ31/nHA	FSP	−1.56	−1.51	0.23	0.11

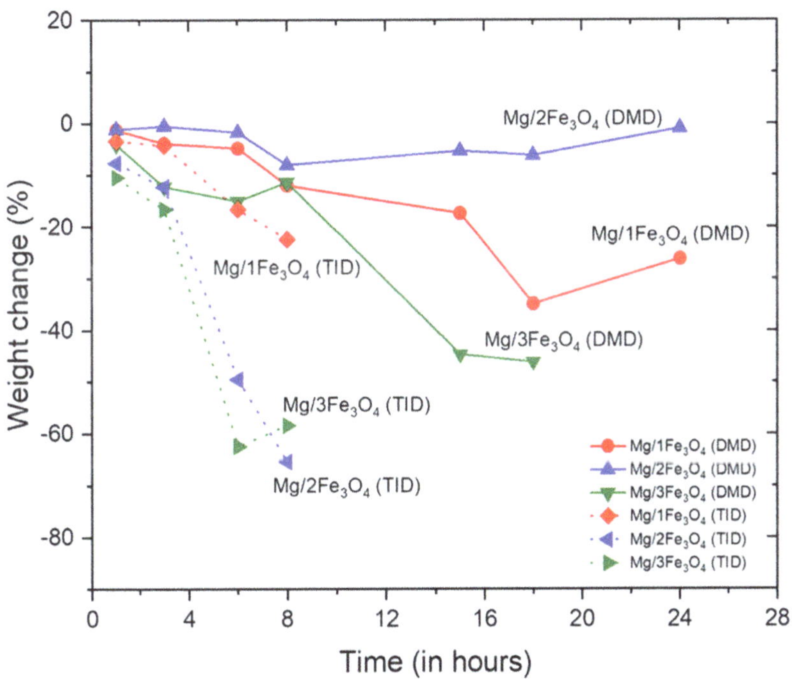

Figure 10.16. Weight change in the Mg/Fe_3O_4 nanocomposites in HBSS solution at 37 °C. Reproduced from [32], with permission from MDPI.

Ma *et al* [33] investigated the effects of processing methods on the degradation behaviour of magnesium nanocomposites reinforced with HA nanoparticles. In this study, the Mg/HA nanocomposites were developed using (i) friction stir processing, (ii) ultrasonic processing, (iii) two-state nano-processing (TSnP), which combines the benefits of liquid state ultrasonic processing and solid state FSP, and (iv) laser surface melting. The developed nanocomposites were tested for their corrosion characteristics in modified-SBF (m-SBF) and the results were compared to those of the as-cast counterparts. From the potentio-dynamic polarization curves of as-cast

Mg, FSP processed Mg, ultrasonic processed Mg/HA, and TSnP processed Mg/HA (B4), it was clear that the TSnP processed Mg/HA nanocomposite exhibited the least corrosion current and hence a great improvement in the corrosion resistance (figure 10.17). A further improvement in corrosion resistance was realized for TSnP processed Mg/HA nanocomposites after laser surface melting.

Magnesium nanocomposites based on AZ91 alloy matrix and fluorapatite nanoparticle reinforcement were produced by ball milling [34–37]. The developed nanocomposites were tested for corrosion in Ringer solution and SBF. From the Tafel plots of electrochemical tests it was evident that the increasing amount of FA nanoparticles reduced the corrosion rate of AZ91 alloy in Ringer solution (table 10.11). Similarly, the results of immersion testing conducted on the nanocomposites in SBF medium showed an increase in the corrosion resistance of AZ91-FA nanocomposites due to the formation of a protective layer on the surface of AZ91/FA nanocomposites.

Mg–Zn–Zr alloy nanocomposites reinforced with different amounts (0.5%, 1.0%, and 1.5%) of nano β-TCP particles were prepared by induction melting and extrusion [38]. In this study, the results of electrochemical testing (table 10.12) conducted in SBF showed enhanced corrosion resistance due to increasing amounts of β-TCP nanoparticles. Similar improvement in corrosion resistance was also reported for Mg/β-TCP nanocomposites with MgO on the surface of β-TCP particles (figure 10.18) [39].

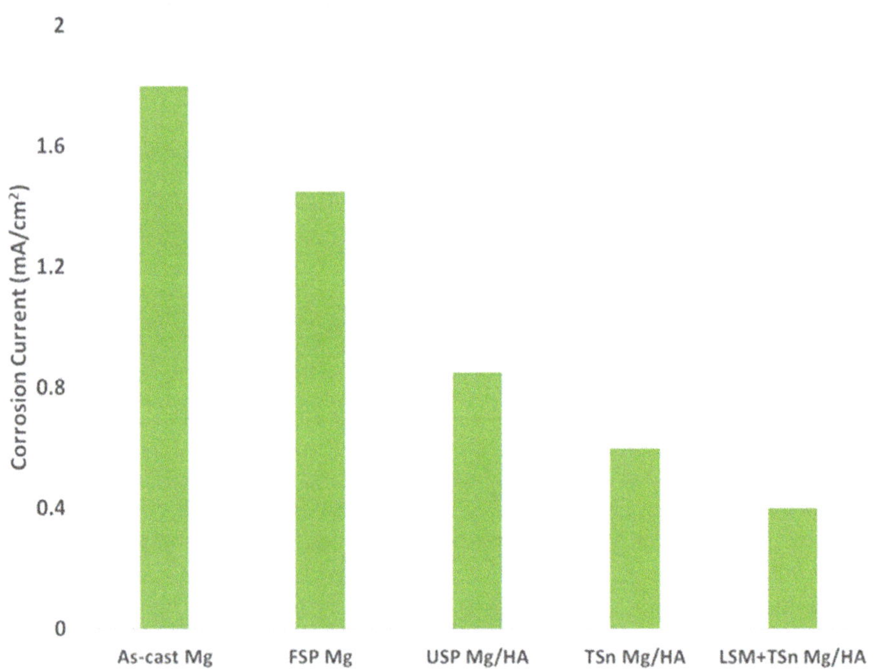

Figure 10.17. The effect of processing methods on the corrosion performance of Mg/HA nanocomposites. Data from [33].

Table 10.11. Corrosion performance of AZ91 alloy and AZ91/FA nanocomposites as reported in [34].

Materials	I_{corr} (A cm^{-2})	E_{corr} (V)	Corrosion rate (mg cm^{-2} h^{-1})
AZ91	8.5×10^{-4}	−1.54	1.3
AZ91/10FA	7.4×10^{-5}	−1.48	1.11
AZ91/20FA	2.3×10^{-6}	−1.43	0.8
AZ91/30FA	3.5×10^{-7}	−1.38	0.44

Table 10.12. Corrosion performance of Mg–3Zn–0.8Zr/β-TCP nanocomposites as reported in [38].

Materials	Corrosion potential (V)	Current density (μA cm^{-2})
Mg–3Zn–0.8Zr	−1.648	1.78
Mg–3Zn–0.8Zr/0.5 β-TCP	−1.630	2.04
Mg–3Zn–0.8Zr/1.0 β-TCP	−1.547	1.20
Mg–3Zn–0.8Zr/1.5 β-TCP	−1.531	1.74

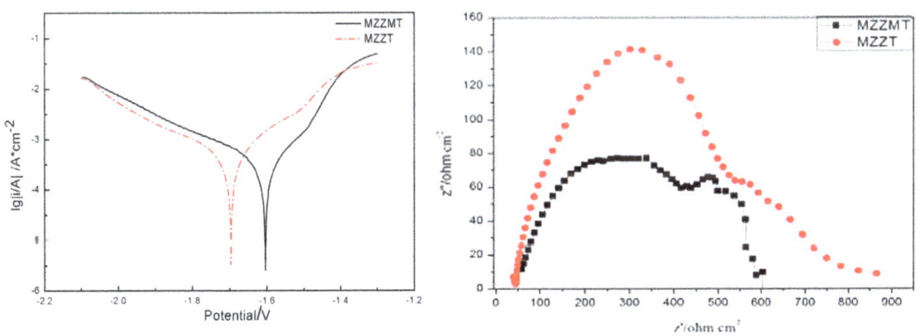

Figure 10.18. Electrochemical corrosion characteristics of Mg nanocomposites reinforced with unmodified TCP nanoparticles (MZZT) and MgO surface modified TCP nanoparticles (MZZMT). From [39], © KeAi.

Mg nanocomposites reinforced with nano-diamond (ND) particles were prepared by the powder metallurgy method and tested for immersion and electrochemical behaviour in SBF [40]. The results of immersion testing were analysed in terms of corrosion rate and pH changes (figures 10.19(a) and (b)). While the corrosion rate of nanocomposites decreased with ND addition, the pH value of the corrosion medium increased steadily until a steady state was reached at the end of 24 h. From the polarization and impedance measurements, a higher corrosion potential and electrical resistance was reported for the Mg/ND nanocomposites with higher ND content (figures 10.19(c) and (d) and table 10.13).

The electrochemical corrosion behaviour of DMD processed Mg nanocomposites reinforced with different amounts of nanoscale TiO_2, TiC, and TiN were evaluated

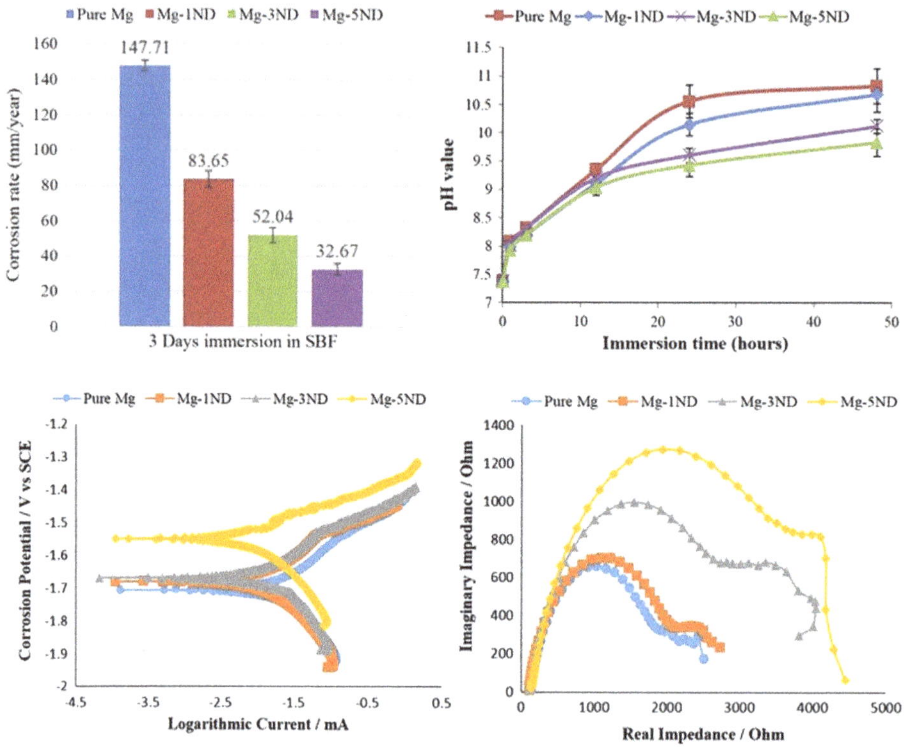

Figure 10.19. Corrosion properties of Mg/ND nanocomposites. Reproduced from [40], with permission © Springer Nature.

in SBF [41]. The results (figure 10.20) showed an improvement in the corrosion resistance due to nano-reinforcement addition, except for Mg/2.5TiC. Among the nanocomposites, Mg/1.98TiC exhibited the lowest corrosion rate of 0.02 mm yr^{-1}.

10.5 The biocompatibility of magnesium nanocomposites

Cytotoxicity or the measure of the cell viability upon exposure to materials can be assessed by *in vitro* methods during which the cultured cells are exposed directly to materials or indirectly using an extraction medium (as given in table 10.3). Normally Mg-based materials are tested for indirect toxicity and the medium is prepared by immersing the sample in an extraction medium for a specific duration and maintained at a humidified temperature of 37 °C and 5% CO_2. Since the concentration of Mg ions varies with respect to the extraction time, the samples are often exposed to extraction media of different concentrations.

Using the method of friction stir processing, nano-HA particles were added to Mg and AZ31 [30, 31]. The cytotoxicity in terms of the cell viability was calculated using the MTT (3-(4,5-dimethylthiazole-2-yl)-2,5-diphenyl tetrazolium bromide) assay by exposing the samples to rat skeletal muscle (L6) cells and the results (figure 10.21) showed no adverse effects due to FSP processing or nano-HA particles and the toxicity was negligible for all the materials. The results of microscopic analysis

Table 10.13. Corrosion properties of Mg composites reinforced with nano-diamond particles as reported in [40].

Properties	Pure Mg	Mg/1ND	Mg/3ND	Mg/5ND
Corrosion potential (V)	−1.55	−1.67	−1.68	−1.70
Resistance (Ω)	2796	2919	4279	5059

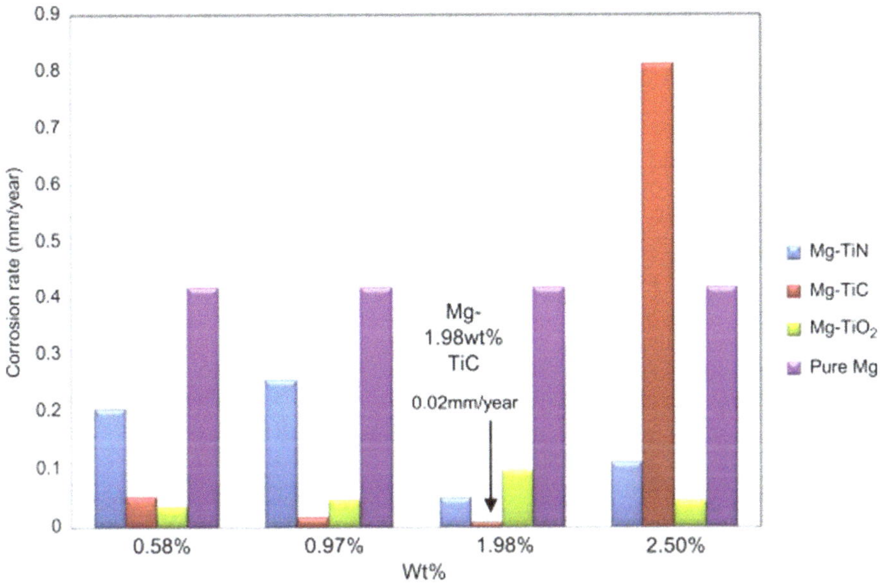

Figure 10.20. Corrosion rates of Mg nanocomposites reinforced with nano-sized TiO_2, TiC, and TiN reinforcements. Reprinted from [41], Copyright 2018, with permission from Elsevier.

conducted on the incubated samples seeded with cells and culture medium showed an improvement in the adhesion behaviour due to better wettability, surface energy, and osseointegration as facilitated by the FSP and nano-HA particles.

Jaiswal *et al* [27] used the powder metallurgy method to synthesize Mg–3Zn/HA with different amounts of nano-HA particles. In this study, the biocompatibility of Mg–3Zn/HA nanocomposites was assessed in terms of cell viability and adhesion characteristics using human foetal osteoblast (hFOB) cells in Dulbecco's modified Eagle's culture medium (DMEM). The results (figure 10.22) showed a 144% and 128% increment in hFOB viability for Mg–3Zn/5HA and Mg–3Zn/10HA compared to Mg–3Zn due to the biologically active apatite layer which enhanced the adsorption of proteins for cell adhesion and proliferation. However, the agglomeration of HA nanoparticles at higher content lowered the cell viability. The adhesion characteristics evaluated by microscopic and wettability tests also showed better results for Mg–3Zn–5HA. While the fluorescent images of hFOB cells stained with DAPI showed more cells adhering to the nanocomposite surfaces (figure 10.23), the

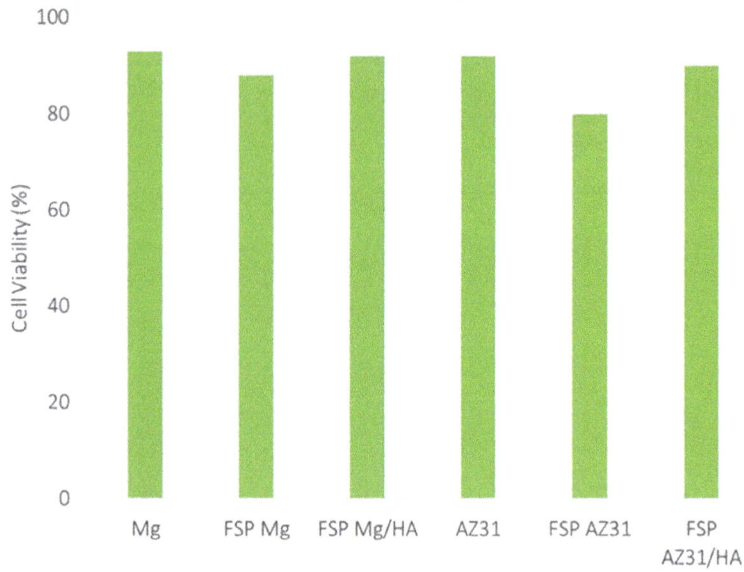

Figure 10.21. The effect of FSP and nano-HA addition on the cytotoxicity Mg and AZ31 alloy as reported by Sunil *et al.* Data from [30, 31].

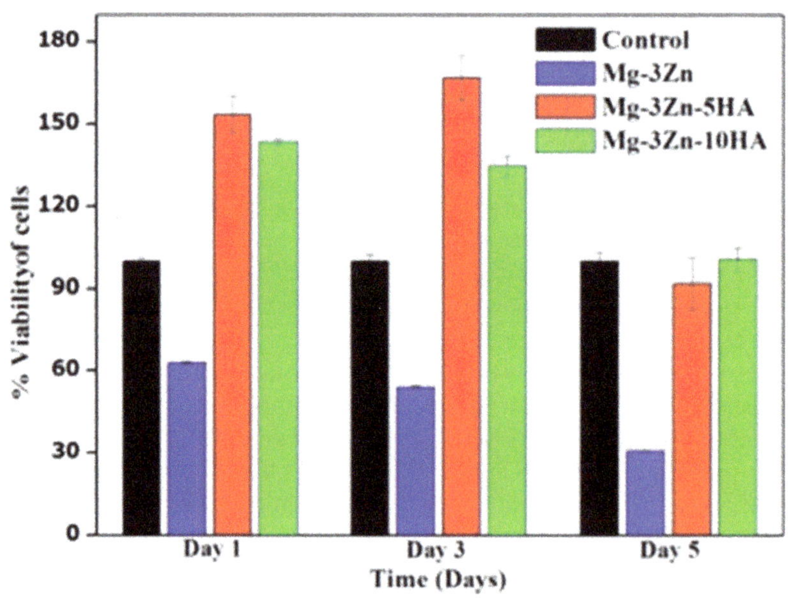

Figure 10.22. Viability of hFOB cells for Mg–3Zn/HA, as reported by Jaiswal *et al.* Reprinted from [27], Copyright 2018, with permission from Elsevier.

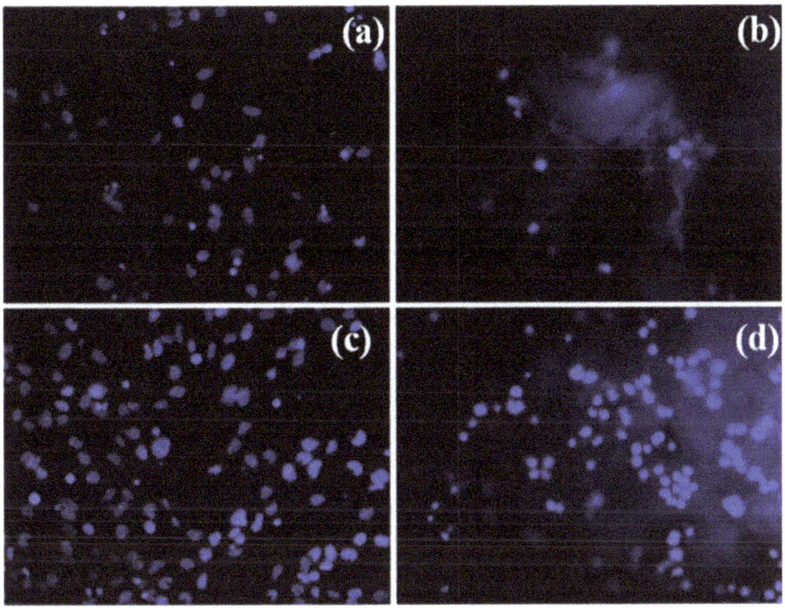

Figure 10.23. Fluorescence images of hFOB cells stained with DAPI after day 1 of seeding on (a) control, (b) Mg–3Zn, (c) Mg–3Zn–5HA, and (d) Mg–3Zn–10HA surfaces, as reported by Jaiswal *et al.* Reprinted from [27], Copyright 2018, with permission from Elsevier.

contact angle of Mg–3Zn/5HA (54.56°±2.9°) was relatively less when compared to Mg–3Zn (79.76°±4.1°) and Mg–3Zn/10HA (59.9°±3.3°), respectively.

The cytotoxicity of Mg–Zn–Si/HA nanocomposites was investigated for osteoblast-like murine MC3T3-E1 cells based on lactate dehydrogenase activity using alpha-minimum essential medium supplemented with 10% FBS and 1% penicillin–streptomycin culture medium [28]. The results (figure 10.24) from this study revealed level 0 cytotoxicity (⩽25% cell death) for Mg–2.5Zn, Mg–2.5Zn–0.5Si, and Mg–2.5Zn–0.5Si–1HA nanocomposite samples after 24 h of exposure. While the addition of Si and HA reduced the cytotoxicity of Mg–2.5Zn, the least cytotoxicity was recorded for the nanocomposite. SEM observations also showed an interconnected network of cells adhered to the surface of Mg–2.5Zn–0.5Si/1HA nanocomposites, with a significantly improved volume of cell attachment and proliferation compared to Mg–2.5Zn and Mg–2.5Zn–0.5Si.

Various amounts of HA and MgO nanoparticles were used to synthesize Mg/HA–MgO nanocomposites by powder metallurgy and tested for human osteoblast (NHOst) cell viability using MTT and ALP differentiation assays [29]. Osteoblast basal medium with the osteoblast growth medium supplemented with 0.1% ascorbic acid, 10% FBS, and 0.1% gentamycin was used for culture and extraction purposes. From the MTT assay results, it was found that the proliferation and differentiation of cells was significantly higher for the MgO modified nanocomposites, which was attributed to the difference in their corrosion behaviour and hydrogen evolution. However, this response was found to be true only until 7 days of cell culture and the

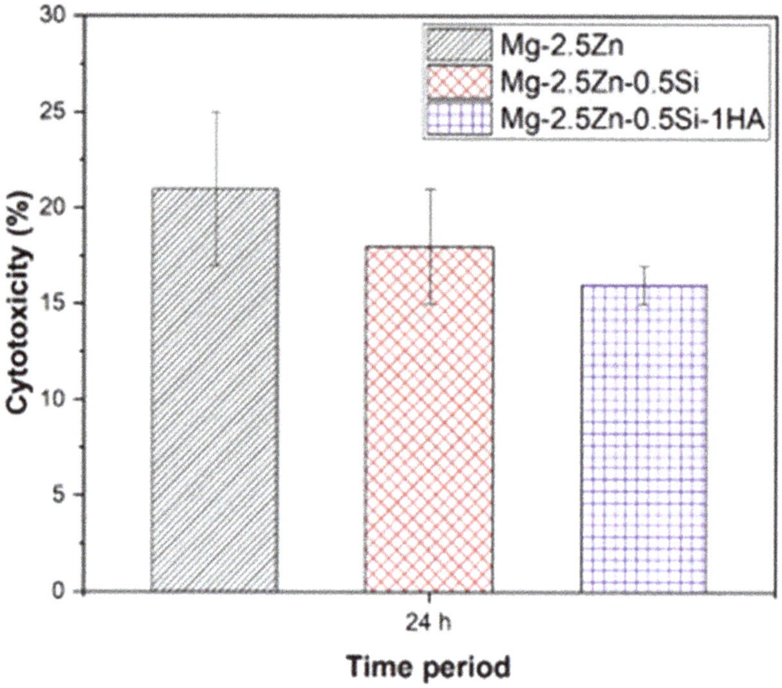

Figure 10.24. The effect of Si and nano-HA addition on the cytotoxicity of Mg–2.5Zn alloy. Reprinted from [28], Copyright 2020, with permission from Elsevier.

Mg/HA nanocomposite showed a higher level of cell density after days, i.e. the Mg/HA and Mg/HA–MgO nanocomposites exhibited time dependent behaviour. Further, it was also reported that the variation in pH value also affects the cell viability significantly. The differentiation of osteoblast cells as evaluated by the ALP activity showed an initial reduction and an increase after 9 days of cell culture which confirmed the enhanced differentiation activity of NHOst cells from days 3–9. Similar improvements in biocompatibility were reported by Zheng *et al* [39] in the case of Mg/HA nanocomposites with MgO on the surface of HA nanoparticles.

Powder metallurgy processed Mg nanocomposites containing GNPs were assessed for cytotoxicity by evaluating the viability of human osteoblast-like cells ($SaOS_2$) [23]. In this study, the extracts of Mg/GNP nanocomposites were prepared using DMEM supplemented with 10% FBS and 1% penicillin–streptomycin. While the results (figure 10.25(a)) showed no significant difference in the cell viability ratio, the Mg/0.1GNP nanocomposite exhibited higher cell viability. In another study, Mg was reinforced with nano-sized diamond particles by powder metallurgy and tested for the metabolic activity of fibroblast (L929) cells using DMEM medium with 10% FBS [40]. From the results (figure 10.25(b)), it was reported that the addition of ND particles did not significantly affect the cell viability and the cells were found effectively adhered to the surfaces of pure Mg and Mg/ND nanocomposites.

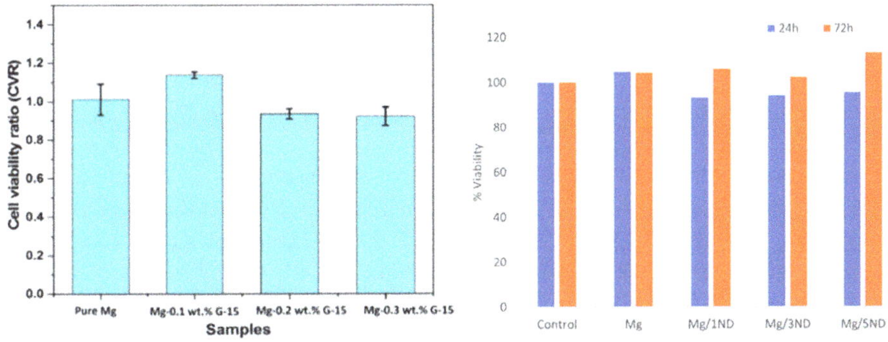

Figure 10.25. Cell viability ratios of Mg/GNP and Mg/ND nanocomposites. Adapted from [23, 40], © KeAi.

10.6 Summary

Magnesium-based nanocomposites with an excellent combination of mechanical properties, wear resistance, fatigue life, and creep resistance are promising for weight-critical applications. With respect to corrosion, the published literature highlights that the corrosion resistance of Mg nanocomposites depends on the type of corrosion medium and the composition of matrix material and nano-reinforcements. The addition of nano-lengthscale reinforcements to Mg improved the surface energy of the nanocomposites and formed a reasonably protective layer on the surface to prevent rampant corrosion. The controlled degradation of Mg nanocomposites is also beneficial for utilization in biodegradable implants. While most nano-sized reinforcements improved the protein adsorption and cell adhesion characteristics to facilitate the cell viability, the actual response differed with reinforcement volume fraction and the type and nature of the cultured cells and extraction medium.

References

[1] Zeng R C, Zhang J, Huang W J, Dietzel W, Kainer K U, Blawert C and Ke W 2006 Review of studies on corrosion of magnesium alloys *Trans. Nonferrous Met. Soc. China* **16** s763–71

[2] Ghali E, Dietzel W and Kainer K U 2013 General and localized corrosion of magnesium alloys: a critical review *J. Mater. Eng. Perform.* **13** 7–23

[3] Esmaily M, Svensson J E, Fajardo S, Birbilis N, Frankel G S, Virtanen S, Arrabal R, Thomas S and Johansson L G 2017 Fundamentals and advances in magnesium alloy corrosion *Prog. Mater. Sci.* **89** 92–193

[4] Sezer N, Evis Z, Kayhan S M, Tahmasebifar A and Koç M 2018 Review of magnesium-based biomaterials and their applications *J. Magnes. Alloy.* **6** 23–43

[5] Prasad K *et al* 2017 Metallic biomaterials: current challenges and opportunities *Materials* **10** 884

[6] Mahyudin F, Widhiyanto L and Hermawan H 2016 Biomaterials in orthopaedics *Adv. Struct. Mater.* **5** 1137–58

[7] Gupta M and Meenashisundaram G K 2015 *Insight into Designing Biocompatible Magnesium Alloys and Composites* (Materials Park, OH: ASM International) p 109

[8] Gu X N and Zheng Y F 2010 A review on magnesium alloys as biodegradable materials *Front. Mater. Sci. China* **4** 111–5

[9] Chen Y, Dou J, Yu H and Chen C 2019 Degradable magnesium-based alloys for biomedical applications: the role of critical alloying elements *J. Biomater. Appl.* **33** 1348–72

[10] Witte F, Hort N, Vogt C, Cohen S, Kainer K U, Willumeit R and Feyerabend F 2008 Degradable biomaterials based on magnesium corrosion *Curr. Opin. Solid State Mater. Sci.* **12** 63–72

[11] Ding Y, Wen C, Hodgson P and Li Y 2014 Effects of alloying elements on the corrosion behavior and biocompatibility of biodegradable magnesium alloys: a review *J. Mater. Chem.* B **2** 1912–33

[12] Bommala V K, Krishna M G and Rao C T 2019 Magnesium matrix composites for biomedical applications: a review *J. Magnes. Alloy* **7** 72–9

[13] Radha R and Sreekanth D 2017 Insight of magnesium alloys and composites for orthopedic implant applications—a review *J. Magnes. Alloy* **5** 286–312

[14] Yau T-L, Testing I and Fundam C 2003 *Corrosion: Fundamentals, Testing, and Protection* **vol 13A** (Materials Park, OH: ASM International)

[15] Song G, Atrens A and John D St 2001 An hydrogen evolution method for the estimation of the corrosion rate of magnesium alloys *Essential Readings in Magnesium Technology* ed S N Mathaudhu *et al* (Cham: Springer)

[16] Gupta M and Meenashisundaram G K 2015 *Insight into Designing Biocompatible Magnesium Alloys and Composites: Processing, Mechanical and Corrosion Characteristics* (Singapore: Springer)

[17] Wong L L, Martin S I and Rebak R B 2006 Methods to calculate corrosion rates for alloy 22 from polarization resistance experiments *Am. Soc. Mech. Eng. Press. Vessel. Pip. Div. PVP*

[18] Madhan Kumar A, Fida Hassan S, Sorour A A, Paramsothy M and Gupta M 2018 Electrochemical corrosion and *in vitro* biocompatibility performance of AZ31Mg/Al_2O_3 nanocomposite in simulated body fluid *J. Mater. Eng. Perform.* **27** 3419–28

[19] Gomes M P, Costa I, Pébère N, Rossi J L, Tribollet B and Vivier V 2019 On the corrosion mechanism of Mg investigated by electrochemical impedance spectroscopy *Electrochim. Acta* **306** 61–70

[20] Kukreja M, Balasubramaniam R, Nguyen Q B and Gupta M 2009 Enhancing corrosion resistance of Mg alloy AZ31B in NaCl solution using alumina reinforcement at nanolength scale *Corros. Eng. Sci. Technol.* **44** 381–3

[21] Aung N N, Zhou W, Goh C S, Nai S M L and Wei J 2010 Effect of carbon nanotubes on corrosion of Mg-CNT composites *Corros. Sci.* **52** 1551–3

[22] Kumar A M, Hassan S F, Sorour A A, Paramsothy M and Gupta M 2019 Investigation on the controlled degradation and *in vitro* mineralization of carbon nanotube reinforced AZ31 nanocomposite in simulated body fluid *Met. Mater. Int.* **25** 105–16

[23] Munir K, Wen C and Li Y 2020 Graphene nanoplatelets-reinforced magnesium metal matrix nanocomposites with superior mechanical and corrosion performance for biomedical applications *J. Magnes. Alloy* **8** 269–90

[24] Ramezanzade S, Ebrahimi G R, Parizi M T and Ezatpour H R 2020 Microstructure and mechanical characterizations of graphene nanoplatelets-reinforced Mg–Sr–Ca alloy as a novel composite in structural and biomedical applications *J. Compos. Mater.* **54** 711–28

[25] Moradi E, Ebrahimian-Hosseinabadi M, Khodaei M and Toghyani S 2019 Magnesium/nano-hydroxyapatite porous biodegradable composite for biomedical applications *Mater. Res. Express.* **6** 075408

[26] Prasadh S, Manakari V, Parande G, Srivatsan T S, Wong R and Gupta M 2019 Bioresorbable nano-hydroxyapatite reinforced magnesium alloplastic bone substitute for

biomedical applications: a study *Nanocomposites VI: Nanoscience and Nanotechnology in Advanced Composites (The Minerals, Metals & Materials Series)* (Cham: Springer) pp 71–82

[27] Jaiswal S, Kumar R M, Gupta P, Kumaraswamy M, Roy P and Lahiri D 2018 Mechanical, corrosion and biocompatibility behaviour of Mg-3Zn-HA biodegradable composites for orthopaedic fixture accessories *J. Mech. Behav. Biomed. Mater.* **78** 442–54

[28] Parande G, Manakari V, Prasadh S, Chauhan D, Rahate S, Wong R and Gupta M 2020 Strength retention, corrosion control and biocompatibility of Mg–Zn–Si/HA nanocomposites *J. Mech. Behav. Biomed. Mater.* **103** 103584

[29] Khalajabadi S Z, Abdul Kadir M R, Izman S and Marvibaigi M 2016 The effect of MgO on the biodegradation, physical properties and biocompatibility of a Mg/HA/MgO nanocomposite manufactured by powder metallurgy method *J. Alloys Compd* **655** 266–80

[30] Ratna Sunil B, Sampath Kumar T S, Chakkingal U, Nandakumar V and Doble M 2014 Friction stir processing of magnesium–nanohydroxyapatite composites with controlled *in vitro* degradation behavior *Mater. Sci. Eng.* C **39** 315–24

[31] Ratna Sunil B, Sampath Kumar T S, Chakkingal U, Nandakumar V and Doble M 2014 Nanohydroxyapatite reinforced AZ31 magnesium alloy by friction stir processing: a solid state processing for biodegradable metal matrix composites *J. Mater. Sci. Mater. Med.* **25** 975–88

[32] Johanes M, Tekumalla S and Gupta M 2019 Fe_3O_4 nanoparticle-reinforced magnesium nanocomposites processed via disintegrated melt deposition and turning-induced deformation techniques *Metals* **9** 1225

[33] Ma C, Chen L, Xu J, Fehrenbacher A, Li Y, Pfefferkorn F E, Duffie N A, Zheng J and Li X 2013 Effect of fabrication and processing technology on the biodegradability of magnesium nanocomposites *J. Biomed. Mater. Res.* B **101B** 870–7

[34] Razavi M, Fathi M H and Meratian M 2010 Microstructure, mechanical properties and bio-corrosion evaluation of biodegradable AZ91-FA nanocomposites for biomedical applications *Mater. Sci. Eng.* A **527** 6938–44

[35] Razavi M, Fathi M H and Meratian M 2010 Fabrication and characterization of magnesium–fluorapatite nanocomposite for biomedical applications *Mater. Charact.* **61** 1363–70

[36] Fathi M H, Meratian M and Razavi M 2011 Novel magnesium nanofluorapatite metal matrix nanocomposite with improved biodegradation behavior *J. Biomed. Nanotechnol.* **7** 441–5

[37] Razavi M, Fathi M H and Meratian M 2010 Bio-corrosion behavior of magnesium-fluorapatite nanocomposite for biomedical applications *Mater. Lett.* **64** 2487–90

[38] He S Y, Sun Y, Chen M F, Liu D B and Ye X Y 2011 Microstructure and properties of biodegradable β-TCP reinforced Mg-Zn-Zr composites *Trans. Nonferrous Met. Soc. China* **21** 814–9

[39] Zheng H R, Li Z, You C, Liu D B and Chen M F 2017 Effects of MgO modified β-TCP nanoparticles on the microstructure and properties of β-TCP/Mg-Zn-Zr composites *Bioact. Mater.* **2** 1–9

[40] Gong H, Anasori B, Dennison C R, Wang K, Kumbur E C, Strich R and Zhou J G 2015 Fabrication, biodegradation behavior and cytotoxicity of Mg-nanodiamond composites for implant application *J. Mater. Sci. Mater. Med.* **26** 110

[41] Sankar M, Vishnu J, Gupta M and Manivasagam G 2018 Magnesium-based alloys and nanocomposites for biomedical application *Appl. Nanocompos. Mater. Orthop.* pp 83–109

Chapter 11

Applications and challenges

Magnesium, being an ultralight metal with interesting attributes, can be used in structural and non-structural applications. It can provide solutions as a sustainable metal (~12 times more abundant than aluminium) for a sustainable planet owing to its capability to mitigate air, water, land, and electromagnetic pollution. For the past two decades, extensive research efforts have been made to bring magnesium into front line applications, and this has been realized progressively with major players emerging from the automobile and electronics industries. The aerospace, space, and sports industries are gearing towards incorporating magnesium-based materials, opening a potential multi-billion-dollar market in the immediate future. In view of the tremendous potential of magnesium in the very near future and beyond, this chapter focuses on the diversity of magnesium-based nanocomposites, showcasing their different attributes for widespread applications and their existing challenges.

11.1 Introduction

Magnesium production was estimated to be significantly high (up to 32 000 tonnes) even prior to World War II (WWII). However, during the war its production was at a peak with over 232 000 tonnes of magnesium produced (most of it from United States) and it was used significantly during the war. However, the demand world-wide for magnesium dropped in 1946 until the 1980s–1990s [1]. Surprisingly, during the early 1990s, the recorded weather patterns showed catastrophic changes and such unexpected weather patterns have only been on the rise ever since [2]. These unusual weather patterns have been attributed to the substantial rise in the emissions of greenhouse gases. In addition, pollution is also on the rise in bodies of water and land, leading to increased toxicity that is now threatening the wellbeing of the whole food chain. With this damage happening at an alarming rate now more than ever, keeping the global temperature increase below 2 °C in comparison to pre-industrial times is the current recommendation and guideline to ensure no irreparable damage is caused to the planet. In addition to such global-scale problems, mankind also

doi:10.1088/978-0-7503-3535-5ch11

faces other challenges, such as those related to electromagnetic pollution. Electromagnetic pollution is an underestimated form of pollution which has become significant in the last three decades due to the excessive use of electronic appliances and may eventually pose a serious health risk. At the same time, the average expected life span of humans is increasing due to advances in science, technology, and medicine. However, with an unprecedented rise in the ageing population and their related health risks, there is an increasing need for biomaterials that can aid in the faster recovery of patients without any side effects such as toxicity, etc. Therefore, there is a need for a paradigm shift in the usage of materials that can help achieve all the above goals. In this regard, magnesium serves as an all-in-one solution as it plays multiple roles and demonstrates multiple beneficial properties, as discussed in the previous chapters. Magnesium was ignored to a large extent for about five decades (1960–2000s), and this is partly touted to be a reason for the worsening climate change and also the significant contamination of water bodies and land masses across the world. With magnesium being one of the most dynamic metals and which is present abundantly in the Earth's crust, sea water, and the human body, it is not too late to adopt magnesium in widespread applications. Some of the key attributes of magnesium are its light weight (~33% lighter than aluminium, 75% lighter than steel), abundance, non-toxicity (it is a nutritional element for both plants and humans [3]), and, most importantly, its recyclability. Further, due to the renewed interest in magnesium over the past 15 years, the most cost-effective options have been explored for its extraction and production. Therefore, there has been a drop in the price of Mg in recent years, which makes it a viable candidate for multiple applications. The multidisciplinary prospects and promise of magnesium, in particular magnesium-based nanocomposites, will be unravelled in this chapter. Further, the challenges pertaining to the application of these magnesium nanocomposites are also discussed.

11.2 Automobile applications of magnesium nanocomposites

With automobiles contributing a significant amount to greenhouse and other gas emissions, it is imperative that magnesium finds its most significant application in automobiles, in particular for reducing weight. In fact, decreasing the weight of an automobile by just 10% leads to a 7% increase in its fuel efficiency [4]. Apart from their light weight, magnesium nanocomposites demonstrate superior creep resistance, specific strength, damping capacity, ignition resistance, and excellent castability, with other advantages as shown in figure 11.1. These make magnesium nanocomposites extremely promising for the automotive industry. Using magnesium as an alternative to the currently used aluminium and steels results in a weight reduction of 22%–70%. Among several magnesium alloys, Mg–Al-based alloy series, i.e. the AZ and AM alloys, demonstrate good strength and ductility at room temperature, along with good resistance to corrosion (salt spray) and superior castability [5]. Hence, they are predominantly being used in the automobile sector as sheets or even engine blocks and in other automobile components, such as steering wheels, boot areas, etc. For elevated-temperature applications such as engine blocks,

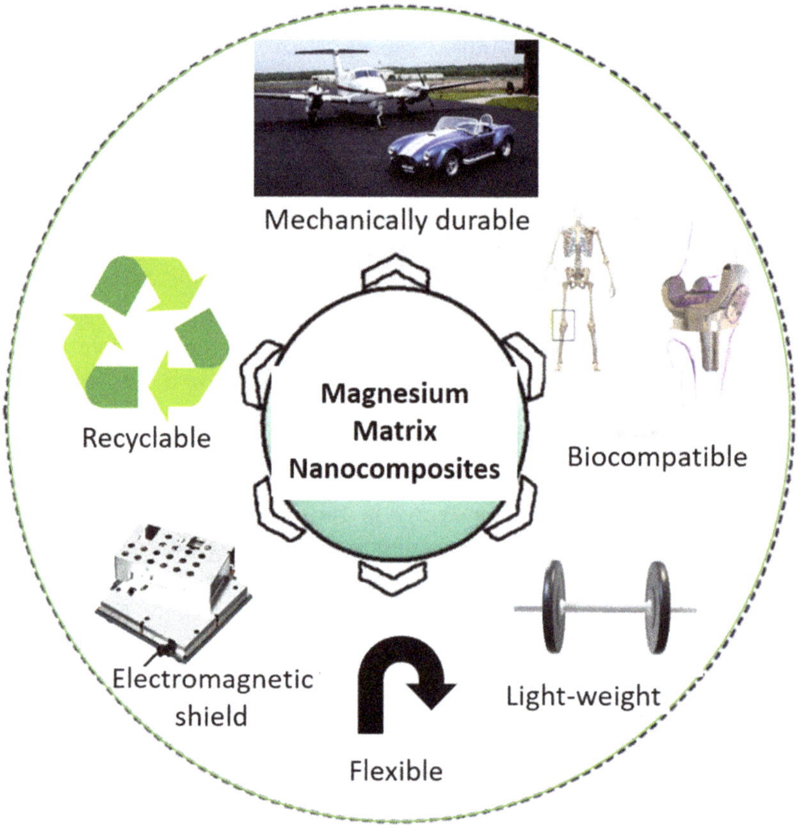

Figure 11.1. Benefits of Mg nanocomposites in the automobile sector. Sourced from: www.recycling.com (Recycling symbol) / Wikipedia, Author: Evan-Amos (Electromagentic sheilding) / Wikipedia, Author: BruceBlaus (Biocompatible) / Pixabay.com, LakeyC (Mechanically durable).

newly developed magnesium–rare earth/calcium alloys are used as they demonstrate good creep resistance. This is where magnesium nanocomposites have tremendous potential as they exhibit high thermal stability in addition to a superior combination of mechanical properties. With the use of magnesium nanocomposites in automotive components, tremendous advancements can be achieved in terms of the design freedom of these parts and fuel efficiency [5].

11.3 Aerospace applications of magnesium nanocomposites

Enabling a total weight saving in aircrafts is a key benefit of magnesium, with its light weight and tremendous potential when it replaces the currently used denser aluminium and other metallic components in the aerospace sector. While Mg was previously used, in particular during the 1940s, the Federal Aviation Administration (FAA) restricted and banned its use in aeroplanes owing to its susceptibility to ignition and its perceived flammability. In fact, magnesium in its finer forms is predominantly used for applications such as flares and fireworks as it is regarded as

Table 11.1. Magnesium versus the most commonly used materials in commercial aircrafts [7].

		% Distribution in commercial aircrafts		
Material	Density	Airbus A380	Boeing 787	Boeing 777
Mg	1.74	—	—	—
Composites	1.6–1.9	22	50	11
Al	2.7	61	20	70
Steels	7.9	—	10	11
Ti	4.5	10	15	7
Others	—	5	5	1

extremely flammable. During WWII magnesium was mainly used in the making of incendiary bombs. Hence, it was always regarded as an unconventional metal in the aircraft industry. However, with the recent upliftment of the ban by FAA, Mg usage is now allowed in interior components of the aircrafts, particularly because it results in large weight savings of up to 33% and 75% when it replaces aluminium and steels, respectively. The commonly commercially used materials in current aircrafts are listed in table 11.1 in terms of their distribution and density. By replacing these existing denser metallic alloys with lighter magnesium, one can achieve reduced weight, fuel consumption, and therefore reduced emissions of CO_2 in the aerospace sector, where the weight of an aircraft is critical for its efficiency. With recent developments and advances in magnesium alloys and composites with excellent mechanical performance as well as ignition resistance, these new-generation magnesium materials are promising for the aircraft industry [6]. Some of the alloys such as Elektron WE43 and Elektron 21 have been approved by the FAA for use in-cabin applications in commercial aircrafts [6]. While there are some concerns pertaining to the corrosion tendencies of magnesium alloys, they are being overcome by adapting advanced coating technologies.

11.4 Electronics applications of magnesium nanocomposites

The flexibility offered by plastic injection moudling in the consumer electronics sector in terms of design freedom, appealing looks, and texture is unparalleled. However, the two important shortcomings of the excessive use of plastic are its recyclability and poor shielding of electromagnetic radiation. This has led to both electromagnetic as well as land pollution from the enormous amount of electronic scrap produced. This calls for an alternative material. The key requisites for a material to suit electronic applications are light weight for portability, and the capability to endure ambient/outdoor temperatures and withstand impact. The obvious choice to meet these criteria is magnesium as its density is similar to that of most plastics and hence there is no compromise in the weight of the components. Magnesium materials are also very lucrative in this regard as they are recyclable, and can be reused and remelted to form other parts. Furthermore, the key advantages to using magnesium in consumer electronics compared to plastics or even the heavier aluminium alloys are its tendency to shield electromagnetic radiation coupled with its excellent damping capacity and specific stiffness. Table 11.2 gives the electromagnetic shielding capacity of magnesium in comparison to aluminium [8]. Although magnesium alloys have been studied for their promising electromagnetic shielding abilities [9], magnesium nanocomposites have not been studied for their electromagnetic shielding capacities. However, micron-sized reinforcements have been added previously [10, 11] to a magnesium matrix and have demonstrated extraordinary potential in mitigating electromagnetic radiation. Recently, magnetic nanoparticles (Fe_3O_4) added to a magnesium matrix showed positive effects on the mechanical behaviour and have been proposed as promising materials with good shielding capabilities owing to the magnetic properties of the nanoparticles [12]. With a careful choice of reinforcements that can inhibit electromagnetic shielding, magnesium nanocomposites are extremely promising for electronics such as mobile phones, laptops, and the housings/enclosures of many electronic devices, and their application can also be extended to radars, space communications, networking, and terrestrial communications, etc. Some of the typical electromagnetic shielding cages used currently to mitigate electromagnetic pollution are given in figure 11.2.

11.5 Biomedical applications of magnesium nanocomposites

For over a century, metals have been used as permanent/temporary implants in the human body because of their excellent mechanical properties such as fatigue strength and toughness. While the currently used materials such as the expensive Ti alloys, Co–Cr alloys, and relatively cheap stainless steel dominate the market for permanent implants, they all have different challenges. For example, the Co–Cr alloys have demonstrated toxicity effects and hence have not been encouraged as implants of late. While Ti and steels do not have challenges with toxicity or mechanical properties, they have an extremely high elastic modulus, almost 5–10 magnitudes higher than the modulus of human cortical bone, resulting in the stress shielding effects between the bone and the implant. In this context, magnesium is a potential substitute as a temporary implant material as it has an elastic modulus (40–45 GPa) matching that of bone, along with other advantages such as its light weight and excellent specific mechanical properties. Further, since bone is a natural composite, magnesium nanocomposites have a natural advantage in behaving similarly to bone. The mechanical properties, such as specific strength and elongation to failure of magnesium alloys and nanocomposites can be tuned to

Table 11.2. Average electromagnetic shielding effectiveness of Mg and Al [2].

Material	0.5–6 GHz	7–13 GHz
Al	59 ± 1	65 ± 6
Mg	65 ± 1	66 ± 4

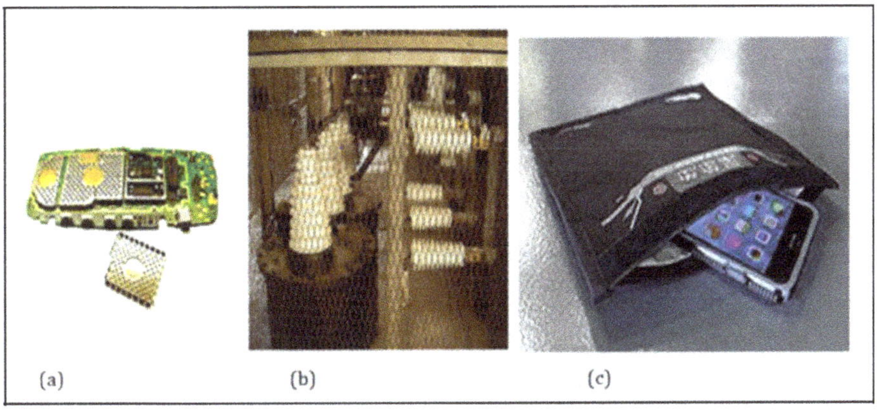

Figure 11.2. Cages used for electromagnetic shielding: (a) a mobile phone, (b) a power plant, and (c) a metallic fabric flexible pouch for protecting digital data from theft and to improve privacy [13], with permission of Jupiner Publishers. The currently used aluminium can easily be replaced by much lighter magnesium, in particular for portable electronic gadgets.

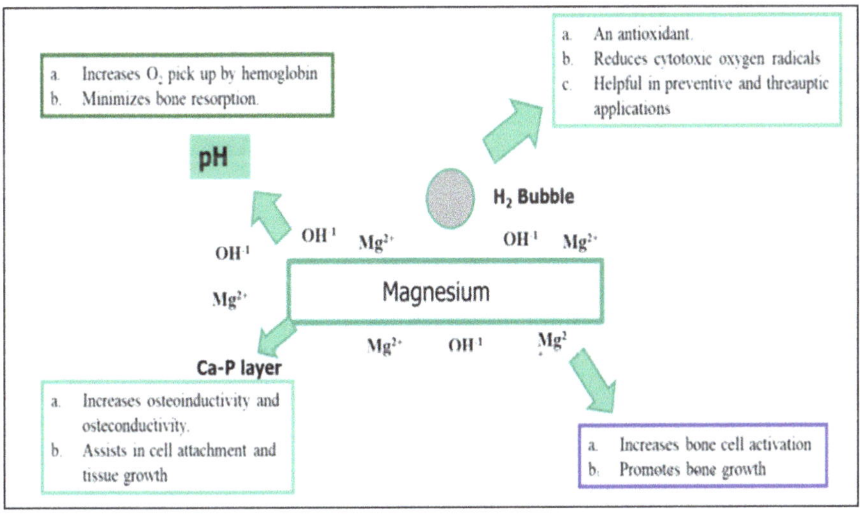

Figure 11.3. The advantages of magnesium in bio-medicine [3], with permission of MedCrave.

suit the needs of bone implants. Further, our bodies require 250–400 mg of Mg a day since Mg is the fourth most abundant cation in the body for assisting functions such as metabolic processes, aiding bone health, etc. In addition to being anti-bacterial, Mg has the additional advantage of being biodegradable in the body, hence is a potential temporary biodegradable implant with no requirement for second/revision surgery which reduces patient trauma, medic hours, and medical expenditure. The reactions of magnesium with simulated body fluids form magnesium ions, bubbles of hydrogen, and a Ca–P layer, in addition to an increased pH, as shown in figure 11.3 [3]. It may be noted that the rate of corrosion of magnesium in the body should be controlled and tailored to match the injury healing time to avoid the detrimental effects of earlier degradation, if any. It may further be noted that magnesium-based stents and screws have already been successfully tried in humans, and components such as surgical plates and clips are other potential applications. Applications of controlled drug delivery can also be explored with the aid of magnesium nanocomposites with special nanoparticles such as Fe_3O_4, etc, which are typically used as drug carriers.

11.6 Sports applications of magnesium nanocomposites

Akin to the requirements of the transportation industry, the sports industry also requires lightweight and high-performance materials. In this regard magnesium is valuable for its use in different sport-related equipment. Its applications include the manufacture of lightweight tennis racquets, golf clubs, archery bow handles, bicycles, wheelchairs, snowboards, etc. In particular, magnesium nanocomposites, with their even better mechanical properties and damping capacities [14], are a good candidate for these applications, particularly for bicycle frames and the chassis of in-line skates, where shock and vibration must be absorbed quickly. In fact, magnesium enables a better user experience as absorption of these shocks and vibrations leads to smoother and more comfortable rides by the users. Further, vaulting poles made of magnesium are also reported to have been produced. Therefore, magnesium nanocomposites have excellent potential in different sport-related applications.

11.7 Challenges of magnesium nanocomposites

Mg nanocomposites with ceramic reinforcements have unparalleled combinations of mechanical properties, such as hardness, tensile and compressive strength, wear, fatigue, and creep properties. The high intrinsic thermal stability of the ceramics makes the nanocomposites stand out in the realm of Mg-based materials. Despite the potential of these lightweight and arguably next-generation materials, there are also certain challenges and limitations to magnesium nanocomposites. Adding to the limitations generally posed by magnesium alloys, such as that of poor corrosion resistance, additional drawbacks persist in the case of magnesium-based nanocomposites [15]. Al and Ti-based alloys and steels are easily fabricated by standard routes involving continuous casting, extrusion, rolling, forging, etc, and are also recyclable. To replace the standard Al, Ti, and steels with Mg, the processing of Mg must be feasible in terms of the economics of cost. In recent times, the fabrication of lightweight magnesium alloys has also been shown to be economically feasible and industrially viable. However, the synthesis of magnesium composites, and in particular nanocomposites, is still not being scaled up to the industrial level despite having been extensively studied in the laboratory scale in the last two decades. One of the major concerns in the industrial viability of nanocomposites is the uniform dispersion of the reinforcement in the matrix and tailoring the interface between the matrix and reinforcement, which dictate the properties of the material. Apart from the limitations in production due to inhomogeneity in the dispersion of nanoparticles, other crucial factors such as reactions between the matrix and reinforcement, agglomeration of the particles, uncontrolled distribution of the particles in the matrix, etc, pose challenges in making these nanocomposites industrially scalable. The performance of these nanocomposites depends on all the above-mentioned factors, such as the distribution of reinforced particles within the matrix, and reactions at the matrix–particle interface or liquid–solid interface [16]. For example, the plausible reactions are debonding of the particles [4], dissolution of the particles in the matrix [17], and the reaction of particles with the matrix [18], which affect the mechanical properties such as the strength and ductility of the nanocomposites. Further, very few articles report thermodynamic reactions, and hence there is a lack of knowledge on the possible interfacial reactions between the magnesium matrix and reinforcements. Since these dictate the properties of the material, when uncontrolled they result in inferior mechanical properties and ultimately may result in premature failure in service.

While several techniques, such as the disintegrated melt deposition technique with optimized stirring and ultrasonic agitation, are applied for uniform dispersion of nanoparticles, these methods are still limited to the laboratory scale due to poor control of large scale synthesis. Although recently one plausible solution to develop sturdy magnesium nanocomposites through *in situ* reinforcement synthesis techniques was proposed as a futuristic approach for the industrial scalability of magnesium nanocomposites, research in this field is still at a very nascent stage and requires further attention before this method can be industrialized. However, on the bright side, this limitation of poor industrial scalability opens a large window for the research community working on the domain of Mg-based materials to devise new approaches and techniques that can aid in industrializing the production of these nanocomposites which will aid in light-weighting the future transportation sector as well as finding applications in the electronic, biomedical, and sports sectors.

11.8 Summary

This chapter makes a genuine attempt to provide a bird's eye view of the diverse and inter-disciplinary potential of magnesium-based nanocomposites in ubiquitous applications. Magnesium nanocomposites have tremendous potential in the automotive industry with several advantages, particularly in view of the worsening state of climate change and environmental conservation. With the aerospace sector being one of the bigger contributors of emissions, light-weighting with strong and flame-resistant magnesium nanocomposites can help reduce emissions and improve the fuel and cost efficiency. Further, their recyclability and efficiency in shielding from electromagnetic radiation render magnesium nanocomposites remarkable materials in the electronics and defence sectors, with primary applications being those that require withstanding shocks and vibrations and the casings and enclosures of electronic equipment. Magnesium-based nanocomposites also offer the advantage of degrading within the body. This helps in minimizing patient trauma, surgery, medical treatment costs, and doctors' time. With an elastic modulus matching that of bone coupled with its non-toxicity, this class of magnesium-based materials makes an excellent choice for non-permanent implants in the human body. With such great potential in all sectors, it is hoped that the existing challenges in the industrial scalability of magnesium nanocomposites will be overcome and magnesium nanocomposites will become the next new weapon in the innovation race.

References

[1] Clow B B 2016 History of primary magnesium since World War II *Essential Readings in Magnesium Technology* (Berlin: Springer) pp 85–7

[2] Gupta M 2018 Global emergence and significance of magnesium/technology *Mater. Sci. Res. India* **15** 107–13

[3] Gupta M 2018 A snapshot of remarkable potential of Mg-based materials as implants *Mater. Sci. Eng. Int. J.* **2** 30–3

[4] Joost W J and Krajewski P E 2017 Towards magnesium alloys for high-volume automotive applications *Scr. Mater.* **128** 107–12

[5] Kulekci M K 2008 Magnesium and its alloys applications in automotive industry *Int. J. Adv. Manuf. Technol.* **39** 851–65

[6] Tekumalla S, Nandigam Y, Bibhanshu N, Rajashekara S, Yang C, Suwas S and Gupta M 2018 A strong and deformable *in-situ* magnesium nanocomposite igniting above 1000 °C *Sci. Rep.* **8** 7038

[7] Gupta M and Gupta N 2017 Utilizing magnesium based materials to reduce green house gas emissions in aerospace sector *Aeronaut. Aerosp. Open Access J.* **1** 1–6

[8] Pandey R, Tekumalla S and Gupta M 2020 EMI shielding of metals, alloys, and composites *Materials for Potential EMI Shielding Applications* (Amsterdam: Elsevier) pp 341–55

[9] Pandey R, Tekumalla S and Gupta M 2018 Effect of defects on electromagnetic interference shielding effectiveness of magnesium *J. Mater. Sci. Mater. Electron.* **29** 9728–39

[10] Pandey R, Tekumalla S and Gupta M 2019 Enhanced (X-band) microwave shielding properties of pure magnesium by addition of diamagnetic titanium micro-particulates *J. Alloys Compd.* **770** 473–82

[11] Pandey R, Tekumalla S and Gupta M 2019 Magnesium–iron micro-composite for enhanced shielding of electromagnetic pollution *Composites* B **163** 150–7

[12] Johanes M, Tekumalla S and Gupta M 2019 Fe_3O_4 nanoparticle-reinforced magnesium nanocomposites processed via disintegrated melt deposition and turning-induced deformation techniques *Metals* **9** 1225

[13] Gupta M 2017 The promise of magnesium based materials in electromagnetic shielding *Juniper Online J. Mater. Sci.* **2** 555598

[14] Tekumalla S, Yang C, Seetharaman S, Wong W L E, Goh C S, Shabadi R and Gupta M 2016 Enhancing overall static/dynamic/damping/ignition response of magnesium through the addition of lower amounts (<2%) of yttrium *J. Alloys Compd.* **689** 350–8

[15] Gupta M and Wong W L E 2015 Magnesium-based nanocomposites: lightweight materials of the future *Mater. Charact.* **105** 30–46

[16] Dieringa H 2018 Processing of magnesium-based metal matrix nanocomposites by ultrasound-assisted particle dispersion: a review *Metals* **8** 431

[17] Song J, She J, Chen D and Pan F 2020 Latest research advances on magnesium and magnesium alloys worldwide *J. Magnes. Alloy* **8** 1–41

[18] Tekumalla S and Gupta M 2017 An insight into ignition factors and mechanisms of magnesium based materials: a review *Mater. Des.* **113** 84–98

Lightning Source UK Ltd.
Milton Keynes UK
UKHW051314091220
374844UK00002B/25

9 780750 335331